WOMEN AND THE WORLD OF WORK

NATO CONFERENCE SERIES

III HUMAN FACTORS

WOMEN
AND THE
WORLD OF WORK

Edited by
Anne Hoiberg

Naval Health Research Center
San Diego, California

SPRINGER SCIENCE+BUSINESS MEDIA, LLC

Library of Congress Cataloging in Publication Data

NATO Symposium on Women and the World of Work (1980: Lisbon, Portugal)
  Women and the world of work.

  (NATO conference series. III, Human factors; v. 18)
  "Proceedings of a NATO Symposium on Women and the World of Work held August
4 – 8, 1980, in Lisbon, Portugal" — T.p. verso.
  Bibliography: p.
  Includes index.
  1. Women — Employment — Congresses. I. Hoiberg, Anne. II. Title. III. Series.
HD6052.N37 1980   305.4   82-5271

ISBN 978-1-4613-3484-2   ISBN 978-1-4613-3482-8 (eBook)
DOI 10.1007/978-1-4613-3482-8
                                                                        AACR2

Proceedings of a NATO Symposium on Women and the World of Work
held August 4 – 8, 1980, in Lisbon, Portugal

© 1982 Springer Science+Business Media New York
Originally published by Plenum Press, New York in 1982
Softcover reprint of the hardcover 1st edition 1982
A Division of Plenum Publishing Corporation
233 Spring Street, New York, N.Y. 10013

This volume is dedicated to the many courageous and tenacious women who have fought or are fighting for equality in the world of work.

. . . the large-scale movement of women into the work force opens up the exciting possibility of creating a much improved society.

Alan Pifer

PREFACE

From August 4 to 8, 1980, the Science Committee of the North Atlantic Treaty Organization (NATO) sponsored the symposium, "Women and the World of Work," which was held at the Hotel Sintra Estoril in the coastal area south of Lisbon, Portugal. This symposium had been "in progress" since 1977 when the idea to prepare a proposal for a NATO-sponsored symposium on the topic of women and the military was first suggested by Dr. Walter Wilkins, then Scientific Director of the Naval Health Research Center, San Diego, California. At that time and during the previous 5 years, increasing numbers of women were being recruited into military service not only in the United States but also in several NATO-allied countries. Few research projects on the utilization of women in the military had been reported in the scientific literature with the exception of work conducted at the University of Chicago and Naval Health Research Center. Several investigators, however, were identified who had recently initiated research in this area—and who expressed interest in participating in the proposed symposium.

A few months after submitting the proposal, Dr. B.A. Bayraktar, Director of the Scientific Affairs Division responded that the members of the Science Committee recommended the deletion of those segments of the tentative program which pertained to women and the military. The reason given for this suggested change was that the topic chosen was similar to the theme of other recent international conferences. The proposal was revised to reflect the Science Committee's growing interest in research on nonmilitary issues and was resubmitted prior to the August deadline in 1978. Upon receipt of official approval in May 1979, the preparations for the "Women and the World of Work" symposium could begin.

The overall aim of this symposium, as well as all other programs supported by the Human Factors Panel of the Science Committee, was to increase the potential value of research within NATO-allied and other countries by providing a forum for the exchange of information and an opportunity for scientists to establish contacts with other researchers. The specific objective of this 5-day program was to examine several areas associated with women's increased participation in the world of work: effects upon the family; women's work, child care, and family wage; effects upon women's physical and mental well-being; economic factors; women as workers; job creation and unemployment; and other related philosophical, psychological, and political issues. Participants represented almost all of the NATO-allied countries and included scientists from each of ten disciplines as well as politicians, government administrators, and policymakers.

The former Prime Minister of Portugal, Maria de Lourdes Pintasilgo, was the Keynote Speaker of the symposium; her thought-provoking presentation, "Women as World Makers," is included in the first section of this volume. Other dignitaries invited to speak during the opening session were: Vaira Vikis-Freibergs, Chairperson of the Human Factors Panel of NATO, who described various scientific research and professional programs sponsored by the Science Committee; Anna Vicente, an official of the Commission on the Status of Women in Portugal, who outlined the achievements and goals of the Commission; and Françoise Latour de Veiga Pinto, Secretary General of CEFRES, who provided historical information and a future perspective on women and the world of work.

At the farewell dinner, the guest of honor was Maria Manuela Aguiar, Portugal's Secretary of State for Emigration. She spoke to the symposium participants about the accomplishments and proposed goals of the Mid-Decade World Conference of the United Nations for Women in Copenhagen at which she served as an official delegate from Portugal. Others who shared their experiences at that conference were Anne Briscoe, Jean Lipman-Blumen, and Hanna Papanek as well as Miet Smet, member of the Belgium Parliament and official conference delegate from Belgium.

This volume, which is a compilation of 19 of the papers presented at the symposium, is divided into four parts. (Unfortunately because of space limitations and other considerations, not all of the presentations could be included; however, the excluded presentations have been summarized.) Each of the four sections begins with an introduction that consists of the summarized presentations, a few brief paragraphs highlighting the main points of the chapters pertinent to the topic of the section, and a conclusion. The titles of the four parts are subsumed under the overall title of Women and the World of Work: Contributions and Progress; Physical, Mental, and Economic Well-Being; Socioeconomic Factors Affecting Women's Increased Labor Force Participation; and Role Integration and Revision.

As editor and convener, I wish to acknowledge those individuals who contributed their time, energy, and expertise to the work of the symposium. To begin, I must thank Jan Moffitt for performing many of presymposium administrative tasks and mail-outs during July and August of 1979. Air Portugal assisted in making the travel arrangements for the participants and attendees while Madalena Goncalves and Candida of Atlántica Tours in Lisbon provided their services in arranging tours and other travel considerations. I feel a special debt of gratitude to Frank Pollock for his assistance both before and during the conference. Mark Zilly also provided administrative support throughout the week of the symposium. For their help in editing this volume, I am indebted to Camille Kim Cook, Larry Palinkas, and Mary Wagner. Final typing was very capably performed by Leslie Griffith-Jacoby of Executive Secretarial in San Diego. Also contributing their professional typing and artistic skills to the preparation of this book were Lucile Cheng, Emily Collins, Bea Neher, and Millie Heeley. Very special thanks also are extended to Camille Kim Cook and Chris Blood who assisted in the proofreading of this volume.

As a final note of appreciation, I am grateful to the Science Committee of NATO for supporting this symposium; special thanks are specifically extended to B.A. Bayraktar, Vaira Vikis-Freibergs, John Nagay, Bert King, and Holger Ursin. Also to be gratefully acknowledged are all of the presenters as well as the following individuals who served as chairpersons or discussants: Frank Pinch, Raquel Ribeiro, Paula Hudis, Joyce Lazar, Alice Cook, Barbara Bergmann, Clair Brown, Judith Stiehm, Nicki Fonda, Freda Paltiel, Anne Briscoe, Martha Darling, Jacob Mincer, Deborah Freedman, Miet Smet, H.M. in't Veld-Langeveld, and Jean Leonard Elliott.

A.H.

CONTENTS

PART I

WOMEN AND THE WORLD OF WORK:

CONTRIBUTIONS AND PROGRESS

INTRODUCTION

Anne Hoiberg

Naval Health Research Center
San Diego, California   U.S.A.

Throughout the ages, women have always worked—both in the home and in the marketplace. With the advent of the Industrial Revolution, women's work force participation extended into the industrial sector, and after the U.S. Civil War they were hired for the first time as federal employees. Since the beginning of this century, women's involvement in the labor force has been cyclical, increasing with each surge in industry's demands for workers and subsiding with each decline in the overall economic situation. During the past 20 years, however, the economic needs of all families as well as unmarried women have spiraled, thereby requiring that a large proportion of women establish a permanent attachment to the labor force and forge an inroad into occupational fields that offer a satisfactory and steady income. Given these increased needs, what progress have women made? To be specific, the most important objective of this section is to examine the progress that women have made during the twentieth century as participants in the "world of work." The introduction centers on women's involvement in the U.S. labor force whereas the summarized presentations and chapters that follow provide the historical perspectives from other countries. Also to be discussed are the behavioral effects attributable to women's increased labor force participation and the recommendations that the symposium authors proffer for enhancing women's contributions to the work force--and to society as a whole.

WOMEN'S LABOR FORCE PARTICIPATION: PAST AND PRESENT

During World Wars I and II, women entered the U.S. labor force in unprecedented numbers in response to the federal government's urgent call to work in wartime industries and in jobs previously held by men. Although assured through a vast advertising campaign during World War I that they would not have to wear trousers or a man's uniform, which

3

suggested that jobs requiring such attire were not considered appropriate occupations for women, they did indeed don men's uniforms to become elevator operators, streetcar conductors, and postal workers.[1] Women in large numbers also were hired to operate drill presses, lathes, millers, and other machines and tools in aircraft plants, shipbuilding yards, and steel mills.

Women officially served for the first time in the military during World War I, although nurses had been recruited into the U.S. Army Nurse Corps for the Spanish-American War; in World War I, approximately 13,000 women volunteered for active duty in the U.S. Navy and 300 for the U.S. Marine Corps. In the latter case, a total of 2,000 women responded to a newspaper article which read that the Marine Corps was seeking applications from "intelligent young women."[2] The 300 women selected were recruited to perform clerical work to free men for combat. More women were enrolled than necessary because officials presumed that it would take about three women to perform the work of two male clerks, which soon proved to be an inaccurate assessment of women's capabilities. To meet our military's needs for clerical workers and telephone operators, General Pershing urged Congress to pass legislation allowing women to serve overseas in the U.S. Army, a request that was vehemently denied.

Women in the labor force during World War II accounted for a third of all workers, approximately 19 million by July 1945.[3] During the early 1940s, many women served overseas as volunteers for various private organizations; one volunteer, Margaret Hughes, served with the ambulance unit of the American Friends of France and was decorated three times by the French government. A total of 350,000 women served on military active duty in capacities ranging from the usual clerical, administrative, and health care specialists to airplane mechanic, parachute rigger, air traffic controller, gunnery and instrument flying instructor, metalsmith, and airplane ferry pilot.

At first, women were reluctant to register for a defense job or to serve in the military; however, through the efforts promulgated in numerous publications, posters, and announcements, they were convinced that the war could be terminated sooner if they joined the labor force. The distinction between a "man's job" and a "woman's job" faded with each feature article or advertisement showing the well-dressed female factory worker attired in a visored cap, slacks, and short-sleeved blouse.

Women worked in all aircraft occupations from welder to test pilot, accounting for 50% of all aviation workers.[4] In the shipbuilding yards, the largest proportions of women were employed as shipfitters (36%), machinists (14.6%), electrical workers (10.2%), and service and maintenance workers (9.6%). In the steelmaking industry, where the work is extremely strenuous, heavy, hot, and dirty, women represented 11% of the steel production work force and performed almost every job except the most strenuous. Women also worked in artillery and ammunition production, as farm workers in the Women's Land Army of the U.S. Crop Corps, in the railroad industry, and as clerical workers. The largest proportion of wartime women employees, however, was in factory work.

To convince defense industry employers that women were productive workers, numerous government studies were conducted comparing women's performance against the standard of the day, which was men's productivity rate. In a 1943 study of 130 firms, which employed a total of 154,587 women and 242,297 men, women's performance was reported as equal to or greater than men's in 88% of the 130 plants. When women's productivity was less, the strength factor was usually involved. Several government researchers also reported that women created many innovative techniques to increase productivity. For example, women in the aviation industry developed a highly efficient method of meeting their work goals by using a team work concept which they in turn were requested to teach their male replacements at the end of the war. These research studies provided overwhelming evidence that women could satisfactorily perform almost all jobs—and with only a minimal amount of training. One male government official recommended that on the basis of their productivity rates and in all fairness to them, women should receive the same pay as men! Unfortunately, this proposal was not implemented.

After the war, the jobs that women held were given back to the returning male veterans although the vast majority of women (75% to 80%) expressed the desire to remain at their jobs. Most women, now unemployed with few prospects for employment in a sluggish economy, resumed their traditional roles in the home. The birthrates in the United States and most European countries during the 1950s reached their highest levels since the early 1900s. Another advertising campaign, although more subtle than that conducted during World War II to entice women into war industries, was initiated which glorified the roles of women as mothers and men as breadwinners. Under the prevailing economic conditions and these social circumstances, most women accepted the view that woman's role was primarily associated with the duties of motherhood and homemaker. The print media of that era supported this view by publishing increasingly more complicated gourmet recipes, by revitalizing the home science of food preservation (canning), and by promoting an interest in such art forms as needlepoint, crewel embroidery, knitting, clothing construction, etc.

The decade of the 1960s heralded a return of many women to the work force, not to resume the factory jobs held in the 1940s, but primarily to enter clerical and service positions, the "women's jobs" described as the "invisible 80%" by Joan Goodin and cited by many authors in this text. At the same time, the housing industry designed and constructed apartments and small tract homes that required less care than the larger homes of previous eras. With the inclusion in these homes of numerous labor-saving devices, such as automatic laundries, dishwashers, freezers, etc., women were freed from many time-consuming tasks to engage in educational pursuits or paid employment. Also during the 1960s, the divorce rate began to climb which has continued its ascendance throughout the past two decades and has resulted in a corresponding increase in employment rates of divorcees.

The question posed for many women during the 1960s centered on the issue of what they should do with their daytime hours to enrich their lives and contribute to the financial support of their families. For most women, the job opportunities fell within a narrow category of clerical and service jobs. It should be noted, however, that the white-collar occupations available to them offered considerably more status than the welding and factory jobs most women held during World War II. During this postwar era, moreover, many college women majored in a 4-year secretarial curriculum—this fact seems incredulous in light of the current climate in which increasingly more women are enrolled in career-oriented programs leading to positions of leadership and power wherein they will be assisted by secretaries or administrative aides.

Between 1960 and 1978, the percentage of women in the U.S. labor force increased from 34.8% to 51.7%, according to government publications cited by Chapman and Chapman. Although women have worked in almost all occupations over the years since World War I, the number of women in fields traditionally associated with men has remained low with a relatively small, gradual increment observed during the past decade. Other changes since 1970 were evidenced in that several occupations such as commercial pilot, telephone repairperson, astronaut, U.S. Supreme Court Justice, and Commanding Officer of a Coast Guard vessel became feasible career choices for women.

The papers and chapters to follow will describe in greater detail the progress that women have made as participants in the labor force, not only in the United States but also in other countries. In the first chapter, an overview of the role that women play in the labor force as well as in the home is presented by Maria de Lourdes Pintasilgo in her keynote address to the symposium. Pintasilgo delineates two possibilities in which women can impact society and the "world of work" in the future: women can attain positions of leadership and power and bring about positive changes including an economic and ideological recognition of the value of "women's work," thereby increasing their own status in society while simultaneously helping to improve it. Or, women can continue to become members of "a new community of slaves"—maintaining a low level of status in society and enduring the exploitation by society. She warns that while considerable progress has been made, the latter circumstance, with all of its dire consequences, still exists. The emphasis in the other presentations is on women's increased participation as members of the professional elite: scientists, engineers, physicians, lawyers, politicians, managers, businesswomen, and other jobs in the decision-making and wealth-producing sectors.

WOMEN'S PARTICIPATION IN THE PROFESSIONS

Scientists and Engineers

According to Lilli Hornig in her chapter, "The Education and Employment of Women Scientists and Engineers in the United States," the proportion of women who have a doctorate in the sciences or engineering has grown substantially during the 1970s, although women represent only 10% of the entire population in these fields. Within the various disciplines, the trend over the decade shows a steady gradual increase, with the most notable gains observed for the social sciences and mathematics/statistics. Hornig states quite emphatically that more women are needed to conduct research and become participants in the important discoveries in these fields. With increases projected for women's educational and employment participation in the sciences and engineering, more women no doubt will contribute to the development of their country's social, economic, and political well-being.

Hornig suggests that our high schools should include advanced science and math, along with English, as required courses. With this proposed change in the curriculum, many high school girls would be less likely to become a "math-anxious technophobe," a label with which some of their mothers can readily identify. In support of her recommendation, Hornig reminds us that almost all jobs, even those considered women's occupations, have become increasingly more mathematically oriented.

Hornig provides evidence to show that women in the scientific professions fare less well financially than men which is reflected by their overall lower mean salaries and the smaller proportions of women in positions of power. Even with these disparities, women have persisted in completing a Ph.D. program and have become members of the professional elite. Hornig concludes that the equal pay laws and other policies have contributed in part to women's attainment of these positions. However, she attributes to women themselves the primary motiviating force for most of their achievements and states that therein lies the basis for future accomplishments.

Physicians and Lawyers

Although not included in this volume, the presentation by Ayse Öncü ("Turkish Women in the Professions: Why So Many?") provides a cohesive explanation for the phenomenon of the relatively high proportion of women in the two most prestigious professions in Turkey, namely law and medicine. One in every five practicing lawyers in Turkey is a woman and one out of six physicians is a woman. By way of contrast, the percentages of women lawyers and physicians in the U.S. for the same year (1970) are 4.7% and 8.9%, respectively. The first woman to register with the Istanbul Bar Association, the oldest and largest in Turkey, did so in 1936. Since then and until the last year of the survey in 1970, the period of greatest growth for women in law and medicine occurred during the 1960s.

Öncü begins to unravel the puzzle posed in her title by first presenting such background information as the fact that only 10% of the labor force in Turkey is composed of women and, moreover, that in rural areas approximately one-half of the women over the age of 15 have not completed primary school. It also has been noted by Kandiyoti in her presentation that the vast majority of Turkish women continue to work on family-owned farms. Other salient facts are the relatively high percentages of women entering the universities who come from civil service and professional backgrounds and live in the metropolitan centers of Turkey. Daughters of farmers and laborers only represent 18% of the female university population. Of those women who complete higher education, 70% are working.

With the establishment of this basic background information, the fairly high proportion of women lawyers and physicians in Turkey is explained in part as a phenomenon that is not unique in many Third World countries and in such intermediate economically developed countries as Mexico, Argentina, Greece, Costa Rica, and India. Women in the higher socioeconomic classes in these countries can exercise a greater number of educational and career options than women in developed countries. One reason for this freedom is the existence of class inequities. Öncü explains that because of the large lower socioeconomic classes, there is a large female labor pool from which to select workers to perform household and child care services. Moreover, child rearing assistance frequently is provided by members of the extended family which, combined with the inexpensive household help, enables upper-middle- and upper-class women to participate in an educational program or a professional career. As contrasted with other occupations, such as the nursing or teaching professions, women lawyers and physicians also can more easily combine the duties of motherhood with a career because of the flexible hours afforded by their occupational choices. Another interesting point made by Öncü is that even if their inexpensive outside help should become scarce, these women of the professional elite have sufficient power and influence (1) to use the political process to generate government action into establishing the needed nursery schools and kindergartens or (2) to stimulate private enterprise into the child care business.

Öncü also presents a historical perspective as an explanation for women's increased participation in law and medicine in both Third World and industrialized countries. In developed countries during the past 150 years, these prestigious occupations have maintained an elite status which has been perpetuated by state regulations and the creation of autonomous organizations to control standards of entry, performance, and conduct. This self-regulation has fostered a restricted entry and a tight-knit unity which has enhanced the status and earning power of individuals in these professions. In developing countries, on the other hand, these professions have evolved into a "trained elite" only during the past 50 years. The government of Turkey, along with other countries, supports an expansion of educational programs to meet the goal of establishing a larger professional class. To meet this objective, an increasing proportion of university students has to be drawn from the manual laborer or peasant segments of society—or from the female population in the upper socioeconomic strata. To men in these professions and educational programs, women of the higher socioeconomic classes are perceived as less threatening and more acceptable than men from the peasant or working class who are likely to be highly competitive and achievement oriented. Another reason that women are accepted quite readily is that these professions have had insufficient time during these past 50 years to become male bastions or to be labeled as "deviant" professions for women. Most of these women also have the social contacts needed to become established in private practice; men from the lower socioeconomic strata may face insurmountable barriers not only in meeting the expenses of their education but also in trying to begin a practice.

On the basis of Öncü's comparisons, therefore, it can be concluded that women represent a relatively high proportion of the prestigious professions in Turkey because of their socioeconomic background, their acceptance, the availability of inexpensive household and child care, the aim of developing countries to expand the professional class, and the flexibility in hours of these professions. As a final point, Öncü cites the results of two recent surveys which show that (1) over one-half of all female lycée students in the city of Izmir aspire to the prestigious professions with medicine as the most preferred choice and (2) 17% of a sample of mothers in Turkey want their daughters to become medical doctors. If these surveys were conducted in the United States, it is conjectured that the percentages would be much lower. Because of the aspirations of Turkish women and the aforementioned considerations, the numbers of Turkish women who select a career in law or medicine no doubt will continue to rise. This projection also seems applicable to women from countries at a similar level of economic development.

Politicians

In political life, as well as in organizational life, numbers are important especially when considering the overall effectiveness of women. In her chapter, "Elected Women: Skewers of the Political System," Judith Hicks Stiehm summarizes Kanter's work on tokenism and emphasizes the importance of increasing the numbers of women in political life if women hope to have an impact on the creation of the laws and policies of any governing body, whether that body is a school board or the U.S. Congress.

In addition to dealing with the credibility problems of being a skewer or token (a member of a 20% or less minority within a group), women elected officials face other unique problems such as their own limited sources of income, a dependency on others for their income, family obligations and pressures (including the double burden), self-presentation, a scarcity of role models, an underestimation of their qualifications for office, and a somewhat limited number of years (commencing with middle age) during which to serve in public office.

Furthermore, the political system with its "head-on" elections, funding, lobbyists, and media appeal tends to be more workable for men than women. Added to these concerns is the problem of discrimination. Thus, Stiehm paints a picture in which women elected officials must address many issues in order to foster their efficacy in office, problems that are unique only to women. If women are to have an impact on all law-making bodies, they have to be willing to serve—and to muster the energy needed to deal not only with the duties of the position but also with each of the extraneous issues just described. To conclude, rather than accepting the media's view that many elected women officials seem to be disinterested in aspiring for higher office, perhaps a more realistic explanation is that many women experience feelings of frustration and futility when they become aware of the many barriers to be overcome in furthering their political careers.

Women in the Decision-Making and Wealth-Producing Sectors

In her chapter, "The Relationship between the Labor Force Employment of Women and the Changing Social Organization in Canada," Lorna Marsden presents a historical perspective of women's increasing labor force participation in Canada, with an emphasis on the extent of women's involvement in the decision-making and wealth-producing sectors. Similar to Stiehm's thesis, Marsden points out that women will have to increase their numbers in these sectors if they are to have an impact on the government's laws and policies.

In comparing data compiled by various governmental agencies with those published by researchers in this area, Marsden reports that not only is there an underrepresentation of women in the decision-making sector, but that the situation is equally unimpressive in the wealth-producing sector. In 1973, for example, only 4% of all lawyers and notaries were women. The author emphasizes that the largest concentration of women in the Canadian work force is in the non-wealth-producing sector: teaching, health care, and clerical work. In Canada, as in Turkey, there also is a growing number of women who have taken over the operation of the family farm because the husband earns an off-farm income.

Marsden observes that although the extent of women's labor force participation has been minimal in the decision-making and wealth-producing sectors, the impact of their overall increased participation has been extensive in the areas of (1) political consciousness (e.g., unionization of 25% of women workers and the formation of the Feminist Party of Canada); (2) public policy such as the establishment of the Human Rights Commission, expanded rights for women, and social welfare considerations; and (3) community life. In an elaboration of the latter, Marsden discusses several thought-provoking issues which are posed in the following questions: As more and more women enter the labor force, who will perform the voluntary work for hospitals, political parties, churches, civic and school organizations, youth clubs, neighborhood associations, and charitable organizations? Also, within the family, who will perform such

activities as food preservation, handicrafts, cultural pursuits, and maintenance of family traditions? These considerations add another dimension
to those described by Pintasilgo in her discussion of women as value givers
and by Papanek in her presentation on family status production.   As
increasing numbers of women enter the labor force, therefore, who will
have the time to perform the aforementioned activities?

BEHAVIORAL AND ATTITUDINAL EFFECTS OF WOMEN'S INCREASED PARTICIPATION IN THE LABOR FORCE

Occupational Behaviors

In Laraine Zappert and Harvey Weinstein's presentation, "Multiple
Role Obligations and the Health Status of Working Women," the authors
propose to determine the overall effect upon women's health status when
women assume more than one major role obligation.   Responses to a
questionnaire are reported for 73 women and 50 men who graduated in
1977 or 1978 with a Master's degree in Business Administration (MBA)
from a prestigious graduate business school. Five domains are assessed on
the questionnaire: Job Tension, Coping Style, Coping Strain, Role
Conflict, and Health Status. The demographic characteristics include
age, marital status, number of children, educational background, and
previous work experience as well as current job information on income,
job level, and income goals.

Results of comparisons show no significant differences on the
demographic variables. Other comparisons indicate that there are no
significant differences between women and men in their ratings of support
from supervisors and cohesion among colleagues. Both men and women
rate their jobs as moderately to highly stressful. They both view their
work as being important; as affording them authority, responsibility, and
autonomy; and as providing an opportunity to set their own limits and to
seek advice if needed. Neither men nor women report any chronic
somatic symptoms and for the most part are in good physical health.

Significant differences, however, are observed for the variable of
income in that men earn an average of $4,000 more per year than women
($29,676 versus $25,688) and men aspire to a higher income level ($42,000)
as contrasted with $31,000 for women MBAs. Men, futhermore, occupy
higher levels in management than women which is reflected by the
percentages of men in executive or mid-management positions (45%)
rather than in junior management or trainee positions (41%) as compared
with 25% and 65% for women in these respective categories. These
differences mirror the entire marketplace which also are reported by
Hornig: men of the same qualifications earn more, are promoted faster
and farther, and have higher expectations for their career goals.

For the comparisons on the five indexes, women have significantly
higher scores than men on all measures except for the nonsignificant
result on the Coping Style Index. The most highly significant differences

among specific items indicate that in comparison with men, women (1) feel more bound by inflexible work schedules (Job Tension Index), (2) report being on the verge of tears more often (Coping Strain Index), (3) are principally responsible for household and child care (Role Conflict Index), (4) respond that work would have a negative effect upon a decision to have a child (Role Conflict Index), (5) more often feel overwhelmed or on the verge of a nervous breakdown (Health Status Index), and (6) report having visited a mental health professional more often in the past 3 years (Health Status Index).

Although at a lower level of statistical significance, the results also show that more men imbibe in alcohol daily while women more often report feeling depressed. In her discussion of these results, Anne Briscoe notes that "If one is paid less, if one's colleagues are given preference in promotions just because they are males and not for superior performance, it is appropriate to be depressed; it is not a sign of less emotional stamina." Given these circumstances, the results of the study are more understandable except for an explanation of why more men than women imbibe in alcohol on a daily basis.

The most important finding from the regression analyses, which were conducted separately for men and women, is that the Coping Strain Index is the most powerful correlate of health status for both sexes. The items in this index pertain to difficulties in controlling one's temper or emotions, impatience, heightened sensitivity to criticism, self-doubt and self-blame, and inaction when confronting problems. The relationship between this index and health status is especially strong for men although it, too, yields the highest correlation for women. To a considerably lesser extent, the Role Conflict Index also contributes significantly to the overall association with health status for both men and women. This scale includes items that reflect an individual's feelings about the integration of his or her work with household and child care responsibilities. Such results clearly point up that work-related factors, as reflected by items on the Coping Strain Index, are significantly more important contributors to an individual's health status than other variables, including role conflicts or multiple role obligations. This conclusion suggests that for these young MBAs, both men and women, the impact upon one's health of several responsibilities is not as powerful a correlate as the difficulties encountered in coping with the stressors in the work setting. These results concerning the minimal impact on health status of multiple role obligations are supportive of the findings by Constance Nathanson to be discussed in the following section.

Power and Political Behaviors

Several authors in this text discuss the importance of knowing the prevailing behaviors and engaging in the strategies pertinent for survival, success, and advancement in organizational and political life. In her chapter, Virginia Schein addresses this issue and comprehensively outlines the roles of power and political behaviors, possible male/female differences in access to power and use of political behaviors in organizations, and recommendations for specific areas of future research.

In describing the key power bases of expertise, control over information, political access, assessed stature, group support, and mobility, Schein reports research results showing that women have acquired expertise (as reflected by the increasing percentages of enrollments in graduate school programs) whereas the other power bases are reported as much less accessible to women than men. Women, however, tend to engage in more frequent use of these and other power bases than men, particularly that of personal appeal. Differences between men and women also are noted for the effective use of various forms of power strategies (e.g., forming alliances or using self-dramatization).

Schein's thesis is that expertise is necessary to achieve managerial success but not sufficient; both men and women have to use other power bases and appropriate strategies. As she states, organizational life consists of rational behaviors such as planning, organizing, directing, and controlling as well as those behaviors associated with the gaining and keeping of power. According to the research cited, women seem to be at a disadvantage in learning and using the prevailing power and political behaviors. The research reported by Zappert and Weinstein provides support to show that for young MBA graduates working in organizations the factor most highly associated with health and well-being is that related to coping with organizational life—for both men and women.

In the future, as more women have an impact on organizational life, there is the possibility that different behaviors and strategies for success and advancement will evolve. As Pintasilgo cautions in her chapter, women should not become carbon copies of men, but should strive to contribute something unique to the workplace that will lead to an improvement in organizational life. Until women increase their numbers, however, it is imperative that they learn and use the current behaviors and strategies.

Achieving Styles

In their presentation, "A Model of Achieving Styles: Implications for Women's Occupational Roles," Jean Lipman-Blumen and her associates provide a comprehensive development of a model of achieving styles, which includes a historical basis of the underlying assumptions, a description of the nine achieving styles, the dimensions of these styles, an inventory to measure an individual's style, and the results of research conducted on subsamples of the 7,000 respondents who completed various versions of the inventory. This model evolved from an interest in achievement satisfaction. The construct under study is conceptualized as falling on a continuum from the active end, wherein individuals satisfy their achievement needs directly and completely by their own efforts, to the passive end in which individuals achieve satisfaction through the accomplishments of others. On the basis of this conceptualization, an

insightful interpretation of Horner's 1968 data[5] is presented which suggests that the well-known "fear of success" motive might be viewed more
accurately as an expression of vicarious achievement.

From this initial interest, various hypotheses were tested which
eventually led to the present effort of examining the similarities and
differences in achieving styles among several occupational groups.
Achieving styles, it was hypothesized, develop through early socialization
and undergo resocialization to meet the demands of an occupational role.
Thus, an important point emphasized by the authors is that an individual's
achieving styles are primarily determined by the occupational role and are
not a reflection of his or her gender.

To test this hypothesis, the authors designed a study comparing the
achieving styles of (1) same gender individuals who are not engaged in
paid employment (female high school students and homemakers), (2)
unemployed and employed women, (3) women in jobs traditionally associated with men (managers) and women who are employed in nonmanagerial
jobs, and (4) women managers and men managers. Results show that
occupational roles indeed do shape the achieving styles of individuals,
which is reflected by the near linearity in mean values of achieving styles
observed from high school students and homemakers to men and women
managers. These findings indicate quite clearly the readiness with which
women are able to acquire the achieving styles needed to meet the
demands of their jobs. Thus, achieving styles of the women managers
more closely resemble their male manager peers than those of their
female nonmanagerial counterparts.

Attitudes toward Occupational Roles

In many of the papers presented, the central issue is the occupational segregation of women in low-paying, low-complexity, low-prestige
jobs, which has been attributed to social structure, sex stratification,
socialization, fear of success, vicarious achievement, or a combination of
these. In many earlier studies, the major theme is that women derive
satisfaction from these undemanding jobs because their most important
role is that of wife and mother. In their presentation, "Perceived Job
Complexity by American Urban Women," Helena Znaniecka Lopata and
her associates at Loyola University of Chicago have designed a study to
verify or dispel this notion. To be specific, the purpose of their study is
to examine the association between women's perception of job complexity
on seven dimensions and their concomitant ratings of satisfaction with
that level of complexity.

Through personal interviews conducted in 1977 and 1978, 996 women
between the ages of 25 and 54 were asked to rate their jobs according to
their perceptions of the job-related constructs. Forty-two percent of this
sample of Chicago residents were full-time homemakers; of the 58% who
were employed outside the home, 46% were in occupations in which
women comprised 75% of the employees, 42% were employed in occupa-

tions which were not heavily dominated by either sex, and 11% were employed in jobs of which less than 25% were women.

The results show that women in the professional category (which consists primarily of the semi-professional occupations of nurse and teacher) rate their jobs as the highest in complexity, followed in order by nonprofessional white-collar workers and blue-collar workers. A more specific breakdown of occupational classifications results in seven groups, each of which differs on ratings of job satisfaction for the variables of creativity, work complexity, control over others, and self-development. Professional and managerial women have the highest ratings of job satisfaction, particularly for the dimensions of independence, creativity, the opportunity to see the end result of their work, responsibility, and work complexity. Homemakers, service workers, and lower-level clerical workers (typists as contrasted with legal secretaries in higher-level clerical jobs) tend to have the lowest ratings of job complexity and the least favorable ratings of job satisfaction. Homemakers are especially displeased with the extent of self-development in their occupation, and lower-level clerical workers have a comparable percentage of respondents (46%) who are dissatisfied with the extent of influence they have over others. Overall, the women in this study have positive perceptions of the complexity of their jobs, as compared with an "average" job, and generally are pleased with that level and with the demands of their jobs. For those women who rate their jobs as below "average" in complexity, they are more likely than others to be dissatisfied with that aspect of their jobs.

Other results of this study show that women who rate themselves as successful and intelligent tend to perceive their jobs as above average in total job complexity. In comparisons with ratings of these jobs by professional analysts, the ratings of the women in this study indicate that they perceive their jobs more favorably. The authors conclude that if there is to be an increase in the number of women in higher-complexity, higher-level, higher-paying occupations, structural changes in the American occupational system will be necessary as well as a redefinition of the job and the self by women. This statement in turn leads to the conclusion that there is a strong need to eliminate or overcome sexist socialization, and society should likewise be willing to provide equal pay for work of comparable value. To obtain these goals, women, too, will have to actively support the efforts to improve their roles in society.

RECOMMENDATIONS AND FUTURE DIRECTIONS: WHERE DO WE BEGIN?

The Need for Research

In her presentation, "Research on Women's Work—Some Missing Elements," which is not included in this volume, Virginia Novarra raises several issues that have an "important bearing on women as people and as workers." Her first consideration, a concern also voiced by others at this conference, is the fact that many women's activities which are considered economically and socially beneficial to the betterment of society tend to be overlooked or minimized in the world of economics because women receive no wage for them.   The most obvious of the activities are reproduction and the rearing of the next generation.   Another important economic issue addressed by Novarra is that of discrimination in the workplace which centers on two questions: (1) What is the explanation for the world-wide phenomenon of unequal pay or why is "women's work" universally undervalued economically?   (2) Besides the well-known economic advantages of employing women, what positive aspects do women bring to the labor force?   Few research projects have been designed to answer these two questions.

Another issue is the absence or paucity of women's presence in educative and training materials; that is, few women are featured in the case studies frequently used in management course work.   Remedying another societal slight centers on how to increase the numbers of women in administrative and leadership positions, particularly in those occupational sectors that either have been pioneered by women or are dominated by them, such as teaching, social services, personnel management, and nursing.   Finally, how can more teen-age girls be encouraged to perform at the highest levels of academic achievement and what is the explanation for the decline in scholastic performance among many teen-age girls?   Closely related to this question is the apparent lack of concern among many parents and educators for the education and training of girls for industrial occupations although women represent at least 40% of all industrialized labor forces.   As a resource, therefore, women have not been adequately channeled, trained, and challenged.

Research projects can be, and need to be, designed to address and resolve these work-related issues.   Novarra states that the position of women in society can be improved through the implementation of recommendations based on the results of effective research efforts--programs that extend beyond the threadbare documentations of women as disadvantaged and provide impetus for direct action.

Changing Sex-Role Patterns

The importance of changing sex-role stereotypes is a topic discussed in the chapter by Anne-Sofie Rosén, "Sex-Role Stereotypes and Personal Attributes within a Developmental Framework."   Rosén states that in Sweden equality has been a cherished principle for many years—and a legal reality since 1920 when legislation was enacted to incorporate sex-neutrality into all aspects of civil rights.   Even with a relatively long history of legislated equality, women's educational and labor force participation still is observed to be primarily in traditionally female fields. For example, Rosén points out that less than 50% of all gifted high school girls in Sweden aspire to a career that requires a university education. High school girls generally prefer "women's jobs" as their occupational choices: hairdresser, nurse, beautician, teacher, etc., occupations where 95% of the total number employed are women.   Further, there continues to be few women in high status jobs, which is reflected by the fact that only 6% of all lawyers in Sweden are women. As Rosén states, "the higher one goes, the fewer women one meets." Thus, despite the sex-neutrality in all aspects of civil rights, the opening of Sweden's universities to women in 1873, and the 1962 school reform to promote equality, these sex-segregated educational and occupational choices have persisted.

In an effort to determine why so little has been accomplished in narrowing this sex-role division, Rosén reports the results of her own and others' research on similarities and differences between women and men in perceived sex-role demands, values, and traits. In the Swedish studies, many similarities between the sexes are noted, particularly for such personality traits as dominance, achieving styles, and the importance of peace as a value. In response to the sex-role demand items, women indicate that they think there are strong demands in today's Sweden for women to be good mothers and at the same time to be self-supporting whereas men report the male role as requiring leadership ability and an achievement orientation.

The author concludes that greater equality between the sexes can be achieved if a concerted effort is made to develop and implement programs designed to change the behaviors of members of society. Thus, rather than continuing to provide funds for educational programs on equality, some of those funds should be allocated for programs that promote occupational shifts and erase existing sex-role patterns. It is only through such endeavors that changes in behavior will occur. Our children, in turn, will then learn through their own observations the true meaning of equality.

CONCLUSION

The chapters and summaries in this section focus on the evolution of women's participation in the world of paid work: the progress that women have made primarily during the past 25 years after having responded to the cyclical demands for their employment during the previous 50 years. In all of the countries represented at this conference, women have increased their participation in the work force not only in "women's jobs" but also as members of the professional elite. The authors report that women are participating in greater numbers in mid- and upper-level management, the prestigious professions of law and medicine, the political arena, and the decision-making and wealth-producing sectors. The authors also remind us, however, that most women work in low-complexity, low-paying jobs that have no career ladders and little status in society. In many developing countries, moreover, women form a new community of slaves in that they receive a very low daily wage from the multinational conglomerates that locate in the countries offering the lowest wage bid. In general, the trend is toward a higher rate of participation in most jobs and an increased proportion of women in the higher reaches of organizational and professional life though the growth rate at these levels continues to be slow.

While the overall gains have been substantial, several specific issues are identified in this section that affect women's participation and contributions in the workplace. These considerations include: socialization, education and training, value of women's work, tokenism or the numbers of women in a specific occupation, organizational behaviors and strategies, and discrimination.

All of the authors and discussants concur that the areas in need of greatest change are sexist socialization and education and training. To bring about changes in sex-role socialization will require the development and implementation of effective educational programs, cooperation from the advertising community, and support from political leaders as well as from men and women themselves. It is only through the results of such efforts that children will have the opportunity to observe both men and women in nontraditional roles and jobs. Rosén emphasizes that the most effective learning experiences for children are those in which they can observe for themselves and draw their own conclusions about the roles of women and men in society.

The second major issue is that of education and training. Included in this area is the need to determine why many teen-age girls have a tendency to "give up" academically, to assume that it is unnecessary for them to make an educational or occupational commitment, and to choose an occupation requiring little educational preparation. Girls should be encouraged, along with boys, to continue achieving in school and to pursue as difficult a scholastic program as possible. Hornig recommends that school officials should consider changing high school curricula to include required advanced courses in math and the sciences. With an increased exposure to these courses, perhaps more girls would become interested in pursuing an educational program leading to a career in a technological field.

Also to be considered is the fact that most high school occupational programs only train girls for such women's jobs as stenographer, typist, and bookkeeper. Many girls may assume that these jobs are their only career options, which suggests that the school programs on educational and career opportunities should be expanded.

For those girls who select a high school vocational training program, they learn upon being hired that their jobs typically have no career ladders and generally offer minimal increases in pay throughout a projected lifetime of work. Even if hired by a large company, most of these young women also learn that they are ineligible for the training programs offered by the organization. Thus, although the vocational training programs in our schools are beneficial for many girls and the organizations that hire them, the actual employment potential associated with this training should be evaluated more realistically—as the springboard to a career that in most cases offers few opportunities for advancement and few possibilities for providing an above average income.

For those women who choose to attend college, the number of career options increases, but most women still select the educational programs typically associated with women: library science, elementary education, nursing, etc. The jobs in these fields pay considerably less than the occupations selected by most male college graduates (e.g., engineer, systems analyst, management trainee). Increases in percentages of women in many of the high-paying occupations have been noted over the years; comparisons of salaries between men and women, however, reveal the widely documented wage differential.

The inequity in pay leads to another important consideration which is the relatively low value our societies place on much of women's work, both in the home and in the workplace. As categorized by Pintasilgo, the work that women perform is primarily concerned with meeting human needs: food provider, health care dispenser, and value giver. In societies that generously reward technologic and business-oriented jobs, women are economically disadvantaged because of their occupational choices.

However, rather than encouraging large numbers of women to abandon their low-paying "women's jobs," a preferable approach for both women and society would be to upgrade these jobs by providing the same wage as paid for jobs of equal value. During recent years, some improvements in pay scales have been achieved, primarily because of union negotiations on behalf of such groups as nurses, supermarket clerks, and civil service employees. Other groups may have to become as assertive in demanding pay increases although their requests no doubt will be countered with pleas to wait until the economies in our countries recover or until any number of future events occurs. Women can no longer afford to allow themselves to be sidetracked from their career objectives but instead must become tenacious in their demands for a higher income—and a higher value for their jobs.

Another important issue is that of numbers. Several authors advise us that until more women are employed in jobs where there are few women, their overall effectiveness will be compromised or lessened. Numbers are important if women's credibility is to be enhanced, whether in a boardroom or onboard a U.S. Navy tugboat. Thus, it is imperative to encourage girls and women to enter nontraditional fields or to choose careers that lead to positions of power and influence—all of those jobs that at present account for relatively low percentages of women.

In addition to the importance of numbers, we also learned that it is essential for women to exhibit the behaviors and strategies required to survive and succeed in organizational life. Although women have attained the credentials and have acquired the achieving styles needed to satis-factorily perform their jobs, they will have to engage in those corporate behaviors that are an integral part of the paths to power. Many authors suggest that women should begin by being more assertive, less accom-modating, and more concerned about self-promotion.

Throughout these chapters a common thread is woven in that women can achieve their career goals although these gains will require considerable energy, fortitude, and emotional stamina to overcome the inevitable obstacles of discrimination. It has become a well-known fact that men from North American and European countries earn more and are promoted faster and farther than women. Even with the enactment of equal pay laws in all NATO-allied countries, women have no guarantee that they will receive a fair wage or one comparable to their male counterparts. Passage of equal pay legislation lays the foundation on which fair practices can be built, but these laws will not be enforced unless women exert the effort to demand such action. Women, therefore, will have to push for equal pay; to establish themselves as winners in the battles for equality, which will serve as examples for future generations; and to provide support for other women who, too, will become involved in this struggle.

NOTES

1. Hewitt, 1974.
2. Ibid.
3. Gregory, 1974.
4. Ibid. See also Glynn, 1942 and Glover, 1943.
5. Horner, 1972.

WOMEN AS WORLD MAKERS

Maria de Lourdes Pintasilgo

Prime Minister of Portugal (1979 to 1980)
Lisbon   Portugal

The topic that I have chosen for my presentation is part of the quotation cited on the program for this symposium: ". . . the large-scale movement of women into the work force opens up the exciting possibility of creating a much improved society." The utopian title of my paper, "Women as World Makers," therefore, can be condensed to this quotation. I only wish that I had the facts, and the inspiration, to give flesh and bones to this exciting concept.

The title I have chosen might be misleading. When we speak of women as world makers, to some ears this statement may sound as if we exclude men. This is not the case, although they have had their opportunity now for quite a long time. It does suppose, however, that for the past few years, and from now on, women are the individuals who have the opportunity to shape up the world, even if we have been unaware of this occurrence.

The title also may appear ambitious, particularly in light of what we see as concrete facts. But it seems to me that unless we consider those facts, not only in historical terms about the past but also in terms of what we want the world to be, the efforts of women of our time, and of those who preceded us, were all wasted.

It is my belief that we are at the end of one type of civilization. We are watchers of the end of the empire era. When we look at history and see what happened with the end of the Ottoman and Roman Empires, which are so near to us in the western hemisphere (not to mention what has been going on in Asia for thousands of years), we may have a slight perception of what historians will say in 200 years about our time. When I studied in grammar school, I learned not only that Portugal ruled on five continents, but that the sun never set on the British Empire.   A completely new order has since developed, and we can see this very clearly when we consider that more than two-thirds of the countries existing today have been "born" since the Second World War. Just those facts, in strictly geographical and political terms, would connote the end of a civilization and the dawn of a new one.

Another important fact is that we were misled by a kind of universal myth in the notion of a technological world. A widely held belief was that as technical civilization spread around the world, different cultures would be in contact with one another which, in turn, would create a unity of nations. As we can see, this unity has not occurred. Rather, what the past empires have given to the new countries are technological tools. It has been through these technological tools that what is often called in United Nations jargon, "nouveau imperialism," has been formed. In spite of the spreading of technological facts and technological culture, divisiveness among nations has been increasing.

When I state that we are at the end of one type of civilization, I have to say very frankly that to discuss women and the world of work in such terms—that women will be nicely and fittingly integrated into a civilization that is fading away—is of no interest to me. Rather, I am deeply concerned with our contribution to the world that is in the making, to the future ahead of us, and with the steps that must be taken to establish our part in it.

EMPLOYMENT IN THE INDUSTRIALIST IDEOLOGY

We speak of the increase of women in the labor force and we speak about the consequences of this increased participation on economic life. In the economic report of U.S. President Ford for the year 1973, one of the five chapters was devoted to the participation of women in economic life.[1] What is remarkable about this fact is that an entire chapter would be devoted to this topic. Therefore, it seems that there is one fact that cannot be ignored anymore in any country of the world: The presence of women in the labor force has profound implications for economic life at both the national and international levels.

During this symposium, many questions will be raised and answered in connection with some already classic topics in this context: the double task of women, the division of labor in society, and the conditions necessary for women to gain their independence and gain participation in the economic cycle. Therefore, when I consider these issues and consider the infrastructures that enable women to cope with a dual role and with the division of labor, I know there is much to be studied in greater depth.

When we speak of work, we speak very often as if work and employment are the same. I noticed this even in the documents of "The Mid-Decade World Conference of the United Nations Decade for Women," just held in Copenhagen. Several outstanding women, however, have attempted to clarify this in a U.N. publication, "Women at Work." [2]

Krishna Patel, for one, very clearly states right from the beginning that "work" and "employment," in actuality, are two very different terms and relate to different conditions. But, very often we speak about women and work in outmoded terms.

My basic assumption in this introduction is that, regardless of the many problems still arising from the perspective of work being equated with employment and the need to study those problems, we can say that there already had been a great leap made in quantitative terms by women in the world of work. So much so that this quantitative leap will give rise to new qualitative questions. Women are so involved in the world of work that new approaches and other outlooks can already be seen.

My premise, therefore, can be formulated in this way: The relation of women to work is not necessarily equated with women who are employed, but rather with women engaged in any activity in which they perform a personal or social function. Such activities spring forth from the personality of the individual and fulfill the many needs of the society in which she lives. In an eloquent statement, Krishna Patel points to this very clearly when she says, "Most women are permanently working, but are not permanently employed in the labor force."[3] And this statement, I think, does show the fundamental contradiction in the ideas attached to the word "work."

We have only to look at the basic difference between the broader meaning of work and the more common notion of work as employment. Employment, of course, is related to paid labor and, therefore, is recognizable and measurable. The term "employment" belongs to the world of defined quantities and, although that definition is sometimes rather downtrodden, it is appropriate. The equation of work with employment is at once a result of what I would call a strictly technical approach to labor and of what might be called a widespread ideology. In fact, the technical concept of "labor force"—a term commonly used—is the direct use of human potential in any process of transforming and processing goods and/or rendering services. This concept of labor force, together with ownership of wealth and property, forms the bulk of what is brought together under the umbrella of economic science. This technical approach, as an interpretation of the process of creating wealth and exchanging it by monetary means, has much to do with what we understand as "economics."

This concept is related to what I term a widespread ideology. It may come under the label of profit, if we are in market economics, or under the label of economic growth which was initially a concept from the planned economies, but has now become a term used by everyone, everywhere. Thus, both profit or economic growth are indeed the by-products of what I call the "ideology of industrialization" or "industrialism." One of the beliefs that this ideology generates is that technology—all kinds of technology—has a magic power which necessarily contributes to progress, well-being, and justice.

The Economics of Technology

I believe we now have many questions connected with technology and the impact of technology on this world of ours. Whatever the political regime may be, this industrialist ideology is always associated with, and presupposes the existence of, a very strong concept of nation-states. Thus, industrialism is something that concerns not only the way in which goods are produced and exchanged, but the way in which people relate to each other and create the social structures within which they live (the production of goods being just one aspect of this relationship).

We can say that the framework for this ideology has been created by what has happened in the world during the last 200 years. We have, on one side, the bureaucratic or centralized state where decisions are always made according to a well-defined hierarchical system and where the state is absolutely overpowering to the individual or any smaller institutions that individuals may form. On the other side, we have several developments that at times are almost absolute. One example comes from the situation in which nuclear energy, an awesome phenomenon that is capable of shaking us to the very basis of life as we know it, can be totally controlled by one single man. I recently read a fascinating book written by Bertrand Goldschmidt, the former Commissioner of the French Atomic Energy Commission. He tells how, in the late 1930s and the beginning of the 1940s, a group of scientists from Canada, France, Britain, and the United States conducted their research. It is very enlightening to read of his own experience at that time of working with two others in a very small research unit. He knew absolutely nothing of what the other scientists were doing, even though they were just next door. Indeed, the only person who knew everything was President Roosevelt. The author does not attempt to prove any particular point in the political field; however, what emerges very distinctly is the tremendous concentration of power held in the hands of one single man and the enormous responsibility he held in relation to the whole world—just because one technology, and the dangers inherent in its development and use, evolved in a particular direction.

We must turn again to the nation-state because it is this state today which has the capacity to use the labor force, to control the ownership of land, to at least guide the concentration of capital, and, finally, to determine if it is possible or not to be in the forefront of technological achievements. When I describe the nation-state in this way, which affirms that all nations are ruled by the industrialist ideology; which stresses that the technical interpretation of work as equated with employment flows from such an ideology; and which combines into an equation the relationship of women, work, and employment, I have to say: "But almost all of the nation-states are governed by men." Therefore, how can we arrive at another interpretation, which is closer to the situation women are in? The employment that is part of the economics of technology, all that is related to the employment of women, is governed and decided upon by men. When we speak of women's employment in this context, we must say that it is an empty expression that "women control their own lives." Employment, economy, control of the nation-state, the power of technology, and women's situation must all be considered together.

The Hidden Unemployment of Women

One related aspect that Kathleen Newland[4] stressed in her book, Global Employment and Economic Justice: The Policy Challenge, is quite interesting. She states that by some perverse logic those who cannot make it into the labor force cannot be considered unemployed. Because women are impeded from entering the labor force, they cannot receive the same benefits as those who are considered unemployed. This example is a very simple way of showing that the machinery has somehow broken down; somehow this logic with regard to women's situation is not what we think it should be. In even more current terms, we can say that economic growth and all the ideological values associated with it are not, and can never be, equated with better employment opportunities for women. But it is my conviction that instead of just correcting the perversity that Kathleen Newland denounces, we also must move toward a healthier way of understanding women's work. For me, this is a very important point. In listening to the comments about the different scientific programs sponsored by NATO, I was struck by the fact that perhaps we are at a point where what we so much hope will happen in the world during the cross-fertilization of different types of sciences may actually occur.

THERMODYNAMIC APPROACH TO THE CONCEPT OF WORK

The concept of work equated with employment is related to our outdated concept of physics. We may not be aware of it, but we are still living in Newtonian times; we are living in an era based on mechanics. Thus, when we speak of work, it is just of employment, just a mechanical interpretation of reality. But nothing in the mechanical approach is said about the subject of work or about the change in the object of work. I think it is all very disconnected from the global process of energy affecting the whole cosmos, which influences all of the changes that are taking place in the world today. In fact, we can say that it is impossible to think of work from other than a thermodynamic approach. Such an approach centers on the exchange of energy: not only the energy that is within a subject, but energy that is internalized in the very process of producing work (in physics, simply called internal energy), and also that energy which is utilized. This thermodynamic approach is really a new concept, which in a way is keeping pace with what happened 200 years ago and is essential for another understanding of work.

The preparatory work of the U.N. conference in Copenhagen made clear that women work, expend energy, and enter into an exchange of energy. However, because of the mechanistic approach in bureaucratic thinking, women are statistically and socially invisible. Although it is true that they are active, most systems of statistics do not take an accurate count of their actions. They are deeply involved in vital processes—vital not only to a few individuals but to individuals in a collective sense—and yet sociologically as well as politically their involvement is totally overlooked as if it did not exist.

In going through the documents from the Copenhagen conference, it is remarkable to observe how the world has changed in the five years since the Mexican conference. The situation certainly is not optimal, but coming from all points there is at least the acknowledgement that <u>women are working</u>. This acknowledgement occurs despite the fact that in many countries rural women, working in family-operated endeavors, are seen as nonactive and are not included in the statistical reports.

This acknowledgement, which has been made in nearly all of the documents available at the international level, is extremely important. Now the important question is: How and where are women working? It is true that women work, but their work is still not included in the category of "employment." They fulfill what might be called "some kind of activity." (I look to the social scientists to find a proper word for what women do.) And yet, despite the inadequacy of words, their activities are the fabric of the existence of all human beings.

These activities have been included under three main categories. In both developing and developed countries, women are food providers: they grow food, gather food, process food, and distribute food. They are value givers: they teach law, science, behavior, and personal and communal history. They are health care dispensers: their knowledge ranges from the most elementary forms of hygiene and nutrition to the various specialized, professional health care levels. In other words, <u>women are at the very core of the most fundamental conditions of living</u>. Unfortunately, these activities are "invisible" when included in conventional calculations, but they are vitally important to the economies of nations.

We know from studies conducted in several European countries that the "invisible" work performed by women, if translated into monetary value, would account for one-fifth to one-third of the Gross National Product (GNP) of each country. This translation, if used, would result in quite an upheaval not only in abstract and figurative monetary value but also in terms of hard currency.

Perceptions of Women's Role in Society

In their activities, women are more involved than men with meeting the basic needs of human beings. If we compare these activities with economics, we perceive that another set of values is at stake. Economics, as such, is no longer the issue: human beings and their needs are. However, the role performed by women is not considered in the different economic systems. Our national planners are neither sensitive to unquantifiable realities nor to basic human needs per se. What women do is then put aside. Thus, when women are channeled into what is called "productive tasks" by national planners, they still must perform their other work, though it continues to remain unrecognized by these same planners.

It is my conviction that one day the satisfaction of basic human needs will become the concern of the nation-state and the powers entrusted with its care. But, as things stand now, it seems very unlikely that this will be so in the near future. The nation-state is far more concerned with its monetary or financial equilibrium than it is with meeting basic human needs. I am not disregarding the importance of the financial system, but this system has to be of service in meeting human needs. (If we speak only in terms of the monetary system, some countries would be almost nonexistent.) Therefore, let us be honest with ourselves and ask: Do we want to use the monetary system as an instrument—a tool in the fulfillment of human needs—or do we make it a goal in itself as many politicians tend to do?

Towards a New Concept of Development

What we need is another concept of the evolution of society and even another concept of development. I purposely use the word "development" because all too frequently we use this word only in reference to poor countries, many of which are not developing. I would like to stress, and the Brandt Report is very enlightening in this respect, that either we all make it together or no one is going to make it at all. Therefore, what we need is another concept of development that is global and all-embracing. I would like to quote the Director of Women's Office of the Lutheran World Federation because she has very clearly expressed my point:

> When we come to formulate a new, future-oriented conceptual framework for development, we should not only ensure that this time women are not excluded, but also that development is defined as human development. By this I mean, the objectives of development have to be: (a) the development of human beings and not that of things; (b) the improvement in the quality of life by satisfying those basic material needs such as food, shelter, clothing, education, medical care, and so on; and (c) the attainment of such basic nonmaterial needs as independence, identity, autonomy, creativity, self-fulfillment, and so on. The development of the structures of institutions, even if they are technological or monetary, should serve only as a means to the above objectives, and never as an end in itself.

I think this is very much the way women are thinking in today's world.

EMPLOYMENT AND ACTIVITY

The issue of society as a dynamic reality is presently at stake. Work performed by women, in terms of meeting human needs, has a much more direct relationship to society as a whole than it does to the nation-state. However, when we place ourselves in the logic of the employment context, we can see that to a great extent much depends upon the political will of the state and its concerns with powers, policies, and plans for implementation. Currently, we are noticing a sudden improvement in women's access to better employment. I cannot say improvements have been made in equal promotion, because this just is not true. We are all fighting for equal pay and for better training in the fields of traditional employment. This situation is a very sad plight, is it not? There are numerous areas for improvement and these need to remain the goal for all of us while, at the same time, we must work toward other concepts of women's roles and work.

Women as a New Community of Slaves

I would like to reemphasize the fallacies concerning employment. Women are not only workers with a double task but, around the world, they are becoming a new type of slave. Women workers in the electronics industry are a striking example of this new slavery. A manufacturing firm, established in my country in 1974, had been forced to close one of its branches in another European country because studies showed that by the age of 30 women employees permanently lost 50% of their eye sight. The turnover of employees was about 50% per year. Of course, the country where this originally took place is a country with great concern for the social well-being of its citizens which resulted in the relocation of the plant. I very much appreciate the efforts of that country's officials to protect workers. The factory moved to Portugal because female labor was very cheap, and the same cycle was repeated. For 9 hours a day, women were required to look at one spot on a television screen or to check one small cord to see that it matched the model exactly. During the 1974 revolution, however, the minimum wage was increased and, when it was no longer highly profitable for that international corporation to operate in Portugal, the plant was established in Thailand and the work is now being done by Thai women, who are being subjected to the same hazards. What is being stressed here is that women around the world form a community of slaves of a new and different type. If one group of women is freed from a difficult chore, the task is certain to be handed over to another group somewhere else in the world. It seems to me that this is just one of the ironies in the employment of women when viewed at the international level. Kathleen Newland makes a very good point in her book when she concludes that the fuel of many economic processes is not oil, but cheap female labor.[5] This statement is not just a good mass media remark, but also leads us to think about what would happen if a revolution similar to the oil situation occurred in the cheap female labor market. If, for instance, women around the world were totally unionized, regardless of national boundaries, and they all immediately demanded equal pay, enterprises suddenly would be forced to function in completely different ways. The redistribution of income and the accompanying increase in buying capacity of more than 40% of the world's population could very possibly lead to far greater inflation than we have at present. And I wonder what one individual would be powerful enough to deal with such a situation—not to mention those who would be the minor stars in the galaxies of power?

The Cultural Impact of Women in the Labor Force

Thus, the activities of women have an enormous potential for changing the world's economic structure. More important, however, is the fact that this increased activity affects culture as well. I think that what we are all looking for is the interdisciplinary approach to life or to any activity that will increase life's value for us. It is rather a platform based upon the general culture, for culture does spring forth from the same basic laws and the same basic inspirations as the different professions.

I think that it is very important to view all activity as work, to see new relationships in work. This concept means that if work is an activity—and not just employment—then work is a source of culture. It enriches the quality of life, adds to what the world already has, and brings about a new interdependence among human beings. It brings about freedom in relation to institutions along with a mobility of mind and body for the individual.

This attitude of activity as work means that an individual does not necessarily have to go up the ladder—one simply does not care about the ladder. Although a person may get to the top; then again he or she may fall. Rather one moves horizontally because it is healthier to move that way, which would eliminate notions of promotion and the importance of status. However, one gets somewhere or something else in return.

Of course there are pitfalls and I do not want to overlook them. We know that people very typically respond to such changes by doing nothing, by losing their own centeredness, or by trying to create some guaranteed sources of security in a kind of cloistered environment.

POLITICAL OUTLOOK

I would like to bring this to a more political level. When we speak of the world of work, are we talking about the world of work which accepts without discussion the North/South divison of labor? Or are we talking about the world of work which takes into account all work performed by women everywhere in those endless activities and which Elise Boulding has not hesitated to call "the 5th World"?[6] Is the world of work comprised of the categories of primary, secondary, and tertiary activities as they are still commonly labeled and still are generally taught, or are we also taking into account what is being called more and more often the "quaternary sector," with activities that belong to a

totally new realm? In this context, there are work problems that have
political implications of their own. From what I have read in the papers
and from what I sensed in the short time I was in Copenhagen, it seems to
me that women have not yet been able to bring to the fore their own
issues as political issues. Rather they have been carried along by the
issues that are important to men, which, of course, all of us have to deal
with, but for which there are already adequate forums. It appears to me
that what women are doing is playing—indeed mimicking—the same songs
that men are playing. Perhaps they are singing an octave higher because
of the range of our voices, but they are doing the same things as men and
as a result are not bringing forth as political considerations those issues
that stem from our experiences as women.

Let us take as an example some of the countries which have
experienced immigration, not at the level of totally unskilled labor, but at
the intermediary level—the level of technicians. In Saudi Arabia and
other countries of the Persian Gulf, and in Venezuela, these immigrants
are not the "sweepers of the streets" that you find in London, Paris,
Amsterdam, Hamburg, and, perhaps, New York. The immigrants to the
countries I have mentioned are at the intermediary level of skill; they are
technicians. When they enter these countries as "guest workers" and are
employed by private enterprise, they ask, and are asking more and more
often, to have key positions. And as these men become more important
personnel sources in strategic industries, the adult women of those
countries are practically excluded from the labor force and from the
intermediary level jobs. Now, to whom does one give priority? This
question is a political issue, the brunt of which is felt by women, although
it is not, in the strictest sense, a woman's issue. It is instead an issue
concerning the division of labor.

Another example is the introduction of appropriate technologies.
We know that they are fundamental to the development process, not only
in the newly emerging nations, but in the developed world as well. I had
an opportunity some weeks ago to visit several industrial complexes,
designed for the future, on the East Coast of the United States. I saw
many wonderful advances in technology being used there which will make
a tremendous difference, not only in the United States itself, but in
developing countries as well. These new changes will be most effective
when we stop thinking that all countries have to go through each of the
various technologic steps of the post-World War II era one by one to arrive
at the current level. Appropriate technologies, I think, are more and
more what is needed to cope with basic human needs but, because these
technologies touch and simplify basic processes that serve humanity, they
tend to attract the interest of women rather than men. For this very
reason, they tend to create new problems. The basic processes and their
attending skills have long been the domain of women. Therefore, if
changes to these technologies are required, implementation of such
changes should be left to women. Unfortunately, there are inherent
dangers to women in this policy: Men will go on with the most advanced
technologies, leaving women in ghettos once again and, this time, without
even their original skills to accompany them.

Let us now tackle the question of women as food providers. I am very impressed by the American book, How the Other Half Dies, by Susan George, published in 1977.[7] In the book there are numerous figures which show that food, when used as a weapon, can be extremely useful—as well as being extremely alarming. If, as statistics show, women are indeed food providers, can women rethink the production of food? Can women change a weapon? I am neither exaggerating nor using a metaphor. I am recalling the address President Ford made to the U.N. General Assembly in 1974 after the oil crisis. At that time he said that, because the oil-producing countries were using oil as a weapon, we may as well use food and wheat as weapons. Therefore, I am not just speaking of a metaphoric reality; I am speaking of something that can occur and is happening in the world today. Can women be the promoters of a totally new redistribution of food?

We can note that the foodstuffs produced by the agricultural industry in France to make pet food for dogs and cats (8 million dogs and 7 million cats) would sustain the entire population of my country.

Can women stop people from eating too much? It would seem that people in the United States cannot eat more than they are eating now. Are women ready to learn how to grow—and use—other foods? I think the relearning process is going to be a major task, and it is not just a humanitarian one: it is a basic task required to attain world peace.

I also have spoken of women as value givers. Can women help pass on values in a political way? This is another political issue that relates to science, codes, behavior, and history. It is a fascinating question that hardly anyone asks. Women hold the majority of teaching positions today; and yet, is there anything new in education? Where is the radical change? Where is the seed of the new world we want to bring about? I think that women in education have the opportunity to introduce some of the radical changes so badly needed. We do not need teachers who merely transmit information—computers can be used in their place along with all of the other forms of information available. Computers serve very well as transmitters of information. We do, however, need women who can be something else: educators and value givers. Has that political revolution even started? I do not think that it has.

FUNDAMENTAL POLITICAL QUESTIONS

On a more global basis, a new international development strategy is in the making. I, of course, am very concerned with political questions. This development will be discussed at the next session of the U.N. General Assembly in September 1980. However, after reading the preparatory documents, I can say that I think people will be disheartened to find that no drastic change has occurred. The same old concept of development is at work. The rich countries will be asked to give approximately 1% of their GNP as aid to developing countries. (It is no longer called aid for development, but something else to make it more palatable.) The coun-

tries who need to increase their wealth are to be stimulated to have their
GNP also increased by a certain amount. These premises are the ones
that have been, and still are, operating. Moreover, I believe that
everything is relegated to the level of negotiations among states. As
positions get stronger, issues get weaker. Ideologies or interests become
the paramount element while the human dimension becomes totally
blurred. I then ask myself, if the dimension of issues negotiated at the
international level is irrelevant, as some politicians say, then tell me what
is relevant in the world? What appears to be lacking is continuity from
domestic national policy to international policy and vice versa. If
North/South negotiations and a new international strategy for development are lacking in human dimension, it is at the domestic level that the
women's movement and women's work can be of vital help. I am stating
this within the context of NATO because I am very much in accord with
the report of the Brandt Commission. Again, I will state that there can
be no solution within the context of the NATO alliance, of which my
country is a member, if there is no solution for the world at large.

Women in the Labor Force in Third World Countries

Indeed, women workers of the Third World make up two-thirds of
the world's female labor force. Their skills overlap with ours to some
extent, although they do not necessarily coincide with one another. By
the year 2000, they will comprise three-fourths of the female labor force.
Do we go on perceiving the situation of this two-thirds (in a few years to
be three-fourths) as marginal and something we have to rise above—as
something outside our interests—or do we make their concerns ours?
Have we as women of the northern hemisphere progressed far enough in
our research to enable us to link our experiences with those of women in
the southern hemisphere without thinking that all women have to go
through all of the same stages we have? I believe that women of the
northern hemisphere are aware of being part of the work force although
we are separated in many other activities. Basically, we do realize that
we cohere together, not necessarily because we are women, but because
we are part of a group that is somehow oppressed. If we can reach out
with a global strategy to women in the southern hemisphere, together we
could make the first steps toward lasting peace.

WORK, PERSONAL FULFILLMENT, AND THE FUTURE

In addition to these political considerations, my other concern is on
a very personal level. My assumption is that work has no meaning if there
is no personal fulfillment and happiness in that work. Work in the
industrialized world, unfortunately, has become for all too many a
substitute, an ersatz, for affective happiness. Prestige and power have
become an integral part of its pattern. I am not denying the need for
self-esteem through the association with others, but knowing and becoming one's self has more to do with personal liberation and with the ability
to make one's own activities. As I pointed out earlier, many women in the
so-called world of work are so utterly carbon copies of their male
counterparts it is not strange that they are considered nonexistent.

But, of course, no woman can change her work just because she is a woman. A long process of probing into oneself must be undertaken along with the outward-oriented and political changes I have mentioned so far. The process may vary from one woman to another, but it is the exploration of one's own roots and one's own aspirations that is the deep adventure which may lead us to another concept and another possibility of women and work.

One of the most important contributions of the women's movement in our time has been the painful, yet revealing, process of self-awareness--not of one's self as a static human being, not as some ideological finding yet to be discovered, certainly not as a self-pitying romantic simply reawakening the past. Instead, the process is an energizing factor of one's own future in that one is always becoming something else, a path to wisdom.

Today women in the world of work have to cope first of all with what they want from their lives—their happiness, their affective lives. Perhaps they should ask themselves if they just want power, as some men do, or if they just want mothering, as some women do, thereby transposing to work the patterns they have known in other spheres. The major question of personal development is, I believe, the path to wisdom, to a liberated self in harmony with a creative world. How else is it possible to work?

I think the idea of mastering the world and mastering nature has disappeared. Ecology has suddenly become of more importance. We are not now considering mastering nature; our ultimate goals are to live in communion with nature. What radical change does this bring to our own fulfillment in our work? We are in a new acknowledgement of others, an acknowledgement in which the feelings of others are sometimes much more important than rationalized statements. I have just come back from a small town in the north of my country where I spoke to a huge crowd packed into a small hall. At the back of the crowd, there was a very young woman, in her early 20s, with a small baby in her arms. At first she could not see because the crowd was really huge, but then she found a table and climbed up on it, and there she stood for more than 2 hours with the baby. Everything that was said which she agreed with was really an acknowledgement. The fact is that this woman may be illiterate, since in my country 26% of the women are still illiterate, but nonetheless there was present this mutual acknowledgement that is transmitted between people, regardless of their positions, which I think is tremendously important in that "pass to wisdom."

CONCLUSION

We need a capacity for wonder before beauty and newness, to enter into what a French writer called "the circle of charm." We ought never to be afraid that charm <u>really will encircle</u> us for it is at that very level of charm and desire that we can meet the fundamental roots of our happiness.   Present in all of this is a tremendous search for meaning because without meaning work may become just a chore.

What I want to say then is that in a symposium such as this one the seeds for the future are deeply rooted in a communion of thought and feelings. I once read in an anthology of poetry that came from Greenwich Village several lines that moved me very deeply: "What are we? We are just men and women of average height orbiting carefully into tomorrow."

NOTES

1.   Economic Report of the President of the United States, 1973.
2.   Patel, 1980.
3.   Ibid.
4.   Newland, 1979.
5.   Ibid.
6.   Boulding, 1977.
7.   George, 1977.

THE EDUCATION AND EMPLOYMENT OF WOMEN SCIENTISTS

AND ENGINEERS IN THE UNITED STATES

Lilli S. Hornig

Higher Education Resource Services
Wellesley College
Wellesley, Massachusetts   U.S.A.

This chapter focuses on a relatively small subset of the intellectual elite: women scientists. These women are products of advanced specialized education and are engaged in responsible and rewarding careers. What can we gain from examining the situation of such a highly distinctive group? The study of these women scientists allows us to delineate the fine structure of sex discrimination with exceptional clarity primarily because their training and work experience are so different from most women's and because they also share the strictly gender-related problems which are common to all women in the paid work force (i.e., household work, child care, occupational segregation, the marriage tax penalty, and others). Further, the scientific professions are characterized by clear standards of achievement and widely shared values; therefore, it is possible to identify quite accurately the ways in which professional outcomes for women depart from the norm of male expectations and the extent to which they do so.

What allows us to examine the details of sex differences in science with great precision is the existence of two unique national data sets in the National Academy of Sciences. One of these is the Doctorate Records File, an annual census of all earned doctorates in the United States which is almost 100% complete. The second is a series of biennial employment surveys of doctoral scientists and engineers, collectively called the Comprehensive Roster, which is based on a stratified sample of the total doctorate population. The most recent survey includes 11.8% of the total of nearly 440,000 Ph.D.'s.[1] Two major reports derived from these data sets and dealing with women scientists and engineers have been published by the National Academy of Sciences' Committee on the Education and Employment of Women in Science and Engineering and these form the primary basis of this paper.[2]

35

Generalizations about women scientists abound not only in conventional wisdom but also in the social science literature; one of the most persistent beliefs is that there are almost no women scientists. Other generalizations concern high attrition rates in graduate school, abnormally long years of training, lack of drive and persistence, and inadequate performance compared to men. An objective analysis, however, reveals that none of these beliefs is grounded in fact.

Indeed, it is true that women are a relatively small proportion—currently about 10%—of the total doctorate labor force in the sciences and engineering. Women also constitute a comparable proportion of most other highly trained professional groups, such as physicians and attorneys. Across the whole spectrum of academic disciplines, only in the humanities are women represented at approximate parity with men. If we consider the numbers of women who choose particular fields, however, we find that in 1979 nearly 4,000 women obtained doctorates in all of the sciences, but only about 1,500 women earned doctorates in the humanities.[3] In comparison with other professional women, women scientists as a group are not the hardy and somewhat peculiar band of pioneers that is the stereotype; what makes them a small proportion of the work force in these fields is simply that men traditionally have made, or been encouraged to make, different educational choices. That is, most professional men enter scientific and technical fields whereas only a very small proportion go into the humanities. The unusual thing about women scientists is not that they are scientists, but that they have chosen a demanding career at all.

What causes men and women to make such different educational choices, determining at an early age that men's work opportunities throughout life will be better, more varied, and more highly rewarded? Have the influences which shape these decisions changed in the last decade or two? Are women's educational and career choice patterns evolving in the direction of greater equality? Have patterns of employment begun to reflect society's increased sensitivity to sex bias? What are the prospects for equity in the future?

These concerns are of increasing importance to the eventual attainment of equality for women because of the fundamental roles that science and technology play in the development of nations and in their social, economic, and political well-being. Unless women in greater numbers have the specialized education and the employment opportunities that will permit them to participate fully in the scientific and technological discoveries which shape the course of national life, they will continue to remain second-class citizens.

Most of the detailed findings to be discussed in this chapter deal with holders of doctoral degrees, both because they are the pacesetters in science and because they are the population on which we have reliable data. Some familiarity with science, however, is increasingly necessary in almost all occupations and at almost all levels. Even some of the most stereotypically female occupations have become far more technical: nursing, clerical work, data processing, and librarianship. An emphasis on broad preparation in science and mathematics must reach well down into the schools if women are to even maintain their status in those fields.

Before examining the status of women scientists and engineers in greater detail, we must note that because formal education in these fields is necessarily sequential, it normally begins in early adolescence and, thus, requires a preparation time of about 15 years or more to produce a professional scientist. Even though the proportion of women doctorates is growing rapidly each year in most scientific fields, the composition of the overall work force has not yet changed dramatically.

EDUCATION

In broad outline, there are few discernible differences in the education of boys and girls before high school. Once in high school, however, their paths begin to diverge. More boys than girls take math throughout school, and more of them elect science courses, especially chemistry and physics, although these differences are narrowing quite rapidly at present. In courses of language and literature, boys generally do less well than girls. However, English remains a required subject in high school thereby more or less forcing boys to acquire some competence in it whether they wish to or not. No corresponding math and science requirements assure a level of competence for girls in these fields.

The choices of college majors and, in fact, the type of college attended differ accordingly between the sexes. Many more men than women attend college in research universities, where both a wider choice of science fields and more advanced work are available than in the institutions typically attended by women (Table 1).

To what degree this unequal access to science education is really a result of free choice, and to what degree it is determined by institutional admissions patterns, remains a subject for investigation. Some of our leading research universities continue to have a stated policy of sex-biased admissions; Harvard, for example, admits men and women in a 60:40 ratio. Women's access even to undergraduate science education, thus, remains limited at the present time, and the numbers of women doctoral scientists will reflect that access barrier for at least another decade.

At the baccalaureate degree level, men and women tend to select different fields of study, with the surprising exception of mathematics where, as shown in Table 2, the percentage for women is approaching 50. In part, this difference occurs because the colleges women typically attend often do not even offer a professional education in some important science fields. These disparities ultimately result in women earning a relatively small proportion of the Ph.D.'s in most of the physical science and engineering fields, but much larger percentages in biological and

TABLE 1

PERCENTAGE DISTRIBUTION OF WOMEN AMONG
POSTSECONDARY INSTITUTIONS FOR FIRST-TIME FRESHMEN,
FALL 1977

First-Time Freshman Women	Percentage
Total in all institutions	52
In public institutions:	
Universities	48
Other 4-year	53
2-year	52
In private institutions:	
Universities	44
Other 4-year	51
2-year	62

SOURCE: Calculated from Tables C and D, p. 24, <u>Fall Enrollment in</u>
<u>Higher Education 1977</u>, National Center for Education Statistics, 1979a.

TABLE 2

RECENT TRENDS IN PERCENTAGE OF WOMEN AMONG BACHELOR'S
DEGREE RECIPIENTS IN SCIENCE AND ENGINEERING

	1971-72	1973-74	1975-76	1976-77	1977-78
Mathematics/statistics	39.1	41.0	40.7	41.6	41.3
Computer sciences	13.6	16.4	19.8	23.9	25.8
Physical sciences	15.1	16.6	19.2	20.1	21.5
Physics[1]	6.9	8.5	10.9	10.5	....
Chemistry[1]	19.4	20.0	22.4	22.8	....
Engineering	1.0	1.6	3.2	4.5	6.7
Agriculture	5.5	9.7	18.3	22.2	24.6
Biological sciences	29.6	31.5	34.8	36.4	38.7
Social sciences	38.5	40.2	42.6	44.4	40.6

[1]Percent statistics were derived from Tables 112-113, <u>Digest of Educa-</u>
<u>tion Statistics 1979</u>, National Center for Education Statistics, 1979b.

social sciences, i.e., the less quantitative fields (see Figure 1). Note, however, that the proportion of women has been growing very rapidly over the last decade in the most nontraditional fields—engineering, agricultural, and computer sciences.

One of the clichés of higher education is that women drop out of school with great frequency. The facts for women scientists in general do not support this view, as shown in Figure 2. Attrition at the graduate level is significant only in the social sciences and mathematics while women baccalaureates in the least traditional fields, such as engineering and agriculture, are actually far more likely than men to complete a Ph.D. program. The high attrition rate in math probably relates to specific employment patterns in this field as well as to sex-biased departmental traditions and climates.[4]

Significant Trends in Higher Education of Women

This brief overview marks a continuation of long-term trends in the higher education of women as well as some significant shifts. The long-term trend is that more and more women are completing degree programs; currently women are a slight majority of college freshmen and a slight minority—about 47%—of baccalaureates. There is also a gradual shift into nontraditional fields, women are almost one-third of all doctorates, although we have seen that in some science fields, their participation is much lower. These shifts, however, are not only in the direction of greater participation in science per se, but also in other nontraditional fields that have quantitative components, such as business and communications. Further, it is important to note that the changes are occurring simultaneously at all degree levels: growth in the proportion of women Ph.D.'s is paralleled almost equally in magnitude at the master's and bachelor's levels. This finding strongly suggests that the reasons for the increase lie in a fundamental shift in social attitudes and women's own aspirations and expectations, rather than in any specific institutional change. If that is true, it is an important policy consideration to which I will return later.

Quality of Women Scientists

Before considering how women scientists fare in the world of work, we need to know how good they are. The fact that they have received the same training as their male counterparts does not necessarily define their level of achievement.

"Quality" of scientists is an elusive charateristic, one not readily measured, and certain proxy measures can be used in addition to such direct ones as grades earned in coursework, where women tend to do better than men. At receipt of the doctorate, women scientists hold a slight edge over men in all of the common proxy measures: They earn their Ph.D.'s as fast as men or even faster; equal or higher proportions of women have completed their work in excellent or outstanding depart-

FIGURE 1

RECENT TRENDS IN PROPORTION OF WOMEN PH.D.'s IN THE SCIENCES

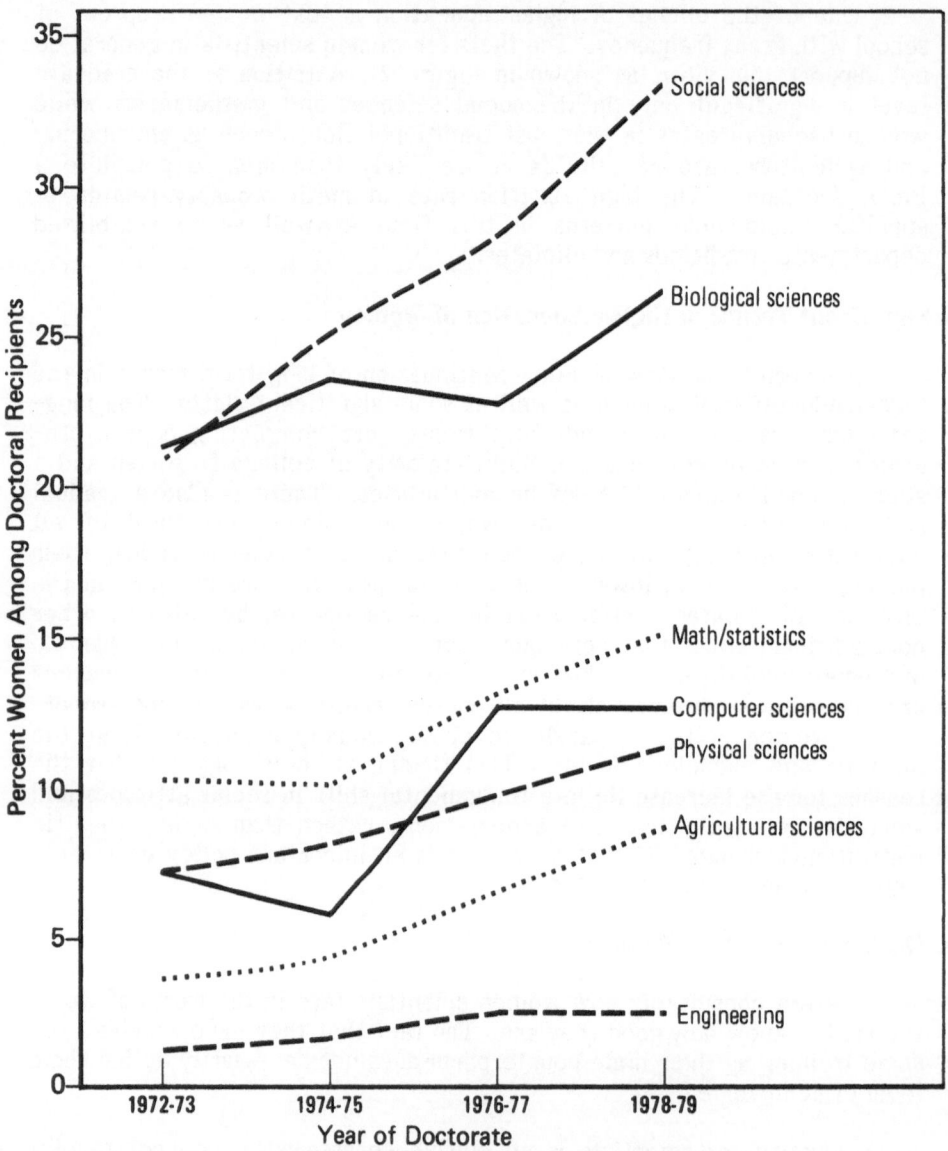

FIGURE 2

PERSISTENCE AND ATTRITION OF WOMEN
IN SCIENCE FIELDS

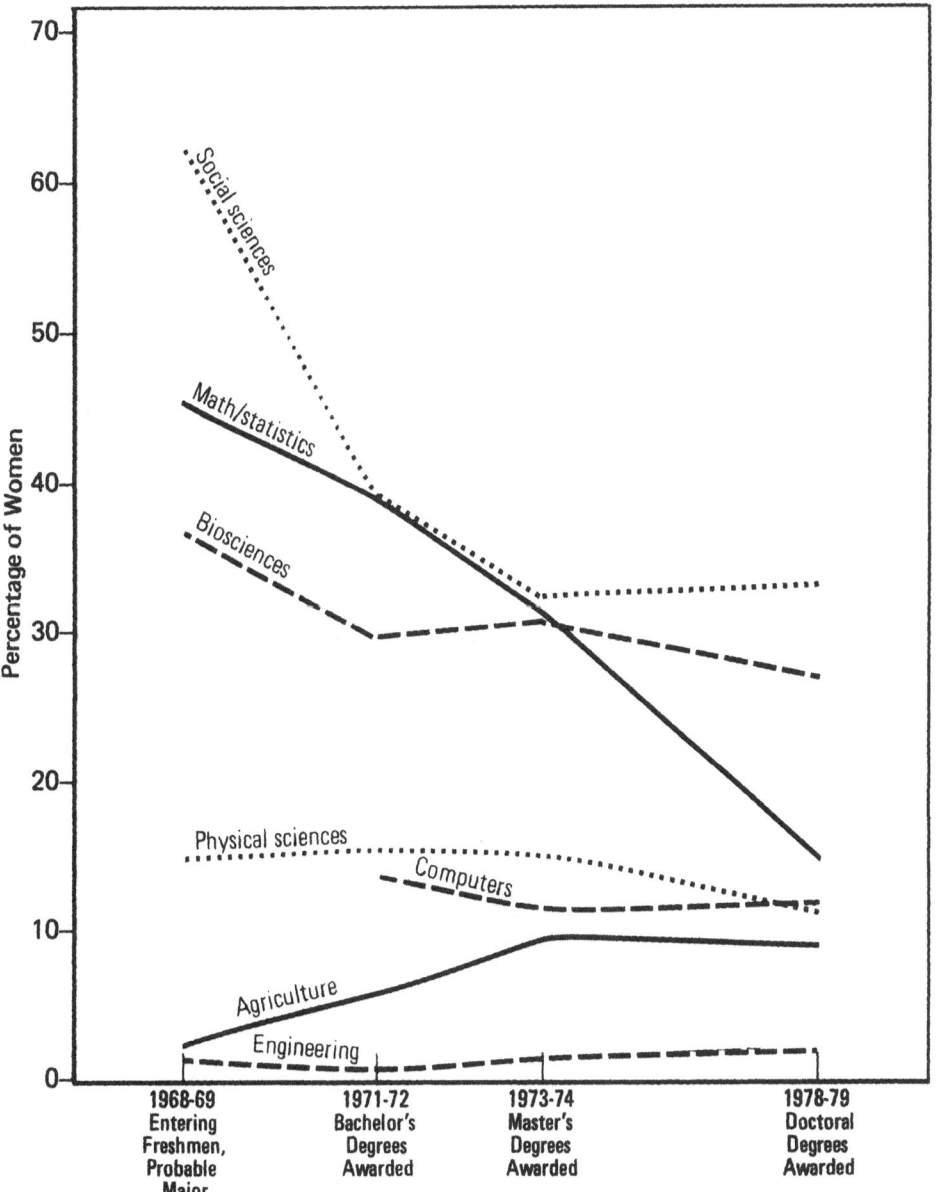

ments; they are slightly younger than men; and they are obviously very highly motivated.[5]

EMPLOYMENT PROSPECTS

Because the training in science or engineering is sufficiently specialized and demanding, few students, male or female, undertake it merely to broaden their intellectual horizons. They have jobs in mind. Do women find the same employment opportunities and rewards as men? By and large the answer is no, but the changes we observe during the past few years are both interesting and significant.

Before comparing the more detailed conditions of employment for male and female scientists, we should note that both the participation rate and level of persistence in the labor force are exceptionally high for women scientists at all degree levels (Table 3) and, at the same time, that their involuntary unemployment and underemployment rates far exceed those of comparatively educated men (Figure 3). These unemployment and underemployment percentages show that although women scientists are excellently trained, willing, and eager workers, they experience greater difficulties in finding suitable work than men do.

TABLE 3

PERCENTAGE OF WOMEN SCIENTISTS AND ENGINEERS IN LABOR FORCE FROM 1975 TO 1978 IN VARIOUS SAMPLES BY DEGREE LEVELS

Degree Year of Sample	Year of Survey	Number in Sample	Percentage		
			Bachelor's	Master's	Ph.D.'s
1974-1975	1976	191,215 *	83.6	87.5	....
1976	1978	116,602 *	84.3	87.2	....
1972	1978	100,623 *	79.9	86.5	....
1965	1975	1,881	63.1	....	....
1962-1977	1978	607	71.0	78.1	96.1
All years	1977	30,470 *	....	....	89.5
All years	1976	1,125	89.0	84.4	92.8

SOURCE: Vetter, Betty M., "Working Women Scientists and Engineers," Science, January 4, 1980.

*Weighted number.

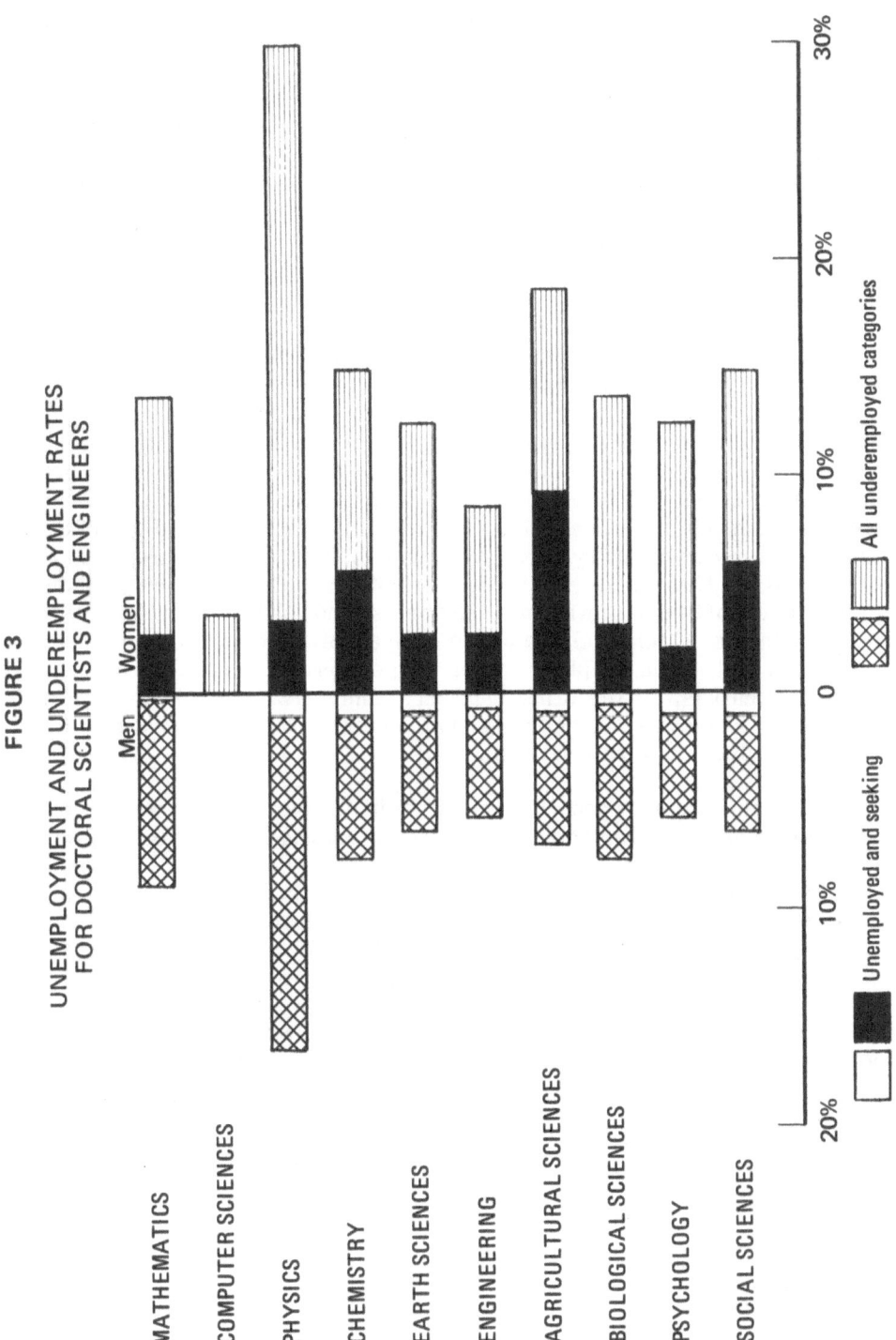

FIGURE 3

UNEMPLOYMENT AND UNDEREMPLOYMENT RATES
FOR DOCTORAL SCIENTISTS AND ENGINEERS

A popular and apparently plausible explanation of women's greater propensity to be unemployed or underemployed is their limited mobility due to the fact that in most two-career families the husband's career takes precedence over the wife's. The ability to move geographically for better opportunities traditionally has been considered essential, particularly to academic careers. It, therefore, comes as a surprise to find that involuntary unemployment and underemployment rates do not differ significantly between married and unmarried women. In fact, other measures of career progress such as rank, salary, and productivity suggest that married women do better.

Employment Sectors

Men and women scientists are distributed differently among employment sectors, and notable shifts in this distribution have occurred in the last decade. As a result, employment prospects for new doctorates have changed quite markedly (Table 4). Despite the academic depression, academic institutions are still by far the largest employers of both men and women scientists, although there are large variations by discipline. Women are much more likely than men to work in academe and much less likely to work in industry. A proportion of this large discrepancy is attributed to their different field distributions, primarily to the small percentages of women who are engineers and physicists. The U.S. government is a relatively minor employer of both male and female scientists, but again employs larger proportions of men than women. Women scientists, on the other hand, are almost twice as likely to work for "Other employers," a category that includes state and local governments, hospitals, and the self-employed.[6]

The changes between 1973 and 1977 in the percentage distributions are notable and occurred because of shifts in employment sectors for new graduates. Academic employment dropped for both sexes, but more drastically for women. On the other hand, the industrial employment rate for new women graduates nearly doubled. The increase in the category of "Other employers" represents the largest percentage increment for women, which is greater than that for men.

Overall, women scientists are markedly overrepresented in the lowest paid and least prestigious employment sectors and are especially underrepresented in industry and government where salaries are the highest. Within the academic world, women scientists are more likely than men to be employed in the least prestigious institutions, overrepresented in small liberal arts colleges and 2-year colleges, and greatly underrepresented on the faculties of leading research universities.

Conditions of Employment: Rank, Salary, and Work Functions

Academic, industrial, and government employment all exhibit similar patterns for each sex, but markedly different ones for the two sexes. In general, men are more likely to hold higher-level positions than women

TABLE 4

DOCTORAL SCIENTISTS AND ENGINEERS
BY EMPLOYMENT SECTOR AND SEX 1973 AND 1977

	1973 Employment		1977 Employment	
	Men	Women	Men	Women
All Ph.D.'s				
Number	185,800	14,700	236,800	24,200
Percentage in:				
Business/industry	24	5	26	7
Academe	57	72	55	68
U.S. Government	9	6	8	5
Other employers	10	17	11	20
New Ph.D.'s[1]				
Number	26,400	3,000	22,500	4,400
Percentage in:				
Business/industry	22	5	25	9
Academe	56	73	52	64
U.S. Government	9	3	8	3
Other employers	13	19	15	24

SOURCE: National Academy of Sciences (1980a), Women Scientists in Industry and Government. Washington, D.C.: Committee on the Education and Employment of Women in Science and Engineering, Commission on Human Resources.

[1]Earned doctorate 1 to 2 years prior to employment survey.

of equal training and experience. In academe, this differential distribution is most clearly visible among academic ranks (Figure 4). About two-thirds of all academically employed male scientists are in the two senior professorial ranks, while as many as three-quarters of the women scientists are below these ranks. The differences are most marked in prestigious institutions, where most of the important research is done. Approximately one-half of all the women scientists employed by these universities do not hold regular faculty rank at all, but are instructors, lecturers, or research associates, whereas only one-fifth of the men are at these levels.

As a consequence of the differential institutional and rank distributions, there are large salary gaps between male and female scientists in academe, amounting to as much as 20%. However, even when rank, field, institutional type, and length of experience are held constant, significant salary differences remain which can only be ascribed to sex bias. The

FIGURE 4

FACULTY RANK DISTRIBUTION
OF DOCTORAL SCIENTISTS AND ENGINEERS
BY R&D EXPENDITURES OF INSTITUTION AND SEX, 1977

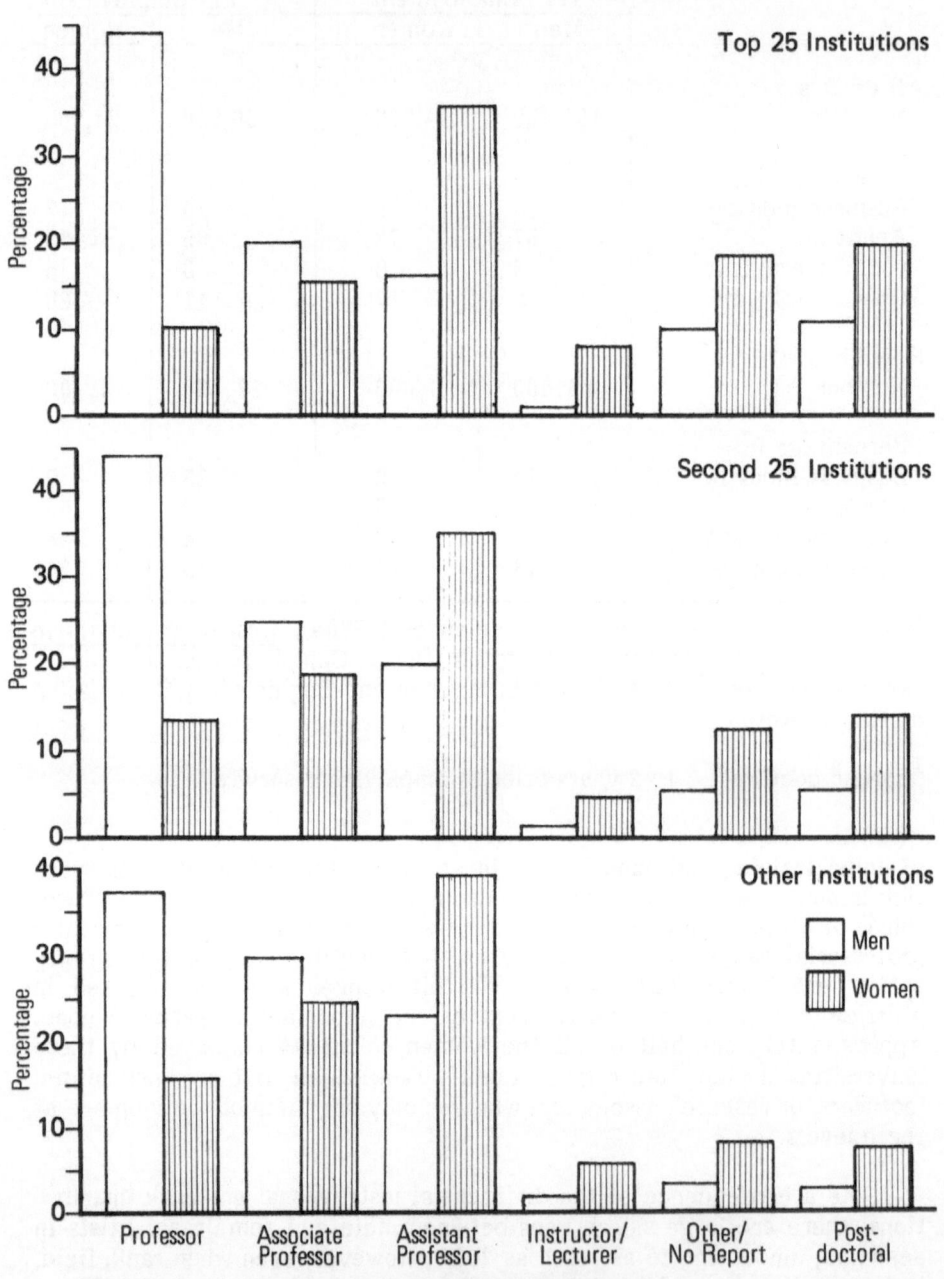

SOURCE: Survey of Doctorate Recipients, National Research Council

disparities are small for new entrants in most fields, but become larger over time.

In industry, there is no simple equivalent of academic rank; each company is likely to have its own internal classification system. For industrial employment, therefore, it is more useful to examine differences in work functions where we note that women are more likely than men to be doing research than managing it, i.e., to perform less highly valued work (Figure 5). They also are markedly overrepresented in such peripheral activities as staff work and technical writing. Median salaries clearly reflect the functional differences and the gap increases with age and experience; but again, even when work activity and experience are held constant, large salary gaps remain (Figure 6). It is of interest to note, though the phenomenon is as yet unexplained, that for those doing research the salary difference increases with experience, while for those who are project managers, the gap narrows suddenly after 15 years experience, but does not disappear.

The employment patterns of scientists in the federal government resemble those in the other sectors, except for the youngest employees. Figure 7 shows the distributions among federal employee grades by sex and age. Overall, far greater proportions of men than women fill the higher levels; the difference is most pronounced with increasing age, and salary differences accurately track the grade distributions.

The sum total of these sex differences in work sector and activity, salary, rank, and associated perquisites adds up to an enormous disadvantage for women scientists and a considerable loss to the quality of science. The large majority of women working in academe are either in pure teaching positions with neither time nor facilities for the research that is the life blood of science, or they hold predominantly subordinate positions in research universities. In these positions, they not only lack the autonomy to initiate and carry out their own research but they have no independent access to the funding, facilities, and graduate students that make research possible. In particular, these cumulative disadvantages make it almost impossible for women scientists to work at the forefront of their field, except possibly as someone else's assistant.

A similar problem is evidenced in industry when we compare growth rates of particular industries with the rates at which women scientists are hired, as shown in Table 5. Increased employment of women Ph.D.'s is occurring primarily in slow-growth industries; the fastest-growing sectors, which are the leaders almost by definition, employ the fewest women and show the slowest growth rates in hiring women.

RECENT CHANGES IN EDUCATION AND EMPLOYMENT

Marked changes have been observed during the 1970s in both the educational and employment patterns of women scientists and engineers. In both areas, the decade began with rather low participation, and the

FIGURE 5

PRIMARY WORK ACTIVITIES OF DOCTORAL SCIENTISTS AND ENGINEERS IN INDUSTRY, 1977

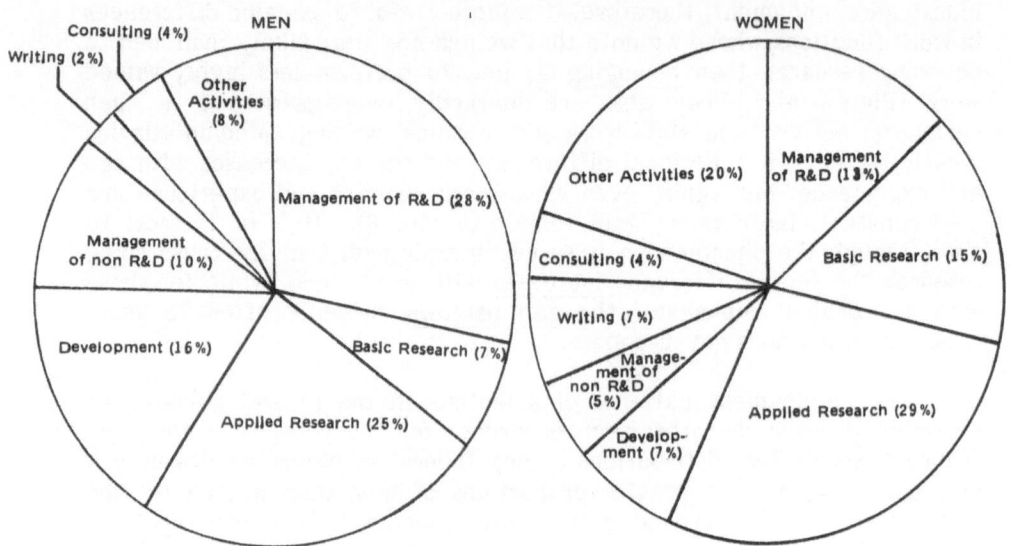

MEDIAN SALARIES OF DOCTORAL SCIENTISTS AND ENGINEERS IN INDUSTRY BY COHORT AND SEX, 1977

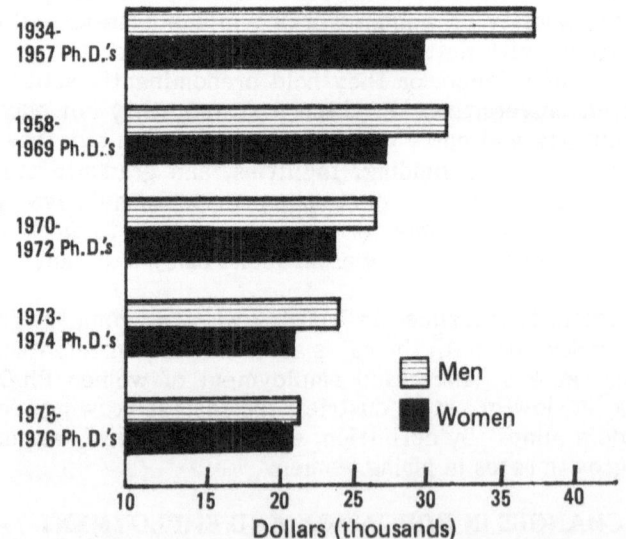

FIGURE 6

MEDIAN SALARIES OF R&D PERSONNEL
BY PRIMARY WORK ACTIVITY AND YEARS OF EXPERIENCE, 1977

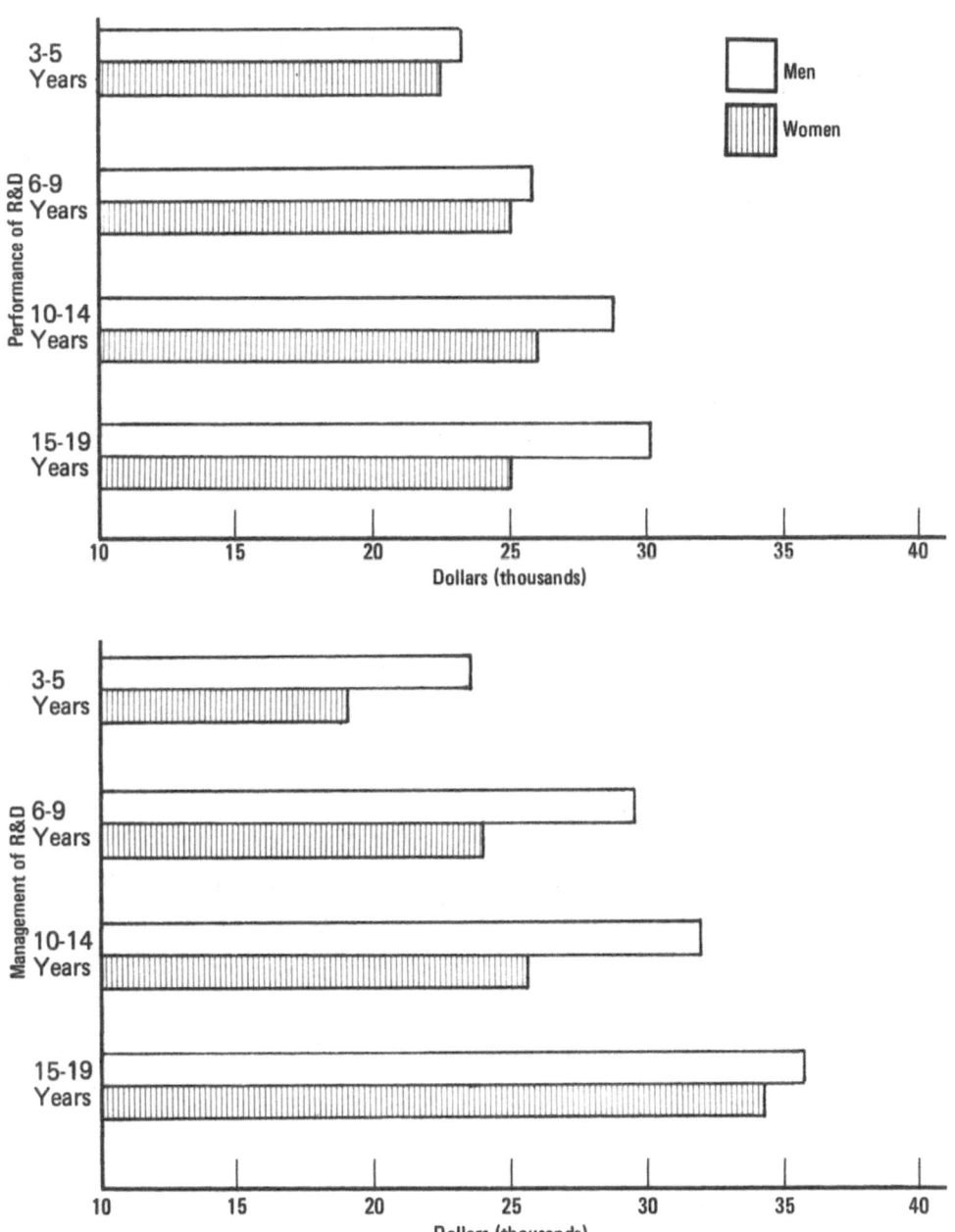

FIGURE 7

PERCENT GRADE DISTRIBUTION OF SCIENTISTS AND ENGINEERS IN THE FEDERAL GOVERNMENT BY AGE AND SEX, 1978

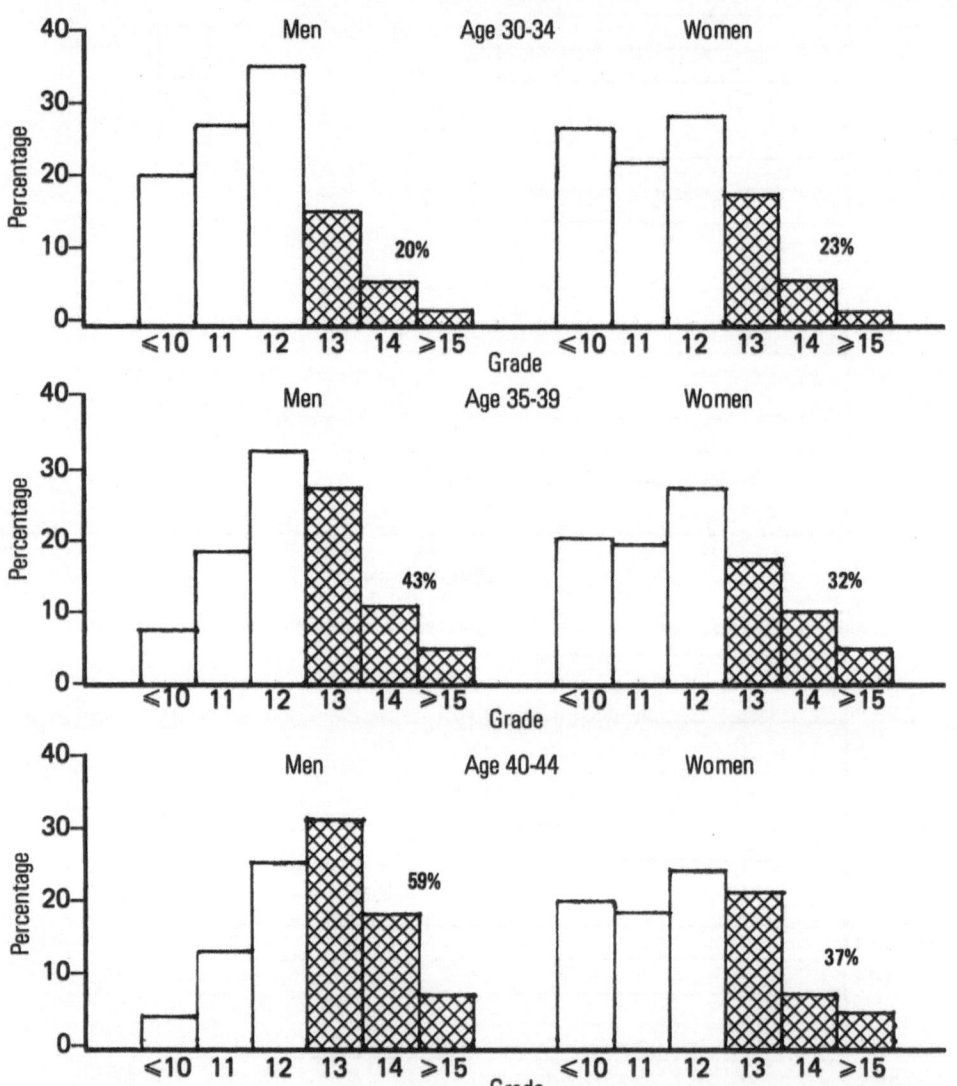

TABLE 5

FOUR-YEAR GROWTH IN R&D PERSONNEL[1] WHO HOLD SCIENCE AND ENGINEERING DOCTORATES BY INDUSTRY GROUPS INCLUDING INCREASE IN NUMBERS OF WOMEN

Industry Group[2]	Doctoral R&D Personnel					
					4-Year Growth	
	1973	1977	Average Annual Growth	Total	Number of Women	Women as % of Increase
Total employed	37,209	46,088	5.5%	8,879	531	6.0%
Classifiable companies	34,974	43,410	5.6	8,436	525	6.2
Manufacturing	32,253	39,603	5.3	7,350	461	6.3
Chemicals	7,751	9,353	4.8	1,602	98	6.1
Electrical equipment	6,085	6,858	3.0	773	86	11.1
Pharmaceuticals	3,206	4,297	7.6	1,091	77	7.1
Petroleum and refining	3,343	3,900	3.9	557	35	6.3
Instruments	2,259	3,118	8.4	859	40	4.7
Other manufacturing	9,609	12,077	5.9	2,468	125	5.1
Services	1,682	2,066	5.3	384	39	10.2
Other non-manufacturing	1,039	1,741	13.8	702	25	3.6
Nonclassifiable companies	2,235	2,678	5.0	443	6	1.3

SOURCE: Survey of Doctorate Recipients, National Research Council.

[1]Includes individuals whose primary work activity is management or performance of research and development.

[2]Standard Industrial Classification.

rates of increase have been considerable in all fields and astonishing in some. Women's share of baccalaureate degrees in engineering, for example, has more than doubled since 1975. Similarly, the industrial employment rate of women scientists has nearly doubled in 4 years.

Simultaneously, we find a greatly increased hiring of women as assistant professors, especially in the research universities where they had been most seriously underrepresented. However, their promotion rates to higher ranks continue to lag behind those for men. Starting positions and salaries in both industry and academe are now almost equal in many fields; in engineering, women hold a very slight edge in starting salaries.

CONCLUSION

Several important conclusions follow from women's increased participation in science education: (1) The educational patterns of men and women are converging, which has become especially evident at the undergraduate level where less time is needed for new trends in career choices to find expression than in advanced training. Striking shifts to nontraditional career aspirations have occurred for women students. If these patterns continue, we can confidently expect women's share of science doctorates to approach parity in the next two decades, provided that overall job opportunities do not deteriorate. (2) Women are now a significant and rapidly growing part of the youngest, most recently trained, scientific work force. It is this segment to which we traditionally look for innovation and work at the cutting edge of research, and women will have an increasing share in these advances. (3) For several years, the total enrollment growth in higher education has been attributed primarily to increasing numbers of women students, particularly at the doctoral level. The higher educational system would have declined drastically during the 1970s without this dramatic growth in women's participation. (4) In some fields of science, serious shortages of highly trained personnel would now be developing if growing numbers of women had not been entering them.

While it is tempting to ascribe the progress that has been made to the implementation of equal opportunity laws, these are at most half the story—especially since they have been enforced capriciously, if at all, on behalf of professional women. In fact, the evolution in educational patterns strongly suggests that the individual choices on which these patterns rest began to change in the 1960s, well before "affirmative action" became a fighting phrase in academe. Perceiving at least the appearance of equal career opportunity ahead, young women began to make educational decisions in response to these perceived opportunities— just as men always have.

The continuing shift of women into the sciences, as into other demanding professional fields such as law or medicine, has been accelerated by the elimination of some formal and informal barriers such as denial of admission to appropriate educational programs. It is less clear that the impact on employment has been comparable, largely because the courts have declined to interfere with the prerogative of faculties to make promotion and tenure decisions. A few recent legal settlements are beginning to question the discriminatory outcomes of such faculty decisions and are cause for cautious optimism.

The most compelling reason for optimism, however, is the clear response from young women that when opportunities become available they will seize them. The scientific professions, for all their pride in being meritocratic, have turned out to be no less discriminatory than others. Setting aside considerations of justice and equity for the moment, however, it remains true that for most people, men or women, sex equality is not the primary consideration in educational and career choices. For women of reasonable talent and high expectations for themselves, the sciences offer excellent and fulfilling career opportunities, certainly more so than many of the more traditionally accessible professions. Women's aspirations have taken a sharp turn upward, and there is cause for hope that the barriers which remain will be overcome.

NOTES

1. Details of sample selection and survey procedures for the Comprehensive Roster may be found in the National Academy of Sciences, 1980a (Science, Engineering, and Humanities Doctorates in the United States: 1979 Profile). The Doctorate Record File is published as an annual report in the National Academy of Sciences, 1980b (Summary Report 1979: Doctorate Recipients from United States Universities).
2. National Academy of Sciences, 1979, 1980c.
3. National Academy of Sciences, 1980b.
4. See, for example, Perrucci, 1975 and Feldman, 1974.
5. National Academy of Sciences, 1979.
6. National Academy of Sciences, 1980c.

ELECTED WOMEN: SKEWERS OF THE POLITICAL SYSTEM

Judith Hicks Stiehm

University of Southern California
Los Angeles, California   U.S.A.

The term "skewer" derives from two sources: the culinary arts and Rosabeth Kanter. As most women know, a skewer is a long, sharp pin stuck into a piece of meat to hold it together. As many social scientists know, Kanter has argued in Men and Women of the Corporation that proportionality has a great deal to do with the interaction of members within a group.[1] Specifically, she says that a group with up to 20% minority members (whether that minority be women, men, black, or white) is a "skewed" group; and only if the minority is over 40% can the group be called "balanced." The elected women of this study are called skewers, because they seem to hold the political system together by keeping it in touch with citizen concerns despite their small numbers. They also are labeled skewers because many of them skew the governing body to which they are elected. Previously those bodies were neither balanced nor tilted; they were homogeneous. The skewing affects the system in that it affects the women officials who may not have had a minority experience prior to taking office.

Kanter calls the individuals who skew a group tokens; she argues that being rare, scarce, or different has important consequences. First, being in a minority means that one is treated differently and, second, that one has a different effect. One is influenced by and influences the group differently. This happens willy-nilly, whether or not the skewer (token) knows it, and in spite of attempts she may make to be assimilated. According to Kanter, a skewer's presence cannot go unnoticed. She has high visibility with attendant pressures to perform and attendant exposure to retaliation.   Further, her abilities may be overshadowed by her appearance. Indeed, her presence causes the majority to heighten cultural boundaries by exaggerating its culture (thus reminding her of her difference), by isolating her, and by testing her loyalty. Finally, her status is reduced, or encapsulated in a special slot or stereotype such as "mother," "seductress," "pet," or "iron maiden." The results are costly. Skewers experience stresses which are not experienced by other members of the group. Further, the presence of skewers generates no pressure for change in the direction of a "balanced" or even a "tilted" group. Instead, the minority's low proportion permits the group to alternatively describe a minority individual's performance as typical of the group or as exceptional. The contradictions and dilemmas of the skewer seem to suggest that there is no winning strategy.

Assuming that Kanter is correct and that skewers are treated differently, this paper will show that women who hold elected office are, in fact, different from the archetypal elected man, and that elected women's differences probably affect both their approach to, and their effect on, the political process.

THE SAMPLE

The subjects of this paper were all women who (1) held public office at the time they were surveyed and (2) were members of an organization called California Elected Women for Education and Research (CEWEAR). Thus, they chose to identify themselves both as women and affiliated with a state organization even though most held only a local office. Much of the data came from CEWEAR's founding year, 1975, with some comparative data from a 1980 resurvey. Continuity during this period was more evident than change. The women elected at the beginning of this country's third century did not seem very different from those elected at the end of the second. The winners showed few signs of consolidating their gains nor was there a dramatic increase in the number of women elected. In fact, there was a decrease in the number of respondents.

The 1975 sample included 184 elected women; the 1980 sample included 150 women of whom 13 were former officeholders. Most of these women held locally elected office and, with the exception of school board members, they were skewers—they were part of a minority of 20% or less.[2] The types of office held by these women are represented in the percentages contained in Table 1.

There were no male controls because the purpose of this paper was to closely examine the women most likely to seek powerful political office. The intended comparison, then, was not with the men who now hold the same office, but against the apparent requirements for winning state or national office. Apart from incumbency and maleness, these seem to include, first, the capacity to win a contested partisan election in which there is a single winner and, second, the ability to meet high campaign expenses by raising large sums of money. In addition, male officials usually have a high income and are white. The only one of these criteria the 1975 CEWEAR women meet is that 180 of 184 are white. Interestingly enough, though, two of the four nonwhite women held two of the highest offices in California: Congresswoman and Secretary of State.

TABLE 1

OFFICES HELD BY CEWEAR MEMBERS, 1975 AND 1980

Office Held	Percentage	
	1975	1980
School Board[1]	31	36
City Council	38	26
Mayor	9	6
County Supervisor	3	9
State Assembly, Senate, U.S. Office	2	2
Judge or other officer	17	21
Total	100	100

[1]Includes both secondary and community school boards.

DIFFERENCES WHICH MAY AFFECT
WOMEN'S APPROACH TO POLITICS

In examining the incomes of officeholders, the 1975 elected women did not have high incomes of their own. Three-fourths of them made less than $10,000 and 90% made less than $20,000. Only 7% of the women earned more than $35,000 annually. On the other hand, the families of the elected women had substantial incomes. Only 11% of these families earned less than $10,000 while 57% earned over $20,000 and 22% earned over $35,000 annually.

The fact that elected women contributed a relatively small part of the family income probably had several effects. First, while the women grasped the need for the use of money, they lacked the direct control over income which elected men possessed. California is a community property state but, even so, women seemed to be inhibited about spending money earned by their husbands as freely as they would spend money earned by themselves. Second, some measure of the psychological impact of making only a limited part of the family income came from a query as to how the women classified themselves economically. The choices were: dependent, independent, or provider. More than three-fourths of the women described themselves as dependent. Some 15% said they were independent. Only 4% assigned to themselves the classic male label of provider.

It seems doubtful that male elected officials would describe themselves as dependent even if they were substantially supported by a wife, inheritance, or political contributions. Further, dependency may not only have an effect on the woman official, but it may also influence other's perceptions of her, thereby indirectly affecting her efficacy. For instance, if she were defined primarily on the basis of her role as someone else's wife or daughter, her effect in office could be quite different (perhaps stronger rather than weaker) than if she were seen as an independent actor.

In the United States, men who are politicians often emphasize their families in their campaigns. Having an attractive, wholesome family is thought to draw votes. Further, wives often campaign for their husbands and participate in ceremonial events with or even in place of them.[3] For elected women, the family's role is more complex. If she is "dependent" on her husband, his attitude toward her office holding becomes very significant. In this study, husbands are described by their elected-official wives as having a wide variety of attitudes.

Even if men have less support from families than they pretend (or believe), a surprising number of elected women's husbands are neutral to negative, as shown in Table 2. It should be remembered that this coolness is directed toward local office holding which is relatively undemanding in terms of the number of hours a week devoted to the office, the number of nights worked, and the work-related travel undertaken without the spouse.

The question is: How would a husband's attitude change if his wife were to run for and win high office? Would he take increased pride in her accomplishment or would he resent her new role? A possible indicator of what might happen is the divorce rate for elected officials. Then one might ask, would the marital status of CEWEAR's women change if they were to run for and win high office?

At the time of the survey, the women officials' marital status was similar to that of most women of their socioeconomic class; 82% were currently married (4% were remarried), 2% were single, and 16% were widowed, separated, or divorced. In spite of the high percentage of married women in this sample, other evidence does suggest that high office takes a toll on marriage. Women in top state and in federal office are less likely to be currently married.[4]

Husbands who do not take an active role in their wives' campaigns are not necessarily opposed to their wives' success. Actually, some may fear that their participation would hurt the candidate's image in that their helpmate assistance could imply either that the candidate is a puppet or that she has a peculiar rather than a traditionally invidious marriage.[5]

Children present a similar problem. Adult children sometimes participate in campaigns but usually have a minor role during the period of office holding. Young children, though, are likely to appear in men's campaign literature as symbols of their father's capacity as a protector. In a woman's campaign, the picture of a young child is likely to spark a different and awkward question: "Who is (or will be) taking care of your children?" Often, elected women have managed this problem by simply not running for office while they still have young children. About 90% of

TABLE 2

ROLE OF CEWEAR MEMBERS' HUSBANDS
IN ELECTION CAMPAIGNS

Husband's Role	Percentage
None	13
Negative	8
Neutral	8
Moral support	31
Active	29
Official help	11
Total	100

the CEWEAR sample had no children under the age of 6, and 60% had none between 6 and 18. Overall, 87% did have children.

Questions about self-presentation were rampant among elected women and those running for office in 1975. Competence and conventionality were understood to go hand in hand for men. For women, the link was thought of as more problematic. In matters of dress, for instance, the question was how to project both seriousness and femininity. One solution was seized upon by so many women that it acquired a name: "The Yellow Blazer Phenomenon." Yellow is dramatic, eye-catching, and, therefore, political. A blazer is severely tailored and, therefore, serious. It is worn with a skirt and, therefore, is feminine. By 1980, some of the questions asked in 1975 seemed ludicrous, i.e., "What physical image should be sought?" Beauty? Sophistication? Nondescript? Other? The success of governors Ella Grasso of Connecticut and Dixie Ray of Washington may have made other women realize that femininity is less important in politics than savvy. It may even be that being too attractive is disadvantageous because it implies to more conventional men and women (the bulk of the electorate) that the candidate official is not just a good politician but is a veritable Wonder Woman who can excel in two wholly different roles at the same time.

Concern for image also stems from women's "double burden," a condition known to anyone who has studied the status of women in any country in the world. The woman who works outside the home almost always continues to perform most of the in-the-home work. Thus, she who labels herself a housewife would seem to place constraints on her public life. Conversely, the woman in public life who is considered by others to be a housewife must be seen by them as constrained. In our 1980 sample, 71% of women holding public office specifically described themselves as housewives. Only a few had full-time domestic help and less than one-quarter had part-time help.

A final image problem needs to be considered. Respondents were asked: "Which nationally prominent woman do you feel most closely represents your political/social views?" This inquiry is a role model question and assumes that adults need models every bit as much as children or adolescents. Perhaps the most shocking finding is that about one-half of the officials gave no response to the question. The problem may lie in the fact that there are so few nationally prominent women, but this lack of response also must reflect elected women's lack of identification with women of national prominence. The responses for the 1975 sample showed that only three women received as many as 5% of the votes. They were Congresswomen Bella Abzug, Yvonne Burke, and Shirley Chisholm. By 1980, two of the three women were no longer Congresswomen; Barbara Jordan was then selected by 11% of the respondents while Congresswomen Pat Schroeder, Shirley Chisholm, and Yvonne Burke and Cabinet member Shirley Hufstedler received 3% or 4% of the votes.

It should be noted here that three of these six women are black although the selectors are overwhelmingly white. As Cynthia Epstein has observed, the double negative may be harmful for most minority women, but for the handful who break into professional life the double negative seems to have a positive effect.[6]

Another variable which suggests that women and men may think differently about themselves and their offices is women's assessment of their qualifications as they relate to the office they hold. In brief, 85% of the women in 1975 rated themselves as qualified when they ran, but 45% rated themselves as overqualified after they gained office. It should be noted that just because women see themselves as overqualified and/or as working harder does not mean that men perceive them in the same way.[7] It may be that women's approach to a hierarchy is step by step; that is, they look about, observe competence less than their own, look up, believe they can do the next job, and then move. The result is slow progress and regular overqualification. Men, on the other hand, may more typically examine the whole hierarchy, select a goal, start as close to it as possible (not at the bottom), and be constantly working toward it even as they fulfill other duties.

Most efforts to correlate family background and adult women's career success have been to little purpose. Having a mother who was a full-time homemaker or who worked outside the home, for instance, has not been explanatory. One study of businesswomen, however, did show that women executives tend to be either only children or the eldest in all-girl families, thereby representing the next generation as a son might ordinarily do.[8] Just about one-half of the women in our 1975 sample have no male siblings, and over one-half have no female siblings. Further analyses will be needed to determine whether or not the pattern of seeing oneself as being responsible for the family's future holds for these women as it does for business executives.

Finally, elected women's occupational perspective is very different from that of men. Most prominent male officials attended law school whereas only 4% of the 1975 CEWEAR women did so.

DIFFERENCES WHICH AFFECT
WOMEN'S EFFECT ON POLITICS

The differences that influence women's effect on politics are not so much differences between them and men holding the same office as they are differences between them and men who hold the state and national offices to which the women may aspire (or to which others may aspire for them). First, there is the matter of party. Some political scientists will argue that the most important thing to know about an individual is his or her party affiliation.   CEWEAR women did identify themselves with parties, but they did not assume leadership roles within their party and did not run for office as partisans. Indeed, 81% of the women in the 1975 sample and 89% in the 1980 sample held offices officially designated as nonpartisan. This is significant because state and national offices are almost always contested by partisans and active party support can substantially increase campaign funds as well as produce volunteer workers. To be apart from the party organization, therefore, is to be seriously disadvantaged.

Second, elected women also may be handicapped by the kind of competition practiced in our political system. There are two kinds of electoral competitions in the United States. One type, in which there is a single winner for a particular geographic district, is termed "head-on." This represents a zero-sum game; one wins what the other loses. The same pattern occurs in seeking employment. The other type is called "at-large." In such an election, there are a number of open seats (perhaps three) and any number of candidates. Thus, an individual can vote for a woman without having to vote for her over all other candidates. In such an election, a woman can win by being one of the best but not necessarily the best; further, she does not have to acquire a majority of the votes. State and national offices are almost always selected in single-member districts. Many local offices are chosen at-large. In fact, about 60% of the women in the two samples were chosen in at-large elections and 64% did not win a majority. There is also some evidence from other countries that women are more likely to win at-large or party-list elections than in single-member districts.[9]

Because at-large elections are rarely held in very large districts, it is difficult to tell to what degree the decrease in women winners is a function of election procedures and to what extent it is simply a function of scale. The fact is that most women officeholders are dealing in units so small that their experience is qualitatively as well as quantitatively different from that of U.S. Congresswomen or a legislator in a large state. In 1975, two-thirds of the surveyed women represented districts of under 50,000 and only 17% represented more than 100,000 people. By 1980, a shift had occurred in that 60% represented less than 50,000 people and 23% more than 100,000 people. A Congressperson, however, represents about 400,000.

The data also showed that average campaign costs were low, although they increased between 1975 and 1980. In 1975, 88% spent under $5,000; only 2% spent over $50,000. A Congressional candidate can easily spend $500,000.

Higher campaign costs would be difficult to absorb because elected women's official salaries are so low. Only 13% of the women in the 1975 survey and 16% of the women in the 1980 survey earned more than $10,000 in their capacity as elected officials. The salary of a Congressperson is $57,000 a year plus a variety of perquisites. These include a substantial budget which they may allocate as they see fit among a large staff. In contrast, few elected women have any full-time, paid staff members.

It is also important to note that the women tended to enter public office only at middle age. This fact suggested that they are not as able to wait for and then seize an opportunity for higher office. Actually, nationally elected men's median age does not differ from that of women, but that median includes numerous bright young men and numerous senior officials who run for re-election term after term. On the other hand, women seem not only to enter late but also to remain in office for fewer terms. Thus, they do not develop experience nor seniority. This may represent a commitment to rotation in office; it also may reflect boredom or a lack of resources for pursuing higher office. Another explanation is that holding office is disappointing--that the women find they cannot be as efficacious as they had anticipated.

In large state and national legislatures, to be effective requires effective use of the committee system. Generally, women legislators report that they are given stereotypical assignments and are left out of important, informal networks. This problem should not arise for CEWEAR women because they typically work in small governing groups which consist of five to seven members. Nevertheless and even though they report that they voted with the majority two-thirds of the time and that they held a leadership role of some kind, CEWEAR women also report they are not full members of their group. In informal discussion they revealed that they fully recognized in their own small groups the subtleties of sex discrimination as it is reported in the research literature.[10] Finally, the new participant has less effect than the professional. A majority of the combined sample were in their first term. While a majority stated they would "probably" run again, only a few expected to run for another office which suggests a lack of the kind of ambition that would lead to higher office.

CONCLUSION

An interesting overarching problem concerns the effect of others' perceptions of elected women in general on the efficacy of particular elected women. If women are seen as short-term, local, and unambitious elected officials, they may not be taken into consideration when party strategy is shaped; lobbyists may not contact them or contribute to their campaigns; journalists may not feature them; and their peers may discount their support or their opposition. What is generally believed affects each individual. If women in general lack credibility, then so does the individual woman. As long as the numbers of elected women are few, they will have a different impact on their peers. The theorems drawn from Kanter do not suggest any reason to expect change simply because the minority performs well. To receive "regular" treatment, the minority must cease to be a minority.

A second point is that we have come to think of all elected women as skewers. In fact, large numbers of them (especially those elected to school boards) are not. They need to be examined separately from those elected to other local office. Further, those who hold local office should not be assumed to be the pool from which candidates for higher office will be chosen. The small-scale, nonpartisan, at-large election, political environment in which these women operate makes them different from state and national officeholders in ways even more fundamental than their gender.

Finally, if women are treated differently and are different as well, it seems fair to assume that their effect in office is different, too. It may well be that holding office is of less worth to women than men. Their presence may raise issues and consciousness, but is it worth raising issues or writing laws when the issues are left unresolved and the laws unenacted? Women do get elected by a heterogeneous electorate and to all levels of government. However, once in office they have to be effective in an overwhelmingly male environment. As Virginia Curry states, even when political women learn the rules of men's politics, they often find that when they try to play by those rules, they "encounter penalties because the rules for women (minorities) in politics are still uncodified, even though they are enforced, cruelly, arbitrarily, unexpectedly."[11]

What would happen if political bodies were as balanced as the electorate? How could that balance be achieved? If we assume that head-on competition between a woman and a man is likely to advantage the latter, could we also assume that two-member districts would give more opportunity to women—and without violating democratic principles about letting the persons with the highest number of votes be the winners? The U.S. Senate now has two members elected from each state. Often these represent very divergent views. (California's Senators currently are Republican S. I. Hayakawa and Democrat Alan Cranston.) The two Senators are elected at different times and are currently thought of as single-member-district winners. But, would not women candidates do better if Senators were perceived as pairs of the state representatives, as at-large representatives from two-member districts? What if election laws governing the House of Representatives were simply rewritten so that they called for two-member districts? Nothing in the Constitution or in democratic theory precludes this.

NOTES

1. Kanter, 1977a:8.
2. In 1975, two out of three of the nonschool board members were skewers (either the only woman or one of a less than 20% minority). However, a majority of the school board members were in "tilted" or "balanced" groups. In some cases, women were 80% of board members. In 1980, five out of eight nonschool board members were skewers, but more than four out of five school board members were not. Indeed, one school board was all women! School boards, it seems, are different.
3. The best account of this is McCarthy, 1972. For further discussion, also see her book, Circles: A Washington Story, 1977.
4. Center for the American Woman and Politics, 1978:13A.
5. Stiehm, 1976a.
6. Epstein, 1978.
7. Center for the American Woman and Politics, 1978:42A-43A.
8. Hennig and Jardin, 1978:100.
9. Stiehm, 1976b.
10. See also, Center for the American Woman and Politics, 1978.
11. Currey, 1977.

THE RELATIONSHIP BETWEEN

THE LABOR FORCE EMPLOYMENT OF WOMEN

AND THE CHANGING SOCIAL ORGANIZATION IN CANADA

Lorna R. Marsden

University of Toronto
Toronto, Ontario   Canada

In the generation of Canadians born about 1950, quite new patterns of adult life are being experienced, especially by adult women. During the past 30 years, Canada has gone through a period of unprecedented economic expansion and growth, and at the same time the patterns of life have undergone a number of changes such that women now in young adulthood (ages 20 to 35) cannot find an accurate model of adult experience in the lives of their mothers. Women who reached maturity during the early 1970s are part of a cohort which has smaller families and a higher level of education than their mothers. After 1959, for example, the average family size (total fertility rate) decreased from 3.9 to 1.8. Further, the entire age distribution of the Canadian population shifted toward older age groups and the proportion of families with children under the age of 6 dropped from 61% in 1951 to 43% in 1971.[1] There also was an increase in the number of two-earner families, one of the major features differentiating economic sufficiency from insufficiency among many households.[2]

Women also have experienced a growing demand for their services in the paid labor force. Tables 1 and 2 show that the Canadian industries and occupations in which most women work expanded very rapidly from 1951 to 1971. This expansion has shown few signs of tailing off despite relatively high rates of unemployment among women and a general economic recession.

Even with this increased demand, full-time women workers earn on an average only about 60% of the wages of men. Because of their educational choices, limited work experience, limited geographical mobility, and sexual discrimination, women generally earn far less than

TABLE 1

INDUSTRIAL DISTRIBUTION OF FEMALE AND TOTAL LABOR FORCE, CANADA,[1] 1951, 1961, AND 1971

INDUSTRY	PERCENTAGE TOTAL GROWTH 1951–71	PERCENTAGE DISTRIBUTION OF TOTAL LABOR FORCE			PERCENTAGE DISTRIBUTION OF FEMALE LABOR FORCE			WOMEN AS PERCENTAGE OF TOTAL LABOR FORCE		
		1951	1961	1971	1951	1961	1971	1951	1961	1971
Agriculture	−41.8	15.6	9.9	5.6	3.0	4.5	3.8	4.2	12.3	23.2
Forestry	−42.8	2.5	1.7	0.9	0.2	0.1	0.1	1.8	2.0	4.5
Fishing and trapping	−50.5	1.0	0.5	0.3	0.0	0.0	0.0	0.8	1.4	3.6
Mining	31.6	2.0	1.8	1.6	0.2	0.3	0.3	2.2	4.0	6.8
Manufacturing	25.4	25.7	21.7	19.8	23.6	17.1	13.7	20.2	21.5	23.7
Construction	53.1	6.6	6.7	6.2	0.5	0.6	0.9	1.7	2.5	4.9
Transportation	44.0	8.8	9.3	7.8	4.6	4.7	3.8	11.7	13.8	17.0
Trade	78.6	13.4	15.3	14.7	18.2	17.1	15.8	29.8	30.4	36.7
Finance	148.4	2.7	3.5	4.2	5.5	5.9	6.2	44.4	45.7	51.4
Community-personal service	163.5	14.6	19.5	23.7	38.4	42.4	39.7	57.8	59.3	57.6
Public administration	109.1	5.8	7.4	7.4	4.6	4.9	5.5	17.8	18.0	25.5
Unspecified	905.4	1.3	2.4	7.9	1.1	2.4	10.2	19.3	26.5	44.2
All industries	62.8	100.0	100.0	100.0	100.0	100.0	100.0	22.0	27.3	34.3

SOURCE: Nakamura, Nakamura, and Cullen, 1979. Calculated from 1951 Census of Canada, Vol. IV, Table 16; 1961 Census of Canada, Vol. III—Part 2, Table 1A; and 1971 Census of Canada, Vol. III—Part 4, Table 2.

[1] Data exclude Yukon and Northwest Territories.

TABLE 2

OCCUPATIONAL DISTRIBUTION OF FEMALE AND TOTAL LABOR FORCE, CANADA,[1] 1951, 1961 AND 1971

OCCUPATION	PERCENTAGE TOTAL GROWTH 1951–71	PERCENTAGE DISTRIBUTION OF TOTAL LABOR FORCE			PERCENTAGE DISTRIBUTION OF FEMALE LABOR FORCE			WOMEN AS PERCENTAGE OF TOTAL LABOR FORCE		
		1951	1961	1971	1951	1961	1971	1951	1961	1971
Managerial	−17.1	8.5	8.8	4.3	3.3	3.3	2.0	8.7	10.4	15.7
Natural sciences, engineering	250.8	1.3	2.1	2.7	0.4	0.4	0.6	6.9	·4.8	7.3
Social Sciences	272.9	0.4	0.6	0.9	0.5	0.7	1.0	27.8	29.4	37.4
Religion	−23.0	0.6	0.5	0.3	1.0	0.6	0.1	39.7	28.9	15.7
Teaching	200.4	2.2	3.1	4.1	6.7	7.3	7.1	67.2	64.4	60.4
Medicine and health	193.9	2.1	3.2	3.8	6.5	8.6	8.2	68.5	72.1	74.3
Artistic	124.1	0.7	1.0	0.9	0.9	1.1	0.7	30.7	31.2	27.2
Clerical	119.6	11.8	13.7	15.9	30.1	30.6	31.8	56.1	61.0	68.4
Sales	165.2	5.8	7.2	9.5	8.8	8.4	8.4	33.3	32.0	30.4
Service	91.8	9.5	11.4	11.2	19.5	19.5	15.1	45.1	46.7	46.2
Farming	−38.3	15.7	10.1	6.0	2.8	4.3	3.6	3.9	11.7	20.9
Other primary	−28.8	4.0	2.8	1.8	0.0	0.0	0.1	0.1	0.3	1.3
Processing	−13.2	7.3	5.4	3.9	4.9	2.7	2.0	14.8	13.7	17.8
Machining, fabricating	32.1	12.5	11.0	10.2	10.2	7.2	5.5	18.0	17.9	18.7
Construction	42.0	7.6	7.0	6.6	0.3	0.2	0.2	1.0	0.8	0.9
Transport	22.2	5.2	4.3	3.9	0.1	0.1	0.3	0.5	0.6	2.4
Other	153.3	3.6	5.4	5.6	2.7	2.7	2.6	16.3	13.6	15.7
Unspecified	1,044.5	1.2	2.6	8.5	1.1	2.4	10.8	20.6	26.0	43.4
All occupations	62.8	100.0	100.0	100.0	100.0	100.0	100.0	22.0	27.3	34.3

SOURCE: Nakamura, Nakamura, and Cullen, 1979. Calculated from 1951 Census of Canada, Vol. IV, Table 4; 1961 Census of Canada, Vol. III—Part I, Table 6; and 1971 Census of Canada, Vol. III—Part 2, Table 8.

[1] Data exclude the Yukon and Northwest Territories.

men and have fewer opportunities for advancement. A large proportion of married women work part time and women occupy 72% of all part-time positions in the Canadian labor force. From this brief historical overview, it is clear that there has been a rapid and visible shift in the lives of women from a generation in which home and family were the most important priorities to a new generation in which labor force attachment is at least on a par with family life.

WOMEN IN THE DECISION-MAKING AND WEALTH-PRODUCING SECTORS

While Tables 1 and 2 indicate the rapid increase in the labor force participation of women in Canada, this increase has had little impact on the power of women in Canadian society. If we can assume that in a capitalist, western-democratic country like Canada, power resides in the decision-making and wealth-producing sectors, what representation do women have? We argue that the top people in these sectors will have a major impact on the general direction of legislation and policy in the country. They will have an impact on the labor force especially through the expansion of their industries.[3] Along with the financial sector, women in decision-making positions within the state and private sectors are the ones to watch as they will be able to take a portion of women's increased power. Furthermore, it is in the decision-making and wealth-producing sectors that women must focus if gains are to be made in the reorganization of social and political life and if power among women is to be protected and increased. The purpose of this paper is to identify changes in women's participation in these sectors and to determine what impact these changes will have upon social organization.

In terms of decision-making sectors, we will look at the representation of women in the highest levels of political office, bureaucratic office, and the judiciary. We will define the wealth-producing sectors by distinguishing them from the service sectors and such industries of internal importance as construction and transportation. The wealth-producing industries include: agriculture, pulp and paper, oil, gas and minerals, fisheries, and manufacturing.

Overall, the picture is unimpressive. In political life, while the number of women in the three most recent Cabinets is a marked improvement over any previous governments, it is still very small. In the 1980 Cabinet, there are 3 women and 31 men; women hold the following positions: the Speaker of the House, the Minister of Health and Welfare, and the Minister of State (Mines). The previous Conservative Government had 1 woman, the Minister of External Affairs, out of a Cabinet of 30 people. In the 1974-79 Government, there were 3 women in a Cabinet of 32. The number of women in the 282 member House of Commons currently is 14, the highest number in Canadian history. In the 10 provincial governments, there are 10 women out of a total of 203 Ministers.

In a study of the civil service, Olsen[4] reports that women comprise only 3% (6 women) of those defined as "the bureaucratic elite." This study, a replication of the 1953 study,[5] shows that no women were included in the bureaucratic elite during 1953. Olsen reports that if women were included in this sector to the same extent as men (i.e., from the same pool of university-educated people), they would comprise 25% of this sector.

In the judiciary, the situation is even more dismal. There has never been a woman among the 9 judges of the Supreme Court of Canada. In the federal courts, of 390 judges identified on the Canada Law list for 1979, 9 are women. In the provincial family and divorce courts where one might expect the appointment of women to be made first, 6 women are found.

In examining the representation of women as leaders in the wealth-producing sectors, the data are very difficult to locate. Time series data are impossible to find. For example, in a study of the Canadian corporate elite, Clement[6] finds that in 1972 only 6 of the 946 elite members he studied are women. This finding is consistent with the Report of the Royal Commission on the Status of Women in Canada,[7] a report of 5,889 directorships and 1,469 executive positions in Canadian corporations. The authors note that women hold only 41 or .7% of the directorships and 8 or .5% of the executive offices. Of the 6 women identified by Clement, 4 are in the media industry. As far as one can tell from his analysis, none of the women is in any of the industries which we have identified as wealth producers. An examination of other sources reveals no evidence that women constitute a significant force in the wealth-producing sector. McGibbon[8] shows that of 79 Canadian manufacturing companies in 1973, only 3 female board members are found among 947 directors. In a study of highly qualified personnel in Canada in 1973 which identifies the occupation (job of longest duration) of university graduates, women constitute 7% of government administrators, 1% of the managers and senior officials, 14% of the administrators of education, and 4% of lawyers and notaries.[9] Not only are these small proportions of the various occupational groups, but they are small in absolute numbers. The managers and senior officials number 140 women. Despite 11 large governments in Canada, only 575 female government administrators are found. There are 65 women civil engineers, which is the largest single group of engineers in Canada. Of all economists in this study, only 585 are women. This scarcity of women limits the scope of linkages among women and shows the small size of the potential pool of women who are promotable to top jobs.

In the wealth-producing sectors, we are interested not only in the top officers, but also in the proportion of women in the industrial sectors. The reason for this interest is that the women discussed thus far have no organized power to exert group pressure on the people in top positions. In the other sectors, however, the withdrawal of the labor of women either through strikes or departure from the labor force would have considerable immediate impact on that sector. That is, wealth will not continue to be created in industry and primary production if women workers strike.

Table 1 indicates that women have grown as a percentage of the total labor force in every industrial sector except community/personal service. We can conclude that because of these increases, women are somewhat more significant as a presence in those sectors which are defined as wealth producing. We should also note that there has been a decline over the 20-year period in the growth of the industrial labor force in agriculture, forestry, and fishing. Therefore, while the overall number of workers in these industries is declining, largely due to mechanization, women have become a larger proportion of this work force. In the same period, there has been a substantial percentage decline in the proportion of the labor force employed in managerial, religious, farming, processing, and primary industries other than farming. As a proportion of all workers, the percentage of women is declining in such expanding sectors as teaching and sales, whereas women represent an increasing proportion of workers in such declining occupations as managerial, processing, farming, and other primary occupations. Women are disproportionately found in such non-wealth-producing occupations as teaching, medical and health care, and clerical work; these rates are considerably higher than those observed for women in the wealth-producing sectors. For example, although agricultural products are a major Canadian export, women, who are an expanding group of workers, account for only 23% of workers in the entire industry and 21% in farming occupations. Of all farm operators, approximately 4% are women. While this seems to be an increase over time, the average income of female farm operators and the increased number of women listed in farm occupations indicate that the majority are operating farms or working in farm labor because their husbands earn off-farm income. Whether or not the major change in the organization of farming has left women holding responsibility for a low-profit business, this still represents a form of control in the industry.[10]

In the fishing industry, so few women are occupied that they do not appear in statistical accounts.[11] However, at least one large fishing concern in Newfoundland is headed by a woman, which is extremely unusual. Women work in various sectors of the fishing industry, but mostly in processing, under which occupational title they would be listed.

The involvement of women in the energy industries, i.e., oil, gas, and hydro, is confined largely to office work.[12] In a recent study of the oil industry, titled quite significantly, The Last of the Free Enterprisers: The Oil Men of Calgary, women are listed as dedicated secretaries, wives, and mothers, but never appear in top jobs. [13]

THE REAL IMPACT OF INCREASED
LABOR FORCE PARTICIPATION

Despite these differences in the labor force experience and power of women in comparison with men, it is possible to observe changes in social organization as a result of the changed pattern of adult female lives, which may be more significant to Canada than the emergence of women into the elite. In addition to identifying some of the changes that are occurring, the purpose of this paper is to suggest the direction or areas in which the impact of women's increased labor force participation is beginning to be felt although it is still not highly visible. These areas include public policy and family and community life. If they occur in the direction indicated at present, it will mean that the generation of women born between 1950 and 1960 is a lead group in the transition to a society organized quite differently from that which we now experience.

Political Consciousness

In terms of public policy, several changes have occurred already. First, it is possible to document that the emergence of women as permanently attached workers in the paid labor force at all levels has led to an increase in political consciousness among women. Elsewhere, I have attributed much of the drive of the contemporary women's movement in Canada to the consciousness-raising aspects of paid labor.[14]

A good example is bank workers. Between 1911 and 1931 (years of the census in Canada), banks moved from being numerically dominated by male clerks and tellers to being dominated by women in these jobs. Although a numerical majority, women did not for many years move into the ranks of managers or into any of the specialized jobs in banking. Most of the women worked in relatively unskilled and low-paid clerical jobs until marriage or childbearing and some returned later in life.[15] The majority of women were young and did not consider themselves permanent workers. While many women in banking still see their work as an interval between school and motherhood, a growing number look with increasing anger at the young men who enter the bank, move very quickly through the training phases, and secure management jobs in the major chartered banks.[16] Second, bank workers have begun to unionize.[17] Unlike other secondary workers in industries where women with low skills have always worked to supplement the family income and where there has been unionization for many years, the bank workers are explicitly feminist in their unionization practices. Indeed, it was a group of feminist bank and clerical workers in British Columbia who first made national headlines by forming a union (SORWUC). Since then, the major unions, such as the Canadian Union of Public Employees and the Steelworkers, have broken the feminist union and organized on their own. Such political consciousness in the workplace is by no means confined to white-collar women workers, although it is most dramatic at that level. The major white-collar union in Canada (Canadian Union of Public Employees) has a majority of women members and is led by women. Overall, about 25% of all women workers and 40% of men are unionized.

Nor is the political consciousness of women as workers confined to the workplace. An increased level of awareness among women of underrepresentation of women at all levels of political jurisdiction has led

to an increased activism in the organized political parties, in the
formation of a Feminist Party of Canada (which, however, has not yet
fielded candidates in an election), and in parapartisan groups such as
Women for Political Action. Political activism is not confined to getting
women elected to municipal, provincial, and federal levels of government. Women have emerged as a major lobby on all levels. This lobbying
activity was spurred on by International Women's Year during which
governments funded women's political activities in many spheres including
the exertion of pressure to review and rewrite textbooks used in the
schools and to break the official and unofficial barriers against the entry
of women to nontraditional jobs in the blue-collar world, the professions,
and education in general.   While much of the political activism was
originally feminist in spirit and intent (however one defines feminism), the
organization of women (the Yvettes) in the province of Quebec during the
referendum to decide whether or not that province would remain in the
federation has been a dramatic statement on the part of more traditional
women.   Associations of women electors, women in church groups, and
women in other voluntary sectors have existed since the fight for the vote
was won in 1918. The Yvette movement in Quebec, however, represents a
new voice for nonradical women. It combines a claim for the dignity and
maintenance of the traditional values of home and family with a call for a
new role and new dignity in the community.   While confined to the
province of Quebec, where the proportion of women in the labor force is
lower than in other parts of the country, it is a significant political force.

Policy Changes

The political consciousness of women has pushed governments
toward an internal examination of their own practices and attitudes.
Equal opportunity programs, affirmative action within the government
service, and the establishment of a Human Rights Commission at the
federal level with investigatory and sanctioning powers on the matters of
opportunity and equal pay for work of equal value have aided an
aggressive attitude on the part of government in comparison to past
practices.   While only 10% of the Canadian labor force works in the
federal jurisdiction, the publicity surrounding claims, disputes, and settlements has led to debate in the provincial jurisdictions.

Many adjustments have been made in pensions and benefits, legal
practices concerning property laws, criminal proceedings, and elsewhere.
The commitment is evident, despite the slow pace of change. The
direction of change is toward expanded rights for women. A new and
major social issue arises directly as a result of the increased attachment
of women to the labor force. There is no family policy in Canada. There
is official and unofficial support of "the family" (whatever that may be).
We deplore, but permit divorce on various grounds; we pay a family
(mother's) allowance of a minimal but symbolically important variety; tax
deductions are possible on the basis of dependents which are defined as
children, dependent parents, etc., but no official policy explains to us
what the family should be.[18]

As the costs of social welfare programs increase and as we move deeper into the current recession facing North America, it is clear that the Canadian government will have to decide whether eligibility for social benefits will be made on the basis of the one-earner or two-earner family. In other words, a specified direction will have to be taken, as in Sweden. This decision is necessary not only because of the outlays for social welfare programs but because of the necessity to collect tax revenues based on some model of compensation.

The two-earner family policy means that child care facilities, lunch and afterschool programs, more flexible pension plans, increased geographical mobility, and changed nepotism rules in work organizations must be developed. Families with only one earner (whether it is a single-parent family or because for some reason one spouse is unable to be in the labor force) will be entitled to more compensation and benefits than others. A single-earner family policy means that sanctions will be applied to the two-earner family in the form of lack of public child care, higher taxes, and a community life organized around the availability of one adult person at home.

In the case of the two-earner family, the pressure to create equality in the workplace in terms of opportunity and earnings may have a depressing effect on the wages of both men and women (the general assumption being that a two-earner family is required financially to maintain the family and, therefore, each earner needs less). If the one-earner family is maintained as the model, the pressure in terms of equal opportunity will be to make it possible for either the father or the mother to stay at home. In either case, maternity and/or parenting "leave" and benefits will require a major overhaul. A major policy decision such as this is a direct consequence of the movement of women into permanent attachment to all levels of the labor force.

Change in Community Life

Much of the community organization of Canadian life has traditionally been based on the volunteer labor of women. Churches, political parties, and various social services to the aged, the handicapped, the imprisoned, and the sick as well as the organized pursuit of traditional crafts have drawn on the labor power of non-labor-force participants. Such organizations and agencies have all had to face reorganization of their focus and activities over the last decade. In many cases, volunteers have been replaced by paid workers. This change is particularly evident in political parties where full office staffs, not only at the provincial and national level but also during election campaigns at the riding level, have replaced the former staffs made up mostly of women volunteers. Churches, too, have hired individuals to perform formerly voluntary duties. Increasingly, corporations volunteer the labor power of executives in community activities. Sports organizations such as the ubiquitous hockey and baseball leagues for young people (boys) and the children's organizations of Brownies, Girl Guides, Cubs, and Scouts as well as Sunday schools find themselves strained for help. Without evaluating the desirability of this change, the impact is fundamental to the nature of community relations and community life. Highly skilled and talented workers are unlikely to be attracted to the low-paying jobs which were formerly performed by volunteer labor. Child care is an important example. The number of complaints, accidents, and community scandals surrounding this problem in Canadian society makes it evident that this type of work for the most part is not attracting high-quality workers. It is more difficult to discover what is happening in other community organizations, although as the churches in Canada have hired professional workers, they have become much more politically conscious. Community pressure groups for environment protection, legal and health care issues, and recreation have sprung up.

The movement of women out of the home and into paid labor has its commercial effect as well. The increased consumption of manufactured goods as well as fast and convenient foods is well documented. The reorganization of the household, however, means that seldom is anyone at home to accept deliveries, admit a variety of utility inspectors, and confer with service workers who repair telephones, appliances, and plumbing facilities. Apartment building superintendents must accept responsibility for fulfilling such domestic arrangements for their tenants. Neighborhoods are deserted during the day. The woman at home with her children finds herself doing a number of favors for her employed neighbors and spending her free time talking only with the retired and the very young. Children's birthday parties, which had been a major feature of afterschool entertainment, are now held on weekends, if at all.

Studies on the division of labor in the household show that, when both parents work, women reduce their leisure time to maintain the household while men hardly increase their domestic work time at all. One of the consequences of this reduction in women's leisure time is that the traditional craft work of women is disappearing. I am not now thinking of handwashed dishes or the preparation of special dishes, but needle skills, dressmaking, gardening crafts, the preservation of food, and musical skills. All of these are either moving into the commercial sphere (handknit clothes created in production line operations are both very expensive and a symbol of high status) or disappearing in the younger generation.

On the other hand, the confining aspects of the neighborhood are reduced by the increased social circle and civic awareness of women in the paid labor force. No longer does the woman meet only her neighbors and her local church congregation. Every day at work, she meets people from a wide range of social situations. Social situations refer not only to the socioeconomic status of co-workers, but, more important in our society, also to people from a wide variety of cultural backgrounds since Canada is not a homogeneous culture. About one-third of Canadians have neither French nor British heritage, and language and religious diversity is characteristic of many communities.

PERSONAL LIFE AND THE NEXT GENERATION

The family is a primary group in which the socialization of the next generation principally occurs. Its fundamental organization is changed when both parents spend a large part of each day in the paid labor force and do so throughout most of the child's life. This reorganization changes the expectations and socializing experience of the child. The debate continues over the issue of whether the children of working mothers are more or less well adjusted, more or less likely to be successful in school, or more or less socially deviant. Very little has been done to assess the impact of both parents being employed outside the home, although this change is as important as the transition from being a predominantly rural society to a predominantly urban society.

Changes will occur in the present generation, not simply in the child's emotional and social development, but also in the relative power of the spouses in the household. It is now understood that as women gain legal and economic equality outside the home, there is a change in the relative power of the parents in the household with respect to one another and to the children. Our culture is built around the notion of the dominance of male values and culture. We turn little boys into big, strong, hardworking fathers while little girls become pretty, submissive, hardworking mothers. It is not at all clear quite what is changing, if anything, on these dimensions. We continue to value marriage and a small family. We seem to be passing that value on to the next generation. We continue to make significant distinctions on the basis of gender in social organization. So far, we are opting out of the problem of defining clearly for little girls and little boys the nature and content of adult roles.

OFFICIAL DATA

Another consequence of this changed organization is the impact on official agencies of data collection. Household studies and census collection have assumed a single-earner household and an adult woman at home. Unemployment figures tend to underestimate female unemployment.

Tax policies have been designed to deal with, at best, a part-time earner among adult married women. Tax inequities are rising. Social programs in pensions and benefits have assumed some sort of family structure which would make it highly unlikely that women would be permanently attached to the labor force. All of these assumptions are changing and with them, difficulties in data collection and the calculation of benefits. The transitional generation of women to which we referred initially is experiencing the squeeze from both sides on this level as well. One response of the official statistical agencies has been to refine estimating procedures, and to produce more standard materials, but not to redo the basic framework.

CONCLUSION

The purpose of this chapter has been to suggest that while the increased rate of labor force participation has not had a major visible impact on the representation of women in the decision-making and wealth-producing sectors of Canadian society, it has had a different impact. The nature of family organization and community life has been changed by the entry of women into a permanent attachment to the labor force, which is much more clearly defined among young women than their mothers. The political consciousness of women who view work as a crucial aspect of their lives with a higher priority than previous cohorts is having an impact on civic life as well as economic organization. The withdrawal of the labor of women from community organizations is forcing a change in their functioning. The governments are faced with the need to adjust social security programs, but are caught in a generational shift which is not fully recognized or understood.

NOTES

1. Nakamura, Nakamura, and Cullen, 1979 b.
2. Love, 1979.
3. Marsden and Harvey, 1979.
4. Olsen, 1980.
5. Porter, 1965.
6. Clement, 1975.
7. Royal Commission on the Status of Women in Canada, 1970.
8. McGibbon, 1975.
9. Ryten, 1975.
10. Shaw, 1979.
11. See, for example, Women's Bureau, Labour Canada, 1978.
12. Smith, 1978.
13. House, 1980.
14. Marsden, 1980.
15. Lowe, 1979.
16. For example, Royal Bank of Canada, 1978.
17. Lowe, 1978.
18. Eichler, 1980.

POWER AND POLITICAL BEHAVIORS IN ORGANIZATIONS:

A NEW FRONTIER FOR RESEARCH ON MALE/FEMALE DIFFERENCES

Virginia E. Schein

Stamford, Connecticut   U.S.A.

Over the last decade, women have been making significant progress in terms of entry and progression in the world of work. In the United States, over 40% of the work force is female, and compared to 1970, the number of women in managerial and professional capacities has risen from 13% to 17%.[1] That women can and want to contribute in the work force is no longer an issue. Despite such progress, women still have not made significant inroads into the upper levels of organizations—into positions of power and influence. Thus, a major problem facing qualified women today is how to gain access to power structures in the organization which employs them. Having a good job is no longer a sufficient entry to these positions; women also must have the same opportunity as men to advance to their highest potential. The acquisition of power and the use of power and political behaviors become essential ingredients in this endeavor.

Although power and political behaviors are important, several researchers[2] have pointed out the neglect of research attention in this area by investigators in the organizational behavior field. Moreover, the limited work that has been done has failed to even consider, much less analyze, the possibility of differential access to power and utilization of power and political behaviors among men and women. The purpose of this paper is to examine the roles of power and political behaviors as they relate to effective managerial functioning in organizations, to consider possible male/female differences in access to power and the utilization of these behaviors defined as "political," and to suggest new avenues for research.

POWER AND POLITICAL BEHAVIORS IN ORGANIZATIONS

One reality of organizational life consists of rational behaviors which involve planning, organizing, directing, and controlling. Underlying these activities is another set of organizational behaviors that revolve around the gaining and keeping of power. Power struggles, alliance formation, and strategic maneuvering may be as endemic to organizational life as planning, organizing, directing, and controlling.

Machiavelli,[3] in giving advice to his Prince, spoke openly of the importance of being political—that the illusion of being honest, compassionate, and generous was important to gaining and maintaining power, yet so, too, was the realistic necessity of breaking one's word, being cruel, and being parsimonious. Being both a lion and a fox was his counsel. Although this advice was directed at those interested in the takeover of principalities, a cursory look at behavior within organizations suggests that Machiavelli is still with us and still doling out advice. His ghostly hand seems to be reflected in memorandums that distort or omit information, meetings held to decide what has already been decided, coalitions formed covertly to block a decision, and rewards promised but never delivered.

If we view an organization as a political environment in which the acquisition of power is the key ingredient, then the similarity in tactics and strategies between managers and the princes to whom Machiavelli was giving advice should not seem strange at all. It is only when we view organizations as rationally structured systems, built on a division of labor, separation of function, hierarchical flow of communications, and formal authority, and operated by individuals motivated by esprit de corps, that these behaviors appear dysfunctional in that they seem to muddy up the waters and prevent the work of organizations from being accomplished. The view presented here is that the latter description is an illusion and that the reality of the way organizations function is far more similar to that of a political arena in which individual managers jockey for power and influence.

Cyert and March[4] have described the executive as a "political broker" operating in a system in which the decisions on the allocation of resources are made by political coalitions, each with potential control over the system. Schein[5] maintains that while some of these power and political behaviors may be for personal aggrandizement, many others are work related and designed to improve employee satisfaction, work unit productivity, or the effective implementation of projects. For example, the powerful department head has the influence to allocate money and promotional slots to reward and motivate a subordinate; the powerful head of production has the bargaining power with the head of supplies to move out production faster; and the powerful head of research has the resources to trade off to get projects implemented. The effective use of power is essential to getting the job done.

Knowledge and use of power acquisition behaviors are essential to another key ingredient of effective managerial functioning: bringing about change in an organization. Creating new programs, developing new policies and procedures, and restructuring units in organizations are important activities for upper-level management. Here, too, power becomes a vital factor, because any proposed program change will threaten the current power distribution. According to Pettigrew, "[People] may see their interests threatened by change, and needs for security or the maintenance of power may provide the impetus for resistance. In all these ways, new political action is released and ultimately the existing distribution of power is endangered."[6] Hence, in order to implement new programs and policies, one must understand power and political behaviors.

As a starting point for research on power and political behaviors in organizations, Schein[7] developed a conceptual framework which focuses on bases of power available to the powerholder and strategies the powerholder uses to bring about a desired outcome. Within this framework, the consideration of male/female differences points to new frontiers for research on equal opportunity in the world of work.

POWER BASES

Power bases are those resources available to an individual which enable him or her to convince another to go along with a proposed plan or idea. Few power bases come with the job. Most need to be developed by the individual and continually maintained lest they be eroded by others seeking to expand their own power and influence.

Key Power Bases

Among the power bases most useful to middle- and upper-level managers are expertise, control over information, political access, assessed stature, group support, and mobility. Expertise, usually acquired through formal education in a particular area such as law, engineering, or accounting, gives an individual the power to influence those lacking knowledge in that area. Related to the base of expertise is that of the ability to control information. Given control over information, the manager can influence others by having more information than the other party, withholding portions of the information, or distorting the information transmitted. Political access refers to the ability to call upon informal networks of relationships within the organization.[8] Through an informal network of relationships, the manager can obtain information pertaining to "what's happening" and has an avenue for influencing other individuals. Assessed stature is perceived competency by others. It allows the manager to influence others based on the general knowledge of his or her prior accomplishments. Another power base is group support from subordinates. In addition to providing a cohesive work unit, influence attempts can be enhanced if subordinates are willing to rally support from their peers in other units. The final power base to be considered is that of mobility, which develops from external recognition. With mobility, if all else fails, the manager can always leave a particular job and get a similar or better job elsewhere.

Power Bases and Male/Female Differences

The development of these and other bases are essential if an individual is to wield influence in an organization. Without such resources, effective power acquisition behaviors are impossible. With regard to women, however, what is unknown is the extent to which women, compared to men, are able to develop the necessary power bases in organizations. Is there differential access for men and women? The question is extremely important, yet there is little, if any, research designed to provide the answer. There are, however, some indirect sources that furnish a partial answer and a background for subsequent research.

Data reported on education suggest that expertise is a power base that women can and are acquiring. For example, women's enrollment in a Master's degree program in Business Administration (MBA) in the United States has increased from 4% in 1972 to 24% in 1978.[9] For both women and minority group members, objective credentialing has been an effective means of overcoming discrimination at the point of entry.

Research findings that are indirectly related to assessed stature and group support, however, show that these bases are not readily accessible to women. Research on performance and performance evaluation suggests that it may be more difficult for women than men to develop a power base of assessed stature. In Deaux and Emswiller's experiment,[10] in which Ss evaluated the identical performance of either a man or a woman on a "masculine" task, the woman's success was attributed to luck, whereas the man's success was attributed to ability. Similarly, Feldman-Sumners and Kiesler[11] found that Ss attributed more motivation to women for the same performance as men while male Ss viewed successful professional women as less capable and/or having an easier task than their male counterparts. Garland and Price[12] found that men who had negative attitudes toward women as managers tended to attribute the woman manager's success to luck or ease of the job. Overall, these results suggest that, despite her successes on the job, a woman will be less likely than her male counterpart to develop a base of assessed stature. Research on subordinate perceptions of leader behavior points up possible male/female differences in obtaining group support. In two separate studies correlating subordinate perceptions of leader behavior and subordinate satisfaction,[13] satisfaction was found to be linked with perceptions of sex-role consistent behavior. Rousell[14] found that high school departments headed by men were rated by teachers high on dimensions of esprit and intimacy while departments headed by women were rated high on the dimension of hindrance. Also, Hansen[15] found that subordinates of both sexes were less satisfied if the superior was a woman. While the studies in this area have not all been clear-cut,[16] they do suggest that women in supervisory positions may be limited in their behavior flexibility and subjected to negative reactions by subordinates for behaving in a non-stereotypical manner. Such behavioral constraints and biased evaluations would make obtaining group support more difficult for a female than a male supervisor.

Control over information and political access also appear to be two power bases more difficult to acquire for female than male managers. According to Mintzberg,[17] managers obtain relevant information by developing contacts outside the formal system and by establishing specific informal communication channels within the organization. Similarly, political access is acquired through the development of an informal network of liaisons outside the formal structure.

Negative perceptions of a woman's managerial abilities may foster exclusion from these informal contacts and networks. Schein,[18] using both a sample of 300 middle-line male managers and a sample of 167 middle-line female managers, found that both successful managers and men in general were perceived to possess the characteristics of leadership ability, competitiveness, self-confidence, objectivity, aggressiveness, forcefulness, motivation to excel, and desire for responsibility. Women were perceived as not possessing these characteristics. The male sample went even further and added emotionally stable, steady, analytical, logical, consistent, and well informed to the list of characteristics possessed by men and successful managers, but not possessed by women. To the extent that to "think manager" is equated with to "think male," women may be less likely to be included in these informal networks and are denied access should they seek them out.

Finally, mobility, the ability to remove oneself from a losing situation, may be a less viable option for women than for men. To the extent that there is no equal job opportunity in the marketplace, a woman's mobility will be less than that of her male counterpart.

Not only is it important to know what bases of power might be differentially accessible to men and women, it is also important to understand the differences with regard to the number and kinds of power bases that are utilized by men and women. Some exploratory research by the author sheds some light on this question. Forty-three middle- and upper-level executives, 32 men and 11 women, were interviewed in depth concerning their attitudes toward and uses of power acquisition behaviors. As part of this interview, they were asked if they "frequently" or "infrequently" used the following power bases: expertise, information, personal appeal, past accomplishments, and support from others. Chi square analyses of the relationship between sex and frequency of use of each power base revealed no differences in usage between the sexes except on personal appeal ($\chi^2 = 4.07$, $p < .05$). Here the women cited frequent use of personal appeal whereas men cited infrequent use. In addition, on only two of the power bases, expertise and information, did more than half of the male sample state frequent use; whereas, more than half of the female sample indicated frequent use of expertise, information, personal appeal, and support from others. Thus, women tended to use more power bases than their male counterparts and, in particular, had a greater tendency to use personal appeal.

Given the exploratory and limited nature of this study, we can only speculate on the interpretation. Are women, who tend to use greater numbers of power bases, actually more powerful than their male counterparts? Or, as an alternative, are they less powerful, with weaker bases of expertise and information power, causing them to seek more power bases from which to operate? Is personal appeal a truly effective power base? Or is it one that women are forced to use because of limited potency of their power bases or limited access to other bases of power? What is needed is more research directed at answering these questions as well as research on the differential access of men and women to power bases.

POWER STRATEGIES

Coupled with research on power bases is the need to examine power strategies. Power bases alone are insufficient unless one uses them astutely. Strategic maneuvers based on the resources available are vital to successful influence attempts. A variety of observational studies of managerial behavior provides some insight into the kinds of strategic maneuvers utilized in organizations. Martin and Sims [19] discuss the tactics practiced by most men whose success rests on the ability to control and direct the actions of others. Among the tactics they report are: taking counsel, forming alliances, maintaining maneuverability, promoting limited communication, compromising, negative timing, using self-dramatization, and exhibiting confidence. In a case analysis of a middle manager's tactics for power expansion, Izreali [20] reports such tactics as neutralizing potential opposition, making a strategic replacement, committing the uncommitted, and forming a winning coalition.

While strategies vary and are usually situationally determined, key behaviors that emerge from the descriptive research are those of coalition and alliance formation.[21] Without such coalitions and alliances, managers appear to be severely handicapped in obtaining important resources, such as budget funds and promotional sales.

All of these liaisons and contacts revolve around an informal network of people built up over time through the development of relationships both on and off the job. Stereotypical thinking with regard to women, however, may foster exclusion from these networks and/or make it more difficult for women to become active participants.

Moreover, such exclusion, based on perceptions of differential characteristics and behaviors, may have no realistic grounds. Lirtzman and Wahba,[22] in their research on the determinants of coalition behavior, found that under certain conditions men and women used the same coalition strategies. More important, their research countered earlier research[23] which did show differential strategies based on sex-role consistent behaviors. Vinacke and Bond and Vinacke[24] reported that men used an "exploitive" strategy in competitive situations whereas females adopted an "accommodative" strategy. Their data suggested that in the organizational context men might behave one way in coalitions and women another. Furthermore, to the extent that women's behavior was contrary to organizational goals of competition and maximization, they would be excluded from these coalitions. In contesting these conclusions, Lirtzman and Wahba maintained that coalition behavior was situationally determined, not sex-role determined. They studied the coalition behavior of 48 women in 16 triads and found that under conditions of high competition and high uncertainty, women, similar to men in a previous study, did not use accommodation strategies. Rather, they used strategies that maximized coalition utility. Women adopted the same coalition strategy as men did when placed in a clearly competitive situation in which outcomes were subject to risk. According to Lirtzman and Wahba, "These findings raise questions about the traditions of business that bar access of women to high organizational positions precisely because it is expected that women will act naturally according to sex-related roles; that is, non-competitively."[25]

CONCLUSION

The countering of sex-role stereotypical thinking, via research, is imperative if women are to have an equal opportunity for managerial success. Exclusion from coalitions and informal networks may impair the manager's ability to function effectively. To the extent that alliances, trading of favors, and influence networks are important for getting work done, a woman's probability for success appears less than that of her male counterpart. And, to the extent that these power acquisition behaviors are rarely acknowledged by organizational members,[26] her lack of effectiveness can then be attributed to her inability to perform technical and administrative managerial skills. A woman may be subject to a very subtle but quite profound form of discrimination by her exclusion from these liaisons. Finally, to the extent that a woman is unaware of the importance of coalitions and liaisons, she may accept the organizational view that her technical and administrative skills are inadequate, in which case she may lower her own aspirations.

The effective use of power and political behaviors are essential to managerial success and advancement. Obstacles to this success, posed by possible differential access to and use of power bases and strategies by men and women, must be overcome. The consideration of male/female differences in this area is a new frontier which demands significant research attention if women are to achieve full equality of opportunity in the world of work.

NOTES

1.   Shaeffer and Axel, 1978.

2.   Porter, 1976; Schein, 1977; Tushman, 1977.
3.   Machiavelli, (1513) 1964.
4.   Cyert and March, 1963.
5.   Schein, 1977.
6.   Pettigrew, 1975:192.
7.   Schein, 1977.
8.   Pettigrew, 1975.
9.   National Network of Business School Women Report, 1979.
10. Deaux and Emswiller, 1974.
11. Feldman-Sumners and Kiesler, 1974.
12. Garland and Price, 1977.
13. Petty and Lee, 1975; Petty and Miles, 1976.
14. Rousell, 1974.
15. Hansen, 1974.
16. Terborg, 1977.
17. Mintzberg, 1973.
18. Schein, 1973, 1975.
19. Martin and Sims, 1956.
20. Izreali, 1975.
21. Sayles, 1964; Mintzberg, 1973.
22. Lirtzman and Wahba, 1972.
23. Wahba and Lirtzman, 1971.
24. Vinacke, 1959; Bond and Vinacke, 1961.
25. Lirtzman and Wahba, 1972: 411.
26. Schein, 1979.

SEX-ROLE STEREOTYPES AND PERSONAL ATTRIBUTES

WITHIN A DEVELOPMENTAL FRAMEWORK

Anne-Sofie Rosén

University of Stockholm
Stockholm   Sweden

Female work participation is high in Sweden today, but not without several problematic features. The problem associated with female work participation is no longer one of whether or not a Swedish woman should work. She does. Rather the issue can be formulated as follows: What kind of work is she doing? What kind of employment, part time or full time, does she seek? What position has she attained within Swedish society? How do young girls today prepare themselves for their future occupations and what are their aspirations and goals? What has been achieved in the years of formal equality between the sexes?

SEXUAL EQUALITY IN SWEDEN

Equality between the sexes has been a cherished principle for a fairly long time in Sweden. Let me briefly sketch recent developments. In 1920, legislation became sex-neutral in all aspects of civil rights. It became possible for women to enter civil service careers on equal grounds with men—with very few exceptions. At that time, the labor market was still highly segregated by sex. Working women were mainly found within a very narrow range of occupations, all in the lower strata. In the 1930s, the economic depression and a very low birth rate led to increasingly hostile attitudes toward married women's employment outside the home. No formal restrictions, however, were introduced. During the 1940s and the Second World War, many women, married or single, were needed in the labor force to substitute for men who were drafted. Many day care centers were started during those years and during the period of economic upswing immediately after the war. But in the 1950s, the Swedish government, in total agreement with the leading unions and the employer organizations, decided to meet the rapidly increasing shortage of labor by recruiting skilled male workers from abroad. These measures were taken instead of expanding day care services and instead of investing in the training or retraining of women to meet the demands of labor. At the same time, the birth rate during the first half of the decade was high.

Still, an increasing number of women sought and obtained employment. By 1960, when most of the children born during the baby boom of the previous decade were entering school or had gone to school for a few years, nearly half of the women in the labor force were married women. By then the stage was set for more serious efforts to be made to bring about greater equality between the sexes in the educational and occupational sectors of life: to change formal equality into actual equality. This decade was characterized by notable changes in attitudes in Sweden toward women's work outside the home and also by greater demands for, and increased building of, day care centers and nursery schools. The educational system was reformed to promote equality. Most important for married women's participation in the work force was the taxation reform at the end of the decade, which finally introduced individual taxation. The two-earner family became the accepted social model. By the end of the 1960s, about 40% of the work force was female. The accumulating pressures from the female half of Swedish society have led to considerable activity among politicians since 1970—for ideological and/or opportunistic reasons. "Real" equality is now the expressed goal of the leading political parties, even if some of our major parties have been extremely slow to effectuate equality between the sexes within their own party ranks. The percentage of women in Parliament is increasing, however, from 3% almost 40 years ago to 26% as of the last election in 1979. Further, sex discrimination in the labor market has become illegal and several special bodies have been created to guard against incidents of sex discrimination within society.

Educational and Occupational Equality

For two decades the Swedish government and several organizations have been active in various attempts to equalize social opportunities and life experiences of women and men. But still, segregation of, and discrimination against, women have been amply documented in a number of studies in the social sciences, history, and literature.[1] The official ideology, in Sweden as elsewhere, does not always find its parallels in the ideology and behaviors of individual members of society. The deliberate attempt to change traditional sex-role patterns in the family, education, government, and the labor market has resulted in what many consider to be a disappointingly slow process of change. About 20 years ago, the basic school system was reformed and the promotion of equality between the sexes became one of the foundations of the entire educational system. Today, the success of the 1962 school reform, as assessed from adolescents' preferences for future occupations, is very disappointing to many. The girls' choices of postcompulsory school education are nearly as traditional as ever. They still prefer the so-called female occupations: nurse, teacher, hairdresser, beautician, etc. Although occupational shifts between the sexes can be noted, there still are typically female occupations where more than 95% of the total number employed are women. At the high school level, less than half of the gifted girls aim at an occupation requiring a higher educational level such as a university degree. For boys, the corresponding figure is much higher.[2] Parenthetically, universities have been open to women since 1873.

The occupational attainments of Swedish women are still generally low. There also is a clear underrepresentation of women in many of the high-status careers when the number of adequately trained women is considered. For instance, more than one-quarter of the students graduating from law school are women. However, only 6% of the lawyers are females. In the judiciary, 8% are women. At present, only a handful of them have reached the very highest levels in the judiciary as compared to some 150 men who have. A similar pattern is clearly discernible in other sectors as well: the higher one gets, the fewer women one meets. This pattern is also true in areas where the proportions of men and women entering an educational or occupational path are equal—and even where women initially dominate in numbers. Educational and occupational choices and careers together with a number of other observations point to the existence of a distinct sex-role division within Swedish society that has been, and continues to be, firmly implanted despite much effort expended on inducing ideological changes.

Sexual Equality and the Socialization Process

In social psychology, a widely held theory states that the existing sex-role division within a society is incorporated into the self-system of an individual as an outcome of the socialization process. In the form of normative expectations of what is appropriate for a woman or man, the societal sex roles will direct many of her or his everyday activities and influence many of the important decisions each individual has to face, such as his or her educational and vocational investments. Since formal equality between the sexes has existed for a fairly long time and all possibilities are open to young girls and women, how can one explain why so little has been achieved? Is it possible that the socializing influences in Sweden work in separate directions for the sexes and that the ambitious effort to equalize opportunities for women and men has not reached beyond a mere surface level of verbally expressed attitudes?

In a series of studies, we have tried to trace possible effects of different socializing experiences of women and men by use of methods that may tell us something about how sex-role demands are perceived, and what differences there are in self-perceptions and value orientations of women and men. In all of the studies to be reported, the age range of our subjects has varied from 20 to 65, with a group median near 35. Although many findings have been reported from studies of similar questions, particularly in the United States, the conclusions may not be valid in the Swedish or Northern European context.

MEASURING PERCEPTIONS OF SEX-ROLE
DEMANDS, VALUES, AND TRAITS

In studies of perceptions of sex-role demands, the subjects have been asked to describe what is expected or demanded of a man or a woman, or to rank order a number of such descriptions in order of their importance in today's Sweden. When a subject of either sex is asked to describe what she or he perceives to be salient normative expectations for both sexes, these descriptions are likely to be contaminated by such influences as memories of recent mass-media debates or a social desirability factor. In order to avoid this confounding, we also have used other methods, where the social desirability factor appears to be less important. In one study, for example, we have applied content analysis to sentence completions. A set of stems have been used that are assumed to elicit thinking along lines influenced by societal sex roles, e.g., "A woman (A man) should always . . . ." "It is difficult to be a woman (a man) today, because . . . ." "If I were back in school . . . ." Such stems are mixed with assumedly neutral ones.

A second problem area concerns value differences between the sexes that may reflect differential socializing processes for men and women. In this research, we have relied upon the work of the social psychologist Milton Rokeach.[3] Values refer to existential states that an individual finds worthwhile to strive for or to personal characteristics she or he would like to have. Rokeach's Value System lists of value items have been used.

Much recent work on the influence of sex roles on the personality makeup of the individual has relied upon the method of self-descriptive trait ratings; that is, subjects are asked to rate the degree to which adjectives, such as kind, warm, or irritable, can be applied to their own typical behavior.[4] In the United States, the average person has been reported to describe herself or himself in a stereotypic fashion, using characteristics that are sex-linked and considered to be more desirable for one's own sex than for the other. Men typically describe themselves as more dominant and achievement oriented, whereas women typically describe themselves as more nurturant and sensitive toward the other person's feelings. The trait descriptors used by these researchers, and many others in the study of masculinity-femininity-androgyny as personality characteristics, have come primarily from what Wiggins [5] denotes as the interpersonal sphere of descriptors in ordinary or everyday language. In our research, we have used trait ratings scales based on the work of Wiggins.[6] Our subjects, then, are asked to rate themselves on a number of scales for when they interreact with others in daily life.

PERCEPTIONS OF SEX-ROLE DEMANDS, VALUES, AND TRAITS
AMONG SWEDISH WOMEN AND MEN

To report some findings, let us start with perceived sex roles. Our female subjects think that there are strong demands in today's Sweden that women become good mothers, are able to support themselves, and become personally independent and self-reliant. They also say that it is expected that women should become more like men in the sense of engaging themselves more in social or political affairs and in their professions and jobs. The male subjects, on the other hand, consider today's female role to be built around demands of being personally independent, but still attractive and feminine. They place less weight on the motherhood demands. As to the male role, the men in our studies report that it requires leadership ability and an achievement orientation, and that men should be able to support a family and not only themselves. Female subjects respond that it is expected of men today not to be wholly absorbed in their jobs and careers and to share the responsibilities for the family on the practical, everyday level. Women also state that it is less important for men to achieve and be successful or show traditional male personal characteristics. Our female subjects seem to have been more influenced than men by the sex-role debate during the 1970s. A similar difference in attitudes has also been noted by other researchers for younger age groups.[7]

A content analysis of sentence completions in another study supports these findings. Female subjects are responsive both to the motherhood mandate and to pressures to become self-supporting and independent as well as to the resultant conflict. Men are well aware of pressures to achieve, take financial responsibility for a family, and show leadership qualities in, what they call, difficult situations. A few content categories are unique for one sex and may provide some further insights into the experienced differential sex-role pressures and their emotional consequences. For women, one category contains responses indicative of experienced frustrations about not being taken seriously. This category has a very high frequency among our female subjects. Another category covers expressed feelings of having been stymied in attempts to achieve a sense of personal freedom. Both categories probably indicate the lower self-regard of women relative to men noted in other studies. For male subjects, a unique category contains expressions of negative emotions, arising from continuous experienced pressure to achieve, to be the competent person on whom others rely, coupled with the wish to be able "to get away from everything," at least for a few moments now and then.

As to value orientations, both sexes report that peace and no war is the most important human goal to strive for, but several differences are noted in the other highly ranked values. To female subjects, the values are, in order: self-respect, mature love and companionship, and equality. For men of our study, however, friendship and comradeship, personal freedom, and self-respect based on achievement are the most important values.

In trait ratings, using adjective descriptors from the interpersonal sphere, we also have observed some sex differences, but none identical to those reported in studies from the United States. Statistically, there is a large variation in the ratings on most trait scales, which generally is not accounted for by the subject's sex. Obviously, Swedish men and women seem less willing to attribute interpersonal characteristics to themselves using sex roles as criteria for deciding whether and to what extent a trait, a positive or negative personality term, applies to them. The same stereotypically based self-descriptions as those noted by Bem, Spence and Helmreich, and others[8] are not obtained in our Swedish groups. The women describe themselves as dominant and achieving as well as efficient and responsible, which is similar to the men's descriptions. The men do not differ from the female subjects in personal warmth and sensitivity to others. Our female subjects, however, describe themselves as significantly more emotional in the sense of weeping more easily.

A few years ago Rebecca, Hefner, and Olehansky[9] suggested a developmental model of sex-role transcendence. Some of the empirical findings reported in the recent literature on the androgynous personality (i.e., one with a happy blend of masculine and feminine characteristics) made us interested in the possibility that the psychological study of sex-role stereotypes also could be approached from a developmental point of view. A structural dimension of the self-system, such as cognitive complexity, is often assumed to be a developmental variable. Higher levels of development become manifest as more complex psychological organization is experienced. This can be observed in patterns of reasoning, beliefs, attitudes, and ways of relating to others. In recent studies, we have included Jan Loevinger's method[10] for assessing ego development in our Swedish version in order to study the relations between perceived sex-role demands, self-descriptive traits, and level of development. From the analyses performed so far, the sex of the subject comes out as a less important variable than level of development in determining self-descriptions and role perceptions. At the intermediate levels of ego development, sex differences similar to Bem's were present in self-ratings and role perceptions. At the higher levels, the self-descriptions did not produce a traditional "male" or "female" pattern. With increasing development, the perceived role demands and the self-attribution of traits may become less based on sex as a criterion as other more personalized standards become more important. As the female and male subjects develop, mature, and reach the highest levels of ego development, their self-descriptions differ and are even more different from those at the lower levels of development.

In interpreting the findings from the series of studies briefly reported here, I think it is necessary to note not only sex-role differences—which we as researchers more or less automatically do—but also to keep in mind the many similarities in how women and men perceive themselves and sex-role pressures. Although they may phrase it differently, both men and women in our groups think that it is expected of them that they strive to become independent, self-reliant, and reliable persons who will take good care of their offspring. It is very important to note that they suffer different frustrations on the road to actualizing these goals. These reports can be assumed to reveal still existing inequalities in life experiences and to reflect how socializing influences differentially restrict the potential development of individuals. It should be mentioned that when sex roles have been debated in Sweden during the past 20 years and changes instigated politically, there has been a conscious effort to change role demands for both sexes. Generally, it has been agreed that the traditional male role is not to be envied or copied in all of its facets.

PERSONALITY DEVELOPMENT
AND THE SOCIALIZATION PROCESS

In a 1973 article, the American psychologist Jeanne Block[11] theorized about the differential effects of the socialization process on the personality development of women and men. Different social influence processes seem to work on the course of development in the early teen years when conformity to rules and roles is the central psychological problem of the individual. Block relies on data from a cross-cultural study when she states that the socialization of adolescent males in the United States and Northwestern Europe seems to enhance experiences that foster the development of the androgynous, or more complex, personality more easily than it does for females. For girls at this age, social influences seem to discourage the acquiring of personal qualities that many have looked upon as conventionally masculine, such as self-assertiveness, achievement orientation, and independence. In Sweden, educational research has revealed that the higher school grades of girls relative to those of boys drop sharply in the early teens and remain lower in central academic fields.[12] One interpretation is that there are firmer social pressures applied to boys to keep on achieving whereas girls are allowed to give up and to develop other interests than academic ones--with consequences for later vocational attainments. It seems possible that the frustrations from sex-role demands, evident in content analysis of responses from our male and female subjects, may reflect consequences of the socializing pressures on boys and girls at a very critical period of personality development. Later in life, women may find themselves without the knowledge and training they then realize are prerequisites to reaching personal goals. Men, on the other hand, may suffer from too much pressure to achieve and succeed.

SOCIALIZATION AND THE CONCEPT OF GENDER IDENTITY

To return to the problem of why little has been achieved to change the social position of Swedish women in the years of formal equality and why they are still lagging behind their potential, it is necessary in a psychological analysis of this problem to consider some of the findings from research on human development and socialization. Fairly early the child will recognize the sex of other people, probably by a simple perceptual dichotomization similar to "big" or "small." This is basic to the acquisition of the concept of gender as part of one's own identity. Around the age of 5, the child is sufficiently cognitively developed to know that another person is either a boy or a girl, that persons are males or females, and that gender is a permanent characteristic of a person. At the same time that the child develops cognitively, he or she also affectively evaluates what is known and learns to attach values, such as "good" and "bad," to what is perceived. When the concept of gender identity is established, it is not something that has to be continuously reestablished. It is maintained as more information is assimilated from the outside world. To uphold the belief that "I am a girl" or "I am a boy," the child is motivated to seek out activities that are congruent with the abstract model or prototype he or she has formed of own-sex behavior and to attend less to other-sex behavior.[13]   In abstracting what is sex-appropriate behavior from observing others, a very important variable is frequency. Thus, the child is motivated to test and engage in activities that are on the average more often performed by people of the same sex just in order to maintain his or her own gender identity.

CONCLUSION

If a child grows up in Sweden, she or he will abstract an image of a society where the division of labor and the distribution of status is not equal for the sexes. Much of what the child is taught about equality from preschool on is clearly contradicted by the child's own observations. For the normal psychological development of an individual, it seems very important that gender identity is firmly established as part of the self-concept. Gender is certainly a variable of profound importance for organizing the experiences in life. This is not to say that existing inequalities in society for women and men are necessary elements in forming the concept of gender as part of the self. What I would like to point out instead is that it seems vital to change the social environment of most children in order to provide each child with a fair chance of forming prototypes of "male" and "female" which are differently composed from those at present. Much money and effort have been spent on new ways of teaching equality to preschool and school children in Sweden in order to erase existing sex-role patterns. Given findings from recent research on social learning in childhood, it seems to me that the process of change would be much faster if some of the money were used to change the actual behavior of people in the child's environment. For instance, more money could be spent on attempts to make occupational shifts between sexes more attractive and common. Growing children, as far as we know, are more likely to rely upon what other people do and to abstract guidelines for their own behavior from these observations than they are to be influenced by what adults tell them about how things should be done. It is not a sophisticated insight of social psychologists but common knowledge, even to children, that verbally expressed attitudes and actual behavior are often incongruent.

NOTES

1. See, for example, Liljeström, Furst Mellström, and Liljeström, 1978.
2. Bergman and Dunér, 1975.
3. Rokeach, 1973.
4. For example, Bem, 1974; Spence and Helmreich, 1978.

5. Wiggins, 1978.
6. Wiggins, 1978; Wiggens and Holzmuller, 1978.
7. See, for example, Wolf-Seibel, 1980.
8. Bem, 1974; Spence and Helmreich, 1978.
9. Rebecca, Hefner, and Olehansky, 1976.
10. Loevinger, Wessler, and Redmore, 1970.
11. Block, 1973.
12. Husén, 1969.
13. Slaby and Frey, 1975.

PART II

WOMEN AND THE WORLD OF WORK:
PHYSICAL, MENTAL, AND ECONOMIC WELL-BEING

INTRODUCTION

Lawrence A. Palinkas

Naval Health Research Center
San Diego, California   U.S.A.

For most women, working outside the home is hardly a matter of choice. Now, more than ever, single and married women enter the labor force because they need the income. This fact is especially relevant for those single mothers (unmarried, separated, divorced, widowed) who constitute their families' sole source of financial support. An increasing proportion of these families live below the poverty level. Even when both parents are present, women are called upon to seek employment outside the home in order to maintain a satisfactory level of living for their families.

Women become members of the labor force to enhance their economic well-being, but it is uncertain as to what extent this well-being is secured. The majority of jobs held by women are low-paying, have limited opportunities for advancement, are often performed in unsafe working environments, and provide inadequate benefits and child care services. Many women endure a "double burden" of working both in the home and outside it. Given these circumstances, the effects of labor force participation on women's health has become a subject of great concern in addition to the efficacy of such participation in meeting women's financial needs.

There is considerable evidence to suggest that work outside the home has a positive effect on women's health. Men, however, who typically have better-paying jobs with greater opportunities and benefits than women, have a higher mortality rate, much of which is attributed to occupational stress.[1] Nathanson[2] observes that the French sociologist, Durkheim, concluded that homemakers live longer than men because they are protected from work-related stressors by their confinement within the family. Yet, Anne Briscoe, the discussant for the conference session on work and women's health, asserts that confinement within the family also exposes women to serious stressors resulting from low self-esteem and vulnerability to financial demands.

It appears, therefore, that labor force participation and the "protective confinement" within the family provide both advantages and disadvantages to women's health. What is the current situation with regard to women's achievement in meeting their financial needs while maintaining a satisfactory health status? The papers in this section address this question by examining the effect of employment outside the home on the overall well-being of women. Well-being is defined as a state of being healthy and able to successfully adapt to the environment. This introduction will review the papers that discuss two major components of that well-being: economic security and mental and physical health.

POVERTY AND ECONOMIC SECURITY

The chapters by Clair (Vickery) Brown, Pat Armstrong and Hugh Armstrong, and Jane Roberts Chapman and Gordon Chapman all point out that most women participate in the labor force "because they need the money." As all of these authors observe, this economic need is particularly acute for a large proportion of women who are the sole source of financial support for their families. Brown reports that families headed by single women now comprise the fastest-growing segment of the poverty population: in 1978, for instance, 49% of all families in poverty in the United States were headed by single women. Further, the National Advisory Council on Economic Opportunity predicts that by the year 2000 the poverty population will consist solely of women and their children.[3] Even in families headed by two parents, according to Armstrong and Armstrong, the woman's financial contribution can mean the difference between economic security and poverty. As financial demands increase, the need to seek a well-paying job outside the home becomes greater.

Unfortunately, single women encounter numerous obstacles when they attempt to provide for their families by working outside the home. The structural causes underlying this dismal circumstance are examined by Brown in her chapter, "Bringing Down the Rear: The Economic Position of Single-Mother Families." Brown maintains that such families are financially at a disadvantage for four reasons. First, they have a high ratio of dependents to able-bodied adults. Second, women usually command low wages; consequently, even a full-time job often is inadequate in enabling a woman to provide for her family. Third, the relative economic position of single-mother families falls in relation to that of husband-wife families when the wife works because the single mother, already working at full capacity, has no comparable options for increasing the number of hours she works. Finally, single women who head families are caught between conflicting policies based on two opposing movements. The first of these, the women's movement, seeks to promote legal equality though women have yet to achieve economic equality. Divorced women, for example, are assuming responsibility for an increasing share of child support even though they may be economically unprepared to do so. The other movement seeks to preserve the integrity of the nuclear family by reducing government funding for such programs as abortion, child care, food stamps, and Aid to Families with Dependent Children (AFDC). Since many women depend upon these programs to meet their families' needs, in addition to employment outside the home, this movement may, in fact, do more to promote the disintegration of single-mother families than to preserve their unity and integrity. Both movements, therefore, seem to constrain the efforts of single women to extricate their families from the grasp of poverty. The social policies generated by these movements have made the labor force participation of single mothers crucial to the existence of their families.

Despite overwhelming evidence to the contrary, such as that given by Brown, the myth that women do not need to work but instead are responsible for the unemployment of "prime age" males is still prevalent. The chapter by Pat Armstrong and Hugh Armstrong begins by attacking this myth. The authors state that men and women tend to be segregated into different occupations and different sectors of the economy so that, in reality, they are seldom competing for the same jobs. Economic need is the primary motivation for most women's job search, as inferred from three observations. First, 54% of widows and formerly married women living alone have incomes below the poverty level in Canada. Second, it is estimated that the number of families living below the poverty line would increase by 51% if the wife did not work outside the home. Third, more women continue to seek employment despite the limited job opportunities, low wages, scarcity of adequate child care, and the "double burden" of housework and outside employment—indicating that they do "need the money."

A key question, then, is to what extent this need is satisfied by labor force participation? Armstrong and Armstrong state that women are more likely than men to be offered and to accept part-time work or low-paying, full-time work. While a large percentage of new jobs are in those occupations and sectors of the economy traditionally dominated by women, female unemployment in Canada has been rising at a faster rate than both male unemployment or female employment. Thus, although there is an increased need to work, the difference between this need and the ability to meet it through employment outside the home is also increasing.

The chapter by Chapman and Chapman provides additional evidence of this disparity and its effects on the well-being of employed women. The authors report that in the United States women have lost ground in their efforts to earn wages at levels comparable to those of men, such that by 1978, the median of women's income is shown to be only 59% of what men earn. The majority of women occupy low-level, low-paying jobs which are concentrated in fewer than 20 occupations and their unemployment rate is rising at a faster pace than that of men. Chapman and Chapman point to six specific factors behind the worsening economic position of women: limited or unstable growth rates in the general economy (which have had a greater negative impact for women than for men), inadequate union representation, the income gap between men and women at all levels within all occupations, occupational segregation and limited mobility, the changing structure of the family, and the failure of the war on poverty.

Chapman and Chapman note that the worsening economic conditions for women in the United States have led to a substantial increase in the number of women and families headed by single women who live in poverty. For these women, the alternatives to paid employment are few. The most important alternative has been an increasing dependence on welfare, even for women already employed. Unfortunately, as government welfare programs are reduced through budget cuts, the viability of this option for women in poverty also is reduced. Another alternative has been "economic crime," although the proportion of women "in crime" is considerably lower than the percentage for men. The authors posit a correlation between increases in the rate of crimes committed by women and periods of diminished job opportunity. Crime is a manifestation of both the inability to attain financial security and the despair associated with this situation, as well as a possibly lucrative source of income.

The inescapable conclusion, then, is that women participate in the labor force because of financial necessity; it is not a matter of "choice" as some would have us believe. However, paradoxically, once these women enter the labor force, there is no guarantee that their economic position will improve appreciably. This dilemma is compounded by the fact that the disparity between wages and financial needs appears to be increasing. A policy-oriented approach should be directed at bringing women's wages in line with their financial needs. Two of the papers, included in the conference program, are concerned with such a policy-oriented approach.

The first paper, presented by Joan Goodin, is titled "The National Commission of Working Women: A New Approach to the Invisible '80%'." The paper addresses the efforts of the Commission to assist the large percentage of women workers in the United States who are concentrated in clerical, sales, service, factory, and plant jobs—the invisible "80%." This segment of the female labor force is comprised of pink- and blue-collar workers in jobs characterized by low wages, little recognition, and almost no mobility.

The paper consists of a brief history of the National Commission of Working Women (NCWW) and an account of its efforts to assist women in these jobs. The author notes that unlike other organizations that are concerned with issues of women's paid employment, the executive staff of the NCWW includes women who are employed in jobs which exemplify those of the invisible "80%." The NCWW also has members from business, government, and the media, each of whom brings necessary talents to promote the efforts of employed women in attaining better-paying jobs and greater occupational mobility.

The NCWW has been involved in four major areas of concern to women in the work force. The first problem area is the low scale of wages and benefits for women. The Commission is striving to bridge the gap between financial need and actual income by promoting a system for rewarding work that is not linked to the gender of the worker. Concomitant with these efforts, the NCWW also is concerned with enhancing the self-esteem of employed women. A third consideration is the need for quality, affordable child care. Because these women earn relatively low wages, the concern over the availablity of low-cost child care is particularly acute. The fourth problem area is the issue of adequate education and training. According to Goodin, access to educational and training opportunities is denied to a vast majority of the "80%" because: (1) of the double burden of job and family responsibility; (2) of the undermining, through the traditional female socialization process, of women's confidence in their ability to participate fully at work and in school; and (3) of employer attitudes and policies.

The current approach of the NCWW is to provide information and advice to both policymakers and employed women. Goodin states that the organization began by designing and implementing regional dialogues throughout the United States, forums for learning from women in the labor force what their perceptions are of the major problems affecting them. These forums are established in conjunction with a public awareness campaign designed to acquaint policymakers and employers with the plight of the "80%." The NCWW also has been involved in the preparation of testimony for legislative hearings and the sponsorship of roundtable discussions which bring together policymakers, employers, and experts on such issues as child care and education.

The paper outlines four specific programs the NCWW has sought to establish for the benefit of employed women. These include: (1) a career development seminar designed to train women office workers to ascertain their needs and familiarize their co-workers with alternatives available to them; (2) a Broadcast Awards Program recognizing quality radio and television programming on women in the labor force; (3) "Project Opportunity" which trains NCWW members to identify their educational and job-related needs and interests; and (4) a leadership skills training program aimed at overcoming internal barriers to success such as low self-esteem, a sense of isolation and alienation, and feelings of powerlessness.

At this stage, the extent to which the NCWW has been effective in bridging the gap between financial need and financial ability is uncertain. No evidence is presented in Goodin's paper that evaluates the effectiveness of the Commission. Nevertheless, the NCWW seeks to improve the working conditions of women employed outside the home and to ensure that women will have the same opportunities as men to provide for themselves and their families.

The second paper, authored by Denise Lecoultre and titled, "Economic Factors of Women's Increased Participation," documents the goals and activities of another association, the Organization for Economic Cooperation and Development (OECD), to promote on an international scale the efforts of employed women to seek better pay, greater occupational mobility, and more flexible working schedules. The paper consists of three parts: (1) the role of women in the economy, (2) women's participation in and the function of labor markets, and (3) a summary of the OECD's high-level conference on the employment of women. Only the first two sections of the paper will be discussed.

The first part reinforces the observation made in several of the chapters, including those by Brown, Armstrong and Armstrong, and Chapman and Chapman, namely, that there has been a great increase in the work force participation rates of women, especially married women, but that occupational segregation and wage discrimination are still major problems for the economic well-being of these women. In France, for example, even after such factors as occupational segregation, level of qualification, and hours of work have been controlled, women still earn only 69.4% of what men earn.

Lecoultre notes several current endeavors which are designed to eliminate occupational segregation and wage discrimination. One example is the effort of governments to reduce this inequality through two European Community Directives concerning equal pay and equal treatment. A second area is in the designing and implementation of different time arrangements which would assist individuals in meeting their obligations both at home and on the job. Such arrangements include: the structuring of work hours to allow parents to work while their children attend school, permanent part-time employment, unpaid leaves of absence, and the relaxation of rules making the age of retirement mandatory. A third area is the provision of adequate child care services for employed parents. Finally, the author discusses the potential benefits of Social Security reform. Divorced or separated women, it is observed, are particularly vulnerable to poverty unless they receive some form of compensation for work in the home. Lecoultre suggests that governments establish family allowances or special benefits payable to women when outside employment is interrupted for family purposes.

The second part of the paper discusses the growth of women's labor force participation and the functioning of labor markets. Lecoultre states that women are motivated to seek employment outside the home by such incentives as the attendant educational or training opportunities and by economic necessity. Occupational segregation is examined and the author concludes that women's overall lower level of education than men and their choice of studies are responsible in part for their segregation in the labor market and, consequently, their lower wages and limited opportunities for advancement.

In addition to this occupational segregation, Lecoultre describes how the introduction of certain technological innovations such as word-processing equipment and electronic cash registers may have a negative effect on employment by eliminating more jobs for women than can be created elsewhere. Efforts are currently under way to retrain women for other, and perhaps more specialized, jobs but the impact of these endeavors has, thus far, been limited. Nevertheless, the author predicts that as traditional occupations become scarce, the integration of women into the labor market will be accelerated as they enter new sectors of the economy and new occupations. This integration, in turn, will promote wage equity and desegregation of the labor market. According to Lecoultre, there is already greater equality for women in newly developing industries than in the older ones.

The paper concludes by stating that married women are particularly vulnerable to the policies and practices of both government and business because such efforts can either facilitate women's occupational endeavors or force them to abandon employment outside the home. Because many married women must work to keep their families financially solvent, the decision to strive for one or the other goal is a crucial one. The OECD has worked for the implementation of antidiscrimination laws, the promotion of education and training, and the design of flexible time arrangements for employed women and men. All of these programs and policies are designed to ensure that the efforts of women to attain economic well-being through outside employment is a successful venture and not one that traps them and their families in an endless cycle of poverty.

HEALTH AND WELL-BEING

The policies advocated by Goodin and Lecoultre may contribute substantially to reducing the disparity between financial need and job income, but can they also protect women from the risks to mental and physical health that are often associated with the labor force participation of men? If the occupational stress associated with the work force is severe enough to contribute to the higher mortality rate of men, will labor force participation have the same effect on women?

Almost all of the authors discuss the "double burden" of employed mothers. Because of this double burden and the stress engendered by poverty or the imminent threat of poverty, mothers who work outside the home are believed to be a high-risk group for mental health problems.[4] What is unclear, however, is whether or not labor force participation creates additional problems, particularly ill health. This issue is examined in the chapter by Deborah Belle and Ruth Tebbets. Using data collected from the Stress and Families Project, they discuss the consequences of outside employment on the mental health of low-income mothers. For these women, according to Belle and Tebbets, unpredictable income, poor housing, dependence on social service agencies for support, and entrapment in low-status, low-paying jobs are only a few of the sources of stress which underlie their relatively high levels of depression. Nevertheless, most of the study respondents have extensive work histories and derive some measure of satisfaction from employment. Belle and Tebbets find that low-income mothers who work exhibit significantly lower levels of depression than unemployed women. With adequate training and education, these low-income mothers believe that they can obtain better-paying jobs which will engender greater confidence, self-esteem, dignity, relief from boredom, and a measure of independence.

In general, the study supports the position that labor force participation of low-income mothers has a substantial positive effect on the mental health aspect of their well-being even though the impact on the economic security component is minimal. More succinctly, a job that is restricted in providing economic security for women is not necessarily limited in conferring important mental health advantages as well.

Belle and Tebbets' findings correspond with the conclusions articulated by the authors of two other papers not included in this volume. The first, authored by Elina Haavio-Mannila, is titled, "Women, Health, and Work." The paper begins with the question of why women typically have higher morbidity rates, but lower mortality rates than men. The answer is believed to lie in one of three hypotheses: (1) there is a greater willingness among women than men to report illness, (2) women have more time than men to adopt a sick role because they have fewer work obligations, and (3) differences in illness do exist with respect to sex.

The author presents data gathered from health statistics and studies in the four Scandinavian countries of Denmark, Sweden, Finland, and Norway. Differences in physical and mental health are investigated and compared across such variable conditions as type of work payment, level of economic activity, and occupational status.

In general, the data show that women have higher rates of illness than men and that hospitalizations for mental disorders occur with greatest frequency for women between the ages of 35 and 65. Men, however, exhibit higher admission rates for alcoholism than do women. Morbidity rates for women vary considerably among the four countries but the results indicate that the higher the proportion of women in the labor force, the lower the proportion of mental and physical health problems.

Haavio-Mannila concedes that her results may be a consequence in part of self-selection—women choosing either to work outside the home because they are healthy or to remain at home because they are not well. This possibility also was raised by Anne Briscoe, the dicussant for these papers. However, Haavio-Mannila also stresses that self-selection could not account for the overall differences in morbidity. She notes that self-selection differs in magnitude from country to country, being less of a factor in countries where the labor force participation of women is small than in countries where it is large. Thus, in Norway, where the tradition of "housewifery" is the strongest in Scandinavia, differences in morbidity between housewives and employed wives are smaller than in Denmark, Finland, and Sweden, where significantly more married women work outside the home.

Although the study does not attempt to measure the effect on physical and mental health of such factors as the socioeconomic status of the worker, hours worked, wages earned, and availability of child care, the data do suggest that an overload of actual tasks, i.e., the "double burden," does not significantly affect women's health. Women in the age range of 35 to 65 have higher morbidity rates than men, but this morbidity tends to decrease as women seek and obtain outside employment.

A similar conclusion is presented in a paper by Constance Nathanson, titled, "Employment and Health Status among Women in the Middle Years."[5] Nathanson argues that the correlation between employment and illness is not a simple one. Traditionally, this correlation has been viewed from two different perspectives. The first, which evolved from Durkheim's classic work on suicide, suggests that homemakers experience fewer stressful situations than their husbands who participate in the labor force. The data, however, largely support the second view, exemplified by the findings of Haavio-Mannila and Belle and Tebbets, which asserts that outside employment is beneficial to the mental and physical health of married women. How can these two disparate views, therefore, be explained?

Nathanson examines illness behavior in an effort to resolve this discrepancy. Illness behavior is defined as "a respondent's behavior in response to perceived symptoms" and is measured "first, by the number of days of activity restriction during the two weeks preceding the interview and, second, by the number of physician visits during the preceding year." Using illness behavior as an indicator of the effect of social and occupational status, Nathanson tests two specific hypotheses: (1) that employed married women (referred to as "women with heavy role obligations") are less likely to do anything about their illness, and (2) that employed married women are more likely to visit a doctor if they decide to do something about their illness while homemakers will treat themselves at home.

These hypotheses are tested using data from the 1974 Health Interview Survey with a sample of almost 13,000 women in the 45 to 64 age group. In addition to the two specific measures of illness behavior defined above, Nathanson utilizes self-health ratings and a report of "chronic conditions" as an objective measure while conceding that only 58% of chronic conditions reported in interviews by women in this particular age are matched by medical records.

The results of Nathanson's analyses confirm both of her hypotheses. She reports that employed married women have significantly better health and display less illness behavior than homemakers. Moreover, employed women have health ratings that are comparable to those of employed men, suggesting that the higher morbidity rates associated with women in this age group are not a function of sex but are attributable to the homemaker status of a majority of adult women. Of those women who are employed outside the home, married women and women with children appraise their health more positively and report less illness behavior than unmarried women with no children, which lends support to the thesis that multiple role obligations such as the "double burden" of housework and labor force particpation do not increase the risk of illness. These results are supportive of those reported by Hoiberg and Ernst[6] who showed that mothers serving on active duty in the Navy have substantially lower hospitalization rates than the women who are not mothers.

On the basis of these results, Nathanson proposes a third hypothesis: employment is significantly more beneficial to the health of unmarried women and those with relatively little education than for married and well-educated women. She reports that this prediction is confirmed by results of analyses of variance and multiple regression analyses. Employment makes less of a difference in the subjective health appraisals of married women than it does for women with fewer role obligations and women with less than a high school education.

Nathanson concludes her study by advancing a third perspective on the relationship between outside employment and the health of married women, one that uses illness behavior rather than actual morbidity as a measure of health status. According to Nathanson, women with a larger number of role obligations display lower levels of illness behavior but are more likely to utilize health services.

Both the Nathanson and Haavio-Mannila papers contribute to our understanding of sex-related differences in morbidity whereas the issue of sex-related differences in mortality is addressed only indirectly. Nathanson reports that employed men and women have similar morbidity rates but her data do not indicate whether the mortality rates of women would increase as more women enter the labor force. However, given the current medical evidence which suggests that women generally have greater resistance to disease than men,[7] this possibility seems unlikely.

Nathanson's data also reveal that the extent to which labor force participation "promotes" health is mediated by the socioeconomic status of the woman. Nathanson's results demonstrate two important facts: (1) that women who usually occupy the lower end of the socioeconomic scale (unmarried women and women with less than a high school education) benefit more in terms of their health status than married women, but (2) their overall appraisal of health is still significantly less than that of married women. Taken together, these two facts indicate that labor force participation is especially important for lower-class women who must struggle against the threat of poverty as well as the stress and illness associated with being poor.[8] Giving these women the opportunity to enjoy the health benefits accrued from employment outside the home requires that the economic security component of their well-being be promoted through a series of specific policies such as those outlined by Goodin and Lecoultre.

CONCLUSION

At the outset of this section, it was suggested that women who work outside the home may be confronted by two major threats to their well-being: the reality of poverty and the risk to mental and physical health associated with labor force participation. Women participate in the labor force to secure some measure of economic well-being for themselves and their families but the low-paying, low-status jobs available to most of them do not provide satisfactory economic security. Rather, the papers document the existence of a growing disparity between women's financial needs and income gained from employment outside the home. Slow economic growth, the wage differential between men and women, limited training or education, job mobility, child care services, and the "double burden" of household responsibilities and labor force participation all contribute to this disparity. Thus, even though more women are seeking jobs outside the home, their chances of securing employment enabling them to adequately meet their families' financial needs are decreasing.

While labor force participation currently has little impact on the economic well-being of women, it does appear to contribute significantly to their mental and physical well-being. Belle and Tebbets, Haavio-Mannila, and Nathanson demonstrate that employment outside the home does not impair the health of women and may, in fact, contribute to its improvement. This finding is particularly relevant for women with multiple role obligations who report better health than women without such obligations.

Outside employment should promote both components of women's well-being. A policy-oriented approach, such as the ones advocated by Goodin and Lecoultre, is needed to ensure that both aspects of women's well-being are enhanced by employment outside the home. Vital to such an endeavor are policies which provide for an expansion of job opportunities for women. This expansion could be facilitated by education and training programs, upgrading of wage scales to levels comparable with male co-workers, child care programs, and greater job mobility. Such an effort might help to overcome the economic disadvantages associated with the "double burden" while retaining and possibly enhancing the health advantages. Employment should promote economic security as well as mental and physical health, for only when both components of well-being are promoted will employment outside the home be a truly profitable venture.

NOTES

1. Waldron, 1976.
2. Nathanson, 1980.
3. See Chapman and Chapman, in this volume, footnote 10.
4. Welch and Booth, 1977; Johnson, 1977; Waldron and Johnston, 1976.
5. Nathanson, 1980.
6. Hoiberg and Ernst, 1980.

7. Marcus and Seeman, 1981.
8. The Framingham study of coronary heart disease found that women
    holding clerical jobs and married to blue-collar workers were at
    the highest risk of developing CHD. See Haynes and Feinleib,
    1980.

BRINGING DOWN THE REAR: THE DECLINE IN THE RELATIVE ECONOMIC POSITION OF SINGLE-MOTHER FAMILIES[1]

Clair (Vickery) Brown

University of California
Berkeley, California   U.S.A

Poverty has gradually become an issue of particular concern to women because families composed of single mothers and their children are the fastest-growing segment in the poverty population. By 1978, 49% of all families in poverty were headed by a single woman, and 56% of all children (under age 18) in poverty were in single-mother families.

In general, the single-mother family is economically disadvantaged because the family's needs are high and its resources low, since the family has a high ratio of dependents to able-bodied adults and adult women usually can command only low wages. The low potential wage rate coupled with the high time demands of raising children and maintaining a home cause these families to be "time poor" as well as "money poor." Single mothers, therefore, are caught in the "double bind" of time and money. They are unable to increase their work hours to supplement their families' incomes because they are already working at full capacity.

While this difficult position of single mothers has remained un-changed, a shift in social custom has resulted in the determination of the relative economic position of single-mother families vis-à-vis husband-wife families. Social custom formerly permitted paid employment for wives and mothers only under special circumstances such as when a husband was incapacitated through injury, illness, or alcoholism, when a husband died, or when a husband provided poorly for his family. Gradually this custom has changed, and the decades of the 1960s and 1970s witnessed an influx of wives and mothers into the labor force. The shift in custom meant that differences in women's work intensities were no longer being used to offset differences in family income.

This paper shows both conceptually and empirically how the relative economic position of single-mother families falls vis-à-vis husband-wife families when the wife becomes employed. In addition to these relative income changes, revisions in government support programs, especially child care and Aid to Families with Dependent Children (AFDC), are examined to see how they have affected the relative resource position of single-mother families.

The political position of single-mother families has also deteriorated during this period as single mothers have become caught in the crossfires of two opposing camps: One is the movement to give women equal rights with men, and the other is the drive to save the nuclear family, neither of which has shown adequate concern for the position of single-mother families. Although unintentionally, single-mother families have been hurt by social programs that advance sexual equality in the law before it is an economic reality (e.g., no-fault divorce and child support laws). Ironically, this premature assumption of equality has improved the position of the absent father at the expense of the single mother.

More consciously, these families also have been hurt by other programs designed to preserve the nuclear family such as a cutoff of government funding for abortions, lack of government-funded child care, and some aspects of AFDC and food stamp programs. The politically conservative and religiously based movement has sought, and has been partially successful, in weakening the supports instituted to accommodate the changes in the structure of the American family in an effort to constrain a fundamental shift in that structure.

The examination of the economic consequences of these social and political factors should be of critical concern to all women, not only single mothers. The economic position of single mothers affects the economic position of all women because being a single mother represents a woman's most vulnerable economic position. The economic vulnerability of the single mother is not an insignificant possibility for today's women, since up to one-half of all children are expected to live in a single-mother family during some period of their childhood.

A STRUCTURAL APPROACH TO FACTORS AFFECTING SINGLE-MOTHER FAMILIES

Two basic assumptions underlie this analysis of single-mother families: (1) a certain amount of housework (i.e., food preparation, clothing care, housecleaning, and child care) is necessary for the maintenance and well-being of all family members, and most of this housework cannot be replaced with market substitutes on a permanent basis;[2] and (2) the family's sense of well-being depends on their relative economic position.[3] Together with the observed changes in the work activities of husband-wife families, in which the wife's increased labor force participation has enabled the family to enhance its share of economic growth achieved in the marketplace, these assumptions point to the conclusion that the economic well-being of single-mother families should have declined over at least the past 20 years.

Single-Mother vs. Husband-Wife Families:
Work, Housework, and Time Allocation

In fact, these assumptions lead us to the inescapable conclusion that a sexual division of labor that puts women primarily in charge of unpaid housework and men primarily in charge of paid market work will result in single-parent families being headed by the mother and in the relative economic position of these families declining since they are severely limited in their ability to switch time from housework to market work.

An example will illustrate how the structural characteristics of work lead to a declining relative position for single-mother families in comparison with families of a husband and wife with two children. Without using any market substitutes for required housework, the single-mother family must perform 61 hours per week of housework and the husband-wife family requires 66 hours of housework.[4] Single mothers are limited by their household responsibilities represented by the 61 hours of required housework (rounded down to 60 hours for the following examples). Assume also that market work activities are constrained in that the husband works 45 hours per week, market goods cannot be substituted for housework, and the single mother works the maximum of 87 hours per week (60 hours at home and 27 hours for pay at a wage rate equal to approximately 60% of the male wage rate). Then the observed earnings ratio of the single-mother family compared to the husband-wife family is .36.[5] If substitution is allowed up to 10 hours per week at a replacement cost equal to the female wage rate (w_f), then the observed earnings ratio rises to .49; however, the ratio net of the cost of the required housework will remain constant at .36.[6] If we compare the family's "full income," defined as earnings plus the value of housework (priced at w_f), then the full income ratio rises to .64.[7] Only when the replacement cost of housework (r) is less than w_f will the single mother be able to increase her full income by working for pay. The more substitution allowed and the greater the difference ($w_f - r$), the greater is the relative income of the single-mother family.

However, the substitution of market goods and services for housework has been found to be empirically unimportant. In addition, the cost per replacement hour of private child care (outside public school) is high relative to w_f, because a large part of child care in the home is performed as a "secondary activity" while other housework is being performed.[8] Therefore, the average of 14 hours of child care (for families with one to three children) in the home will usually have to be replaced with 15 to 40 hours of substitute child care (paid or unpaid) per child when the mother is employed.

More generally, if each adult is working a maximum of 87 hours per week, the single mother has 26 hours after housework for all other work and leisure activities; together the husband and wife have 108 hours after housework for other work and leisure activities. The required housework for these two families is almost equal, but the time available for all work activities differs by a factor of two. The degree of economic inequality between these two family types depends on two additional factors: the amount of housework that can be replaced by market items along with the replacement cost, and the wage rate each adult can command in the marketplace. In order for the single-mother family to have the same economic resources as the husband-wife family, two unrealistic requirements must be met: the single mother's market wage (for up to 87 hours per week) equals the combined wages of the husband and wife, and required housework can be completely substituted with market goods and services at a replacement cost that is less than or equal to the lowest wage of the three adults.

More realistically, if we assume that the female wage rate (w_f) is 60% of the male wage rate (w_m), then the ratio of <u>maximum money incomes</u> of the two families (i.e., single-mother earnings divided by husband-wife earnings) will depend on the allowable substitution of market items for housework ranging from .16 (no substitution possible) to .38 (complete substitution possible).[9] However, if the earnings are the net of the costs of paying for replaced housework ($r = w_f$), then the net income ratio is always .16 in this example. The ratio of full incomes, which includes earnings <u>plus</u> the implicit value of housework (calculated at the hourly price, w_f), is always .38 in this example. If we add constraints on market work, in that no adult can work more than 45 hours per week for pay, then the ratio of maximum money incomes will vary from .22 (no substitution) to .38 (full substitution). In any event, the ratio of family incomes net of costs for required housework is .22. The ratio of full income is .48.[10] The constraint on market work has raised the relative income position of the single-mother family because the husband's market wage advantage is diminished (and any housework he does above market work hours is valued at w_f). In addition, the two-fold advantage of the husband-wife family in total work hours is also diminished because the 24 free hours they have left over after their 90 market work hours and their 60 housework hours have been assigned zero value.[11]

Single-Mother vs. Husband-Wife Families:
Their Relative Economic Positions

The relative position of the single-mother family vis-à-vis the husband-wife family in these different examples without income support programs is summarized in Table 1. If the wife does not work outside the home while the single mother does, then the single-mother family has 36% of the money income of the husband-homemaker family. However, if the wife is also employed, then the single-mother family has only 22% of the money income (net of the replacement cost of housework) of the husband-wife family. If housework can be fully substituted with market replacements (so a mother can work 45 hours for pay) and if earnings are not adjusted for the replacement costs of housework time, then the single-mother family does appear to have 60% of the earnings of the husband-homemaker family and 38% when the wife is also employed. However, these adjusted figures are deceptive, since the net income figures in Case

TABLE 1

RELATIVE ECONOMIC POSITION OF SINGLE-MOTHER FAMILIES
COMPARED TO HUSBAND-WIFE FAMILIES[1]

	SINGLE-MOTHER	HUSBAND-HOMEMAKER	HUSBAND-WIFE, WIFE-EMPLOYED
Ratio of earned income[2]			
Net of required housework	1.00	.36	.22
Gross of required housework	1.00	.60	.38
Ratio of full income[3]	1.00	.64	.48
Earned income minus poverty line (weekly)[4]			
$w_m = \$4.70; w_f = \$2.82$	−$17	$94	$220
$w_m = \$7.30; w_f = \$4.38$	$25	$210	$408
Ratio of above-poverty income ($w_m = \$7.30$)	1.00	.12	.06
Ratio of above-poverty income to poverty line ($w_m = \$7.30$)	.27	1.78	3.46
"Free" time available (weekly)	0 hr.	69 hr.	24 hr.

[1] In all cases, the single mother works 87 hours at housework and paid work. Women's wages (w_f) are equal to .6 men's wages (w_m). Market work is constrained to 45 hours per week for any one person. Required housework is assumed to be 60 hours per week for each family.

[2] For case 1, required housework cannot be substituted or has a replacement cost of w_f and has been netted out.

[3] Earnings plus required housework time valued at w_f.

[4] Federal poverty threshold (1977) was $93 weekly for a single mother with two children and $118 weekly for a husband and wife with two children. The $4.70 male wage rate makes the implied female wage rate of $2.82 approximately equal to the minimum wage. The $7.30 male wage rate is the median full-time, year-round male wage rate in 1977. The comparable median wage for women was $4.30, or 59% of the male rate.

1 hold unless the replacement costs of housework fall below w_f. If housework valued at w_f is added to earnings (for full income), the single-mother family's relative income is much higher: 64% (wife at home) and 48% (wife employed).

However, since only required housework is counted, and since only a small portion of it can actually be replaced with substitutes, a correct comparison of the well-being of these two family types should exclude required housework time (as well as required money income) and should include the money income and the time available <u>above</u> the required levels. These net figures for time and money show how close to the poverty line a family is actually living and how much flexibility they have in their economic decisions.

If we compare these families using the median wage rates for 1977 (w_m = $7.30), then the single-mother family has $25 per week above the poverty line or 27% more than poverty income, the husband-homemaker family has $210 above the poverty line or 2 times the poverty line, and the husband-wife family with wife employed is $408 above the poverty line or $3\frac{1}{2}$ times the poverty line. The income of the husband-homemaker family is considerably above poverty in contrast to that of the single-mother family, and the margin increases substantially when the wife is employed. The single mother's "extra earnings" above poverty is 12% of the "extra" earnings for the husband-homemaker family (and only 6% when the wife is employed). In addition, the husband and wife have more "free" time after paid work and required housework: 24 hours if the wife is employed or 69 hours if she is not. The single mother has no free time.

These dismal numbers hold only for those single mothers fortunate enough to be at the middle of the wage scale. If they are minimum wage workers (w_m = $4.70), then they earn $17 below the poverty level (compared to the husband-wife family earning $220 weekly above the poverty level when the wife works.).

Single-mother families have always been economically far behind husband-wife families because of the absence of adult male work hours at the higher male wage. Their monetary disadvantage has increased as wives have taken paid jobs. Rising market wages have meant that the value of time above required housework has been enhanced, but this has meant little to the single mother who has little (if any) free time to transfer to paid work in contrast to homemaking wives who do.

These hypothetical conclusions have been made for families with two children. A family's time and monetary needs depend on household composition, and both components display large economies of scale. Therefore, the <u>relative</u> economic position of single-mother families is also affected by changes in the household structure of the population. Let us look at how the economic position of single-mother families has actually changed over the past 25 years and see how it reflects shifts in work patterns, household composition, and government programs.

DOWN THE UP ESCALATOR

Changes in the Economic Position of
Single-Female Families since 1950

Single-female families[12] shared in the tremendous economic growth experienced by the United States during the post-World War II period, but their standard of living was less improved than was the standard for husband-wife families. In absolute terms, the incomes of single-female families improved considerably. In 1954, 49% of these families had an income of less than $4,000 and 14% of these families had incomes of more than $10,000; by 1974, the low-income group had fallen to 30% and the higher-income group had risen to 28% (all figures in constant 1974 dollars).[13] At the same time, however, the relative economic position of single-female families was steadily declining. Their median income was 56% of the median income of husband-wife families in 1950; in 1977, it was only 44% (see Table 2).

The typical family headed by a single woman experienced a deterioration in economic position between 1960 and 1977 relative to the typical husband-wife family whether or not the wife worked; however, the decline was steeper in the case where the wife was employed. If the median income ratios of the three family types had remained the same in 1977 as in 1960 and only the participation rates of wives had increased, the income ratio of the single-female family to all husband-wife families would have fallen to .49 (instead of .44).[14]

These income figures include all income (not just earnings) for all family sizes; thus, they differ from the previous hypothetical comparisons of earnings for families with two children. In particular, they include any government support payments, such as AFDC, which should increase the income of single-mother families. A comparison of Tables 1 and 2 shows that the actual income ratios of all single-female families to husband-wife families with wife employed (.38) or with wife at home (.52) are less divergent than in the hypothetical example. However, this divergence is to be expected because of the inclusion of aid payments to single-mother families and because the wife is not necessarily working full time when she is employed. The ratios still indicate wide disparities in money income. Furthermore, the ratios in Table 2 more closely approximate Case 2 in Table 1, where earnings have not been adjusted for replaced housework time. The empirical data, therefore, are upward biased to the extent that single mothers paid for any housework time replacements (especially child care).

If we look at the entire income distribution, we find that single-female families have become increasingly more <u>overrepresented</u> at the low end of the scale and <u>underrepresented</u> at the top end of the scale (Table 3). The predicted female-headed families as a percentage of all families in each income category, which assumes that the relative income position of single-female families remains the same in 1974 as in 1954, is

TABLE 2

MEDIAN INCOME OF FAMILIES, 1950 TO 1977

	1950	1960	1970	1977
Median income of families with female head	$1,922	$2,968	$5,093	$7,765
Percentage of all families	10	10	12	14
Median income of husband-wife families	$3,446	$5,873	$10,516	$17,616
Wife at home	--	5,520	9,304	15,063
Wife employed	--	6,900	12,276	20,268
Percentage of husband-wife families with wife employed	23	30	39	46
Ratio between female head and husband-wife families	.56	.51	.48	.44
Wife at home	--	.54	.55	.52
Wife employed	--	.43	.41	.38

SOURCE: U. S. Department of Commerce, Bureau of the Census, *Current Population Reports,* Series P-60, Nos. 118, 99, 80, 37, 9.

considerably below the actual percentages for the three lowest-income groups. In the two highest-income groups, the predicted percentage is considerably above the actual value.

Demographic Changes in Single-Mother Families since 1950

The single-female family is not a fixed group, and since 1950 this family type has undergone two important changes. First, the presence of children has increased: 45% of these families had children under 18 in 1950, and 62% had children in 1976.[15] While the group of families headed by an older woman, who had lost her husband and whose children were grown, declined in prevalence, the group of families headed by a single mother with young children increased. Second, the single-family is increasingly living as a separate household (i.e., it is not living with other relatives or friends). In 1950, 17% of single-female families were part of larger households headed by someone "outside" the single-female family; by 1979, 7.5% were part of a larger household.[16] Although single-female families may still be aided by relatives and friends, this trend toward fragmentation of multiple-family units means that the everyday sharing of two basics--meals and shelter--has declined.

From the viewpoint of the economic well-being of children, these trends have meant that more children under 18 are living with only their mother for some period of their childhood: 7% of all children in 1950 and

TABLE 3

PERCENTAGE DISTRIBUTION OF FAMILIES BY MONEY INCOME, 1954 AND 1974 (Constant 1974 dollars)

	TOTAL		$4000		$4000 TO $6999		$7000 TO $9999		$10000 TO $14999		$15000 TO $24999		$25000 & OVER	
	1974	1954	1974	1954	1974	1953	1974	1954	1974	1954	1974	1954	1974	1954
Percent of all families	100	100	9	22	13	22	14	24	24	20	28	9	12	2
Female head as percent of all families:														
Actual	13	10	43	22	25	10	16	6	9	4	5	5	2	5
Predicted[1]	---	---	40	---	18	---	11	---	8	---	9	---	9	---

SOURCE: U. S. Department of Commerce, Bureau of the Census, *Current Population Reports*, Series P-60, No. 101, Tables 19, 20.

[1] Predicted female-headed families as percentage of all families in each income category is calculated by assuming the 1954 relative income distribution of female-headed families holds in 1974 and adjusting for the distribution of higher incomes in 1974 and for the relative growth of female-headed families.

17% in 1978. For those children living with only their mother in 1978, 50% were officially classified as poor.[17]

Differences in the Economic Position between Black and White Families

Table 4 shows the relative economic position of children in husband-wife and single-mother families for whites and blacks in 1977. For all children, the ratio of the median income (comparing children in single-mother families to those in husband-wife families) is .37 for whites and .38 for blacks. When the single mother is in the labor force and the wife is at home, the ratio of median incomes is .50 for whites and .59 for blacks; when both mothers are in the labor force, the ratio falls to .44 for whites and .41 for blacks.[18] The participation rates for mothers of black and white children are quite different. Black children are more likely to have an employed mother when the family has two parents; white children are more likely to have an employed mother when the family has one parent.

The work status of the wife makes less difference for whites than for blacks when comparing incomes of the single-mother and husband-wife families. When the wife is not in the labor force, the income ratio for whites is lower than in the hypothetical example for earnings (.60). These ratios indicate that white single-mother families have 50% of the income of husband-wife families with the wife not in the labor force; the ratio is .59 for blacks. Many of the single mothers work only part year or part time, although many receive some government payments in addition to their earnings.

When the wife is in the labor force, the income ratio is higher, especially for whites, than in the hypothetical example with earnings only (.38). This higher ratio probably reflects the addition of government support payments to income, since both single mothers and wives include women who work part year and part time.

In any event, these ratios overstate the economic position of single-mother families relative to husband-homemaker families, if the employed mother must pay for child care. However, additional payments for child care by employed single mothers are not very large. In fact, regular child care by nonfamily members (other than in schools) is small in terms of actual hours and does not differ much according to the mother's work or marital status.

CHILD CARE PATTERNS

Even though almost all children under 6 years receive some regular child care by someone other than their parents, parental care continues to be the primary form of child care. Less than one in three children under 6 years receives more than 10 hours a week of nonparental care and less than one in six receives 30 hours or more of nonparental care.[19]

TABLE 4

COMPARISON OF SINGLE-MOTHER AND HUSBAND-WIFE FAMILIES
WITH CHILDREN UNDER 18 YEARS, BY MEDIAN INCOME, 1977

	Husband-wife Families		Single-mother Families	
	WHITE	BLACK	WHITE	BLACK
Children under age 18	$18,869	$14,194	$6,981	$5,357
Mother in labor force	20,357	16,102	8,856	6,541
Mother not in labor force	17,620	11,017	4,614	4,029
% of children under age 18 with employed mothers	47	64	64	54
% of children under age 6 with employed mothers	37	59	53	47

Ratios of Median Income	WHITE	BLACK
Children of single-mother to husband-wife families	.37	.38
Single mother in labor force; wife at home	.50	.59
Both mothers in labor force	.44	.41
Neither mother in labor force	.26	.37

SOURCE: Elizabeth Waldman, Allison Sherman Grossman, Howard Hayghe, and Beverly L. Johnson, "Working Mothers in the 1970s: A Look at the Statistics," *Special Labor Force Report 233,* U. S. Department of Labor, Bureau of Labor Statistics, 1979: Table 3. Income figures are for 1977; family status is for March, 1978.

The major change in child care patterns over the past decade has been the rising proportion of children enrolled in preschool. From 1967 to 1976, the percentage of 3-year-olds in preschool tripled from 7% to 20%; the percentage of 4-year-olds in preschool doubled from 21% to 42%; and the percentage of 5-year-olds in preschool increased from 65% to 81%.[20] The enrollment levels vary according to the mother's work status for 3- and 4-year-olds, but not for 5-year-olds. When the mother is in the labor force, 25% of the 3-year-olds and 45% of the 4-year-olds are enrolled in preschool; when the mother is not in the labor force, the preschool enrollment rates are 17% and 40%, respectively. Furthermore, the children aged 3 and 4 with employed mothers are in nursery school for more hours: 44% of those with employed mothers attend a full-day program compared with only 8% of those with mothers not in the labor force.

When children aged 3 to 6 years are not in school, they are primarily cared for by a relative (usually the mother at home). This is true for one-half of the children aged 3 to 6 years even when the mother is employed full time. This percentage rises to 80% when the mother is employed part time and 97% when the mother is not in the labor force. Among children aged 3 to 6 years whose mothers are employed, only 21% are cared for by nonrelatives outside of school. This percentage falls to 10% for children aged 7 to 13 years.[21]

The ratios in Table 5 show that employed mothers of young children use more substitute child care than nonemployed mothers. Outside the school system, they rely upon other relatives and, to a lesser extent, nonrelatives. However, employed mothers, especially of young children, remain the primary source of child care. Although substitute child care increases when the mother is employed, there is nowhere near an hour for hour replacement in terms of paid child care.

The growth in kindergarten enrollments, of which the great majority are in public schools, has been accompanied by an expansion of other public child care programs such as Head Start and Title XX day care (from state funds for low-income children). The federally subsidized child care for low-income children has primarily taken the form of direct services.[22] In contrast, federal subsidies to moderate-income families for child care primarily take the form of reimbursement through tax credits for employed mothers which was initiated in the 1970s. In 1977, the child care tax credit was applied to 4 million children and cost $500 million or 22% of all federal expenses for child care.

The amount of child care actually paid for by employed mothers is fairly small, especially for low-income families. This statement is tautological in that mothers with low earnings cannot afford to work unless substitute child care is inexpensive or without cost. Given our lack of knowledge about possibly unmet demands for low-cost, satisfactory child care, we cannot make any generalizations about the probable work patterns of single mothers if the system of subsidized child care were expanded. However, in the absence of low-cost, satisfactory child care, the time requirements of maintaining a home coupled with low earnings potential preclude many single mothers from engaging in paid work, especially full time.

GOVERNMENT PROGRAMS FOR SINGLE-MOTHER FAMILIES

The other side of the double bind of time and money squeezing single mothers is that without adequate monetary assistance from the government the single mother is forced to find a way to increase the family's income while providing the required housework services. Government programs for single-mother families in the United States are largely determined by state regulation, and the variation across states is well known. The government programs have historically been torn between providing adequate support for the children in single-mother

TABLE 5

RATIOS BETWEEN EMPLOYED AND NONEMPLOYED MOTHERS
BY DAYTIME CARE OF CHILD WHEN NOT IN SCHOOL, 1976

	IN HOME BY RELATIVE	CARED FOR BY NONRELATIVE[1]
Children 3 to 6 years, mother in labor force	.65	.21
In nursery school (16%)[2]	.75	.14
In kindergarten or school (40%)	.70	.20
Not in school (43%)	.58	.26
Children 7 to 13 years, mother in labor force	.76	.10

SOURCE: U. S. Bureau of the Census. *Current Population Reports,* Series P-20, No. 318, 1978, and Series P-20, No. 298, 1976.

[1] Includes care by nonrelative in own home or in someone else's home, child cares for self, and not reported.

[2] Sixty-six percent of these children are in private (as opposed to public) nursery schools.

families and providing economic "incentives" for the single mother to raise the family out of poverty either by her own employment or by remarriage. In reality, however, most of these families have little control over such events. In the first instance, the wage rate required by single mothers to meet the minimal time and money requirements to be nonpoor is considerably above their earning power in the labor market, especially when the children are young.[23] In the second instance, many single mothers rarely face a viable option of marrying a man who can financially care for the family. For these reasons, policies that center on providing economic incentives for work and/or remarriage at the expense of providing adequate income support may result in many single-mother families being below the poverty line for extended periods of time.[24]

In the United States, the average AFDC monthly payment per recipient in constant dollars rose 34% between 1960 and 1970 and increased only slightly (2%) between 1970 and 1978. However, payments to families (instead of recipients) actually declined 20% in the 1970s. The average number of recipients per family was approximately the same (3.8 persons) in 1960 and 1970; therefore, the percentage increase in payments to recipients and families was the same. However, in 1978 the average number of recipients per family had fallen to 3.0. Because money needs reflect economies of scale (for example, a mother with 3 children needed 27% more income in 1977 to be nonpoor than a mother with 2 children), the decline in the average number of recipients per family meant that the average recipient payment had to rise in order for the family's standard of living not to fall. If the 1970 average family payment is adjusted for the decline in average number of recipients, then the 1978 payment should have decreased to only $136, which is 4% higher than the actual value of

$131. By 1978, the average family payment was 63% of the official poverty threshold. Of course, this average obscures the enormous variation across states but does not mask the decline in the average payment's actual purchasing power long before the average payment reached the poverty threshold.[25]

When these pieces of the puzzle are placed together, they show a picture of declining economic status for the single-mother family as the husband-wife family increases its number of earners; as all families take advantage of greater use of preschool child care with government subsidies for direct child care for low-income families and tax credits for other families; and as the purchasing power of average AFDC payments stagnate or decline slightly. Although we might have expected the women's liberation movement to improve the relative economic position of single-mother families, in fact the opposite has occurred. Why has this happened? I suggest that the single-mother family has been caught in the crossfire of two opposing movements: (1) the feminist struggle for sexual equality that initially centered on helping women (especially professionals) gain access to jobs while essentially ignoring the problems posed by their household responsibilities, and (2) the counterattack by the profamily forces to maintain the sexual division of labor and women's primary role as mothers. Neither one of these movements addressed the problems of the single mother, who is the sole provider of the time and money needed by her family.

CAUGHT IN THE CROSSFIRE

The single mother does not have the option of focusing her complete attention either on a market job or on running the home. She is constrained by her family's minimum requirements of time and money, which to a large degree are not substitutable, and by government income support programs, which usually do not meet basic needs. The single mother has always been in a precarious economic position, one made even more tenuous by certain recent policy trends.

Impact of the Women's Movement

The movement for sexual equality has thus far resulted primarily in wives' increasing their paid labor time, but not in husbands' increasing their housework time or decreasing their paid work time. Nor has wage equality been achieved. As this paper has documented, one result has been the relative decline in the economic position compared with men. In fact, in some ways their position has deteriorated because the myth of equality, which has been ahead of the reality, has added more to women's legal responsibilities than to their resources. This deterioration can be seen in the changing practices of granting child support and alimony when a marriage is dissolved. Although a wife in the past had seldom shared equally in a husband's income after the dissolution of the marriage, the social presumption had been that the husband had some obligation to help support his former wife while she continued in her role as caretaker of the children. Usually, she (with the children) kept the house when there was one.

Although the increased labor market participation of women has diminished their absolute economic dependence on men, the positive aspects of this trend have been offset by a decline in government-enforced economic support by the husband of wives and children. The social presumption today is that after a divorce or separation women are expected to work.[26] Even those wives who have spent most of their adult lives as full-time homemakers are expected to become wage earners. The husband may be expected to support his former wife while she makes the "transition" into the labor market, and he is still expected to contribute to the support of his children.[27]

In neither period has the father's contribution been based on the "maintenance of life style" of the mother and the children. Rather, it has usually been based on his "ability to pay" after his needs, which reflect his occupational status, have been taken into account. This approach has meant that women and their children have usually suffered more from divorce than men. The latest moves toward presumed equality have worsened the position of the divorced woman, who is now expected to raise the children (with perhaps more visiting from the father) and earn a paycheck. The deteriorating position results from the application of a "lopsided" equality, i.e., the expectation that women and men both have a job without requiring an equalization of male and female earnings and without requiring equal sharing of child rearing. The single mother is caught in the rise of the "new equality" which means an increased workload for her in the absence of real equality.

An increased social awareness of the work overload for employed mothers, both married and single, has brought forth proposed solutions which are centered around lightening this burden, primarily through increasing the social services for children (especially day care programs) and through increasing the part-time work options for women. Both of these approaches will help alleviate the workload of employed mothers, and should be applauded on those grounds. However, the latter solution of decreasing the paid work week for women (and not for men) should be viewed with great skepticism by proponents of sexual equality. While it would give women the time needed to maintain their households without increased help from their husbands, reducing the work week for women's jobs will keep these jobs separate from and unequal to men's jobs. Family life as the woman's domain will continue, yet women will still be a source of well-disciplined, highly educated, and low-paid labor. Economic equality for women, especially during any period a mother is raising children alone, becomes a cruel myth under these circumstance.

Impact of Forces Promoting the Nuclear Family

If single-mother families have been little helped so far by the gains made by the feminist movement, they also have not been aided by the counter movement to "save the family," which refers to the promotion of the intact nuclear family. Because this movement has a politically conservative and religious base, many of their demands center around reducing government social programs or using government programs to punish people who stray from the socially prescribed path. This approach makes two major, controversial assumptions: people have control over the formation and evolution of family structure, and people respond to economic incentives in making decisions about family composition. When the social norm is the intact nuclear family, this approach results in policies that are economically punitive for other family types. One example of this policy in action is Nixon's 1971 veto of the Comprehensive Child Development Bill, which would have expanded government funding and provision of child care. Another more recent development is the cutoff of federal funds for abortions for low-income women (the Hyde Amendment, a law that was ruled constitutional by the Supreme Court). Other examples include the Food Stamp provision which stipulates that recipients must not share meal preparation with others outside the government-defined recipient unit, and the AFDC provision in many states that reduces a family's monthly payment by the amount of reduced rental costs achieved by sharing living quarters with people outside the government-defined recipient unit.[28] These policies, which reinforce sexual inequality and promote women's role as primary caretaker of young children and as housekeeper, reduce the already meager options and resources available to the single-mother family. In terms of resources needed in comparison with those available, the single-mother family remains the most jeopardized group in our economy. At the same time that the movement of wives into the labor force has worsened the relative economic position of single-mother families, government policies that provide income support are stagnant or declining and becoming more punitive.

CONCLUSION

In a fundamental sense, the lowly position of the single-mother family reflects the true position of women in our society, because this situation forms the basic economic ground from which the adult woman builds in her work and family life. This worst possible case for women in America underscores their weak bargaining position in the home and in the marketplace. The deteriorating position of the single-mother family undermines the notion that the position of women vis-à-vis men improved in the United States during the 1970s. Our "bringing down the rear" in the United States must concern everyone who cares about the position of women as well as the lives of the ten million children of single mothers.

NOTES

1. Special thanks are extended to Sean Flaherty for his assistance in the research for this chapter and for his comments on an earlier draft. Support was provided by the Institute of Industrial Relations, University of California, Berkeley.

2. See Vickery, 1979.
3. See Easterlin, 1973.
4. Vickery, 1977: Table 4. The exact hours of housework vary by the age of the youngest child. In the examples used below, adults are assumed to be able to work a maximum of 87 hours per week.
5. For simplicity, the assumption is made that required housework is the same number of hours for each family. This assumption, further simplified so that required housework is 60 hours per week, will be used in all the examples that follow. The earnings ratio is w_m

$$\frac{27(.6w_m)}{45w_m} = \frac{16.2}{45} = .36$$

where the female wage rate (w_f) is .6 the male wage rate (w_m).
6. If 10 hours of housework can be replaced at a replacement cost of w_f, then the earnings ratio rises to

$$\frac{37(.6w_m)}{45w_m} = \frac{22.2}{45} = .49$$

7. The full income ratio is

$$\frac{27(.6w_m) + 60(.6w_m)}{45w_m + 60(.6w_m)} = \frac{52.2}{81}$$
$$= .644$$

This ignores adding in a value for the adult's time outside paid work and required housework. If the value of leisure time is included, the ratio will be lower.
8. Vickery, 1979.
9. Whenever the assumption is made that substitution of market goods and services for housework time is allowed, the assumption also is made that the replacement cost per hour of time (r) is less than or equal to w_f. Here there is no limitation on the market hours worked, and all adults work 87 hours per week in paid and/or unpaid work. Then with no substitution, the ratio of single-mother earnings to husband-wife earnings is

$$\frac{27(.6w_m)}{87w_m + 27(.6w_m)} = .157$$

With full substitution, the ratio is

$$\frac{87(.6w_m)}{87w_m + 87(.6w_m)} = \frac{.6}{1.6} = .375$$

Whenever we assume $w_f = .6w_m$ represents the value of female and male work time, and all adults work 87 hours per week, the the ratio of the value of adult's time in the two families is .375.
10. If market work is limited to 45 hours and 60 hours of housework is

required (without substitutes), the ratio of maximum earnings of the single mother to the husband-wife is

$$\frac{27(.6w_m)}{45w_m + 45(.6w_m)} = .225$$

since the single mother works 27 hours for pay and 60 hours at home, the husband and wife each work 45 hours for pay and together work 60 hours at home (with 42 hours of housework the maximum either one could do). If the value of the 60 hours of housework at price w_f is added, the ratio is

$$\frac{27(.6w_m) + 60(.6w_m)}{45w_m + 45(.6w_m) + 60(.6w_m)} = .483$$

11. If the 24 "free" hours of the husband and wife are valued at w_f and added to full income, then the full income ratio falls to .426.

12. The term "single-female family" is used to denote a family (i.e., two or more related individuals) headed by an unmarried female adult. Not all such families have children, a point that will be covered under the section on family composition.

13. U.S. Bureau of the Census, 1976a:Table 20.

14. The computation is $.49 = .46(.43) + .54(.54)$. As the participation rates of wives increased, wives of husbands with higher earnings entered the labor market in increasing numbers. This raised the median family income for employed wives by an amount even greater than the wife's earnings, thus causing the single-female to husband-wife (employed) ratio to fall even more. At the same time the ratio of the median incomes of families with non-employed wives compared to families with employed wives fell from .83 in 1950 to .74 in 1977 (See Table 2).

15. U.S. Bureau of the Census, 1951, 1977. About 55% of the husband-wife families had children under 18 in 1950 and 1976.

16. U.S. Bureau of the Census, 1979f:Table 3. The figure given is the proportion of (primary, plus secondary, plus subfamilies) that are female headed. A detailed analysis of the changing household structure and its importance for understanding poverty is given in Vickery, 1978.

17. U.S. Bureau of the Census, 1950, 1976, 1979c, 1979d. The poverty rate of children in all families is 16%.

18. The ratios in Table 4 are fairly close to the ratios in Table 2. The ratios in Table 2 are lower than they would be if the single-female families had been broken down by the women's labor force status so that the income of the employed single female would have been higher.

19. Rodes and Moore, 1975.

20. U.S. Bureau of the Census, 1978a.

21. U.S. Bureau of the Census, 1976c.

22. Head Start and Title XX involve direct provision of child care. Other child care expenses for AFDC employed mothers are

reimbursed. Title XX grants served 800,000 children in 1977 and cost $809 million; Head Start served 350,000 children in 1977 and cost $448 million. The Congressional Budget Office estimated that 1.2 million children were eligible but did not participate in Head Start. See U.S. Congress, Congressional Budget Office, 1978.

23. See Vickery, 1977: Tables 2 and 3. These tables show that about one-half of all employed women would not be able to support two to three children as a single parent.

24. This discussion has not included any mention of the child support issue because policies aimed at increasing child support for low-income families seem to be an avenue of limited return in terms of improving the well-being of the families involved. For a discussion of the issues, see Vickery, 1978.

25. The equivalent 1970 rate in 1978 is $X(.2) + (1.27X).8 = 163$. AFDC figures are from U.S. Department of Commerce, 1979: Table No. 566. Poverty threshold figures (1977) are from the U.S. Bureau of the Census, 1979b:Table A-2.

26. This discussion leaves out the group of women who have children outside of marriage. This phenomenon is especially important among the low-income population; in 1973, 32% of the children in the AFDC program had mothers who were not married to the father, U.S. Department of Health, Education, and Welfare, Social and Rehabilitation Service, 1974.

27. These findings on divorce settlements are taken from Weitzman and Dixon, 1980.

28. See Vickery, 1978.

JOB CREATION AND UNEMPLOYMENT FOR CANADIAN WOMEN[1]

Pat Armstrong and Hugh Armstrong

Vanier College
Montreal, Quebec   Canada

In recent years, Canadian women have been flooding into the labor market, joining the ranks of the employed and unemployed. While the steadily rising participation rate of women has been carefully documented and discussed, the more dramatic increase in the female unemployment rate has been largely ignored or dismissed as unimportant. The growing number of women searching for paid work has been explained, if at all, primarily in terms of changing female aspirations and preferences. The search has been viewed by some as dangerous, as the main cause of the increase in both male and female unemployment. More effort has been directed toward dismissing women's unemployment as insignificant--because they do not need the work, because they are secondary workers, and because they claim unemployment primarily to gain eligibility for benefits--than toward investigating the economic conditions which give rise to these massive changes in women's labor force behavior.

The purpose of this paper is to show that women's behavior in the labor market should be understood in terms of structural factors rather than personal characteristics, in terms of women's economic needs and employer demands rather than individual skills and aspirations, in terms of employment opportunities rather than unemployment, and in terms of jobs rather than people. Through an examination of recent trends in the labor force, job creation, and income distribution, we argue that the growth in the number of women working or looking for work can be explained primarily in terms of changing economic conditions. More of the jobs available are "women's jobs" and more women are searching for work because they need the money.

LABOR FORCE TRENDS

In the last 20 years, Canadian women, like women in the United States, Australia, and Sweden, have been steadily increasing their labor

129

force participation.   According to the U.S. Department of Labor's comparison of data for nine countries,[2] the Canadian female labor force participation rate had increased more since 1960 than that of any other country and by 1976 was the fourth highest, tied with Japan. In that year, the female unemployment rate in Canada was second only to that of the United States and almost two percentage points higher than the rate in France, which had the third highest rate.   Compared to these eight countries, then, an increasingly greater proportion of Canadian women searched for work but, with the sole exception of the United States, a large proportion failed to find jobs. The paper begins with an examination of these dramatic shifts in Canadian women's labor force behavior; the final sections investigate the factors which contribute to these unprecedented changes in women's employment and unemployment.

Since 1946, Canadian women's labor force participation rate has almost doubled, from 24.7% to 48.9% in 1979, and their unemployment rate has more than tripled, from 2.4% to 8.8%.[3] As Table 1 indicates, while just over one-third of Canadian women were working or looking for work in 1966, almost half of them were doing so by 1979. Furthermore, although women's share of employment increased by 7% between 1966 and 1979, their share of unemployment increased by 14.3%. Throughout the late 1970s, almost one in ten of the women in the Canadian labor force was unemployed. Even the official unemployment figures (and it is clear that such data greatly underestimate the actual number of people who want work) indicate that women disproportionately suffer from unemployment.[4]

As Table 1 also shows, women in 1979 made up only 38.3% of the employed, but they accounted for 46.1% of the unemployed. Moreover, the frequency and duration of female unemployment has dramatically increased while the reverse pattern is evident for men. In 1979, 11.3% of unemployed men and 21.3% of unemployed women had been without work for a year or more. By contrast, 35.0% of unemployed men and 26.0% of unemployed women had been without a job for less than 3 months at the time of the survey.[5]

That women's share of unemployment had increased more than their share of employment suggests that too many women entering the labor force are not, as many have argued, the only reason for the startling rise in female unemployment rates. The coincidence, as shown in Table 1, of the fluctuations in the female percentage of the unemployed and in the female unemployment rate with the consistent rise in female labor force participation also suggests that women's rising labor force participation rate is not simply caused by too many women seeking work.

EFFECTS OF WOMEN'S LABOR FORCE PARTICIPATION ON MEN'S JOBS

Women's rising labor force participation rate has been considered dangerous, in part at least, because it is feared that women cause the real

TABLE 1

LABOR FORCE PARTICIPATION
EMPLOYMENT AND UNEMPLOYMENT
CANADA, 1966-1979

Year	Participation Rate Women	Men	Women as % of Employed	Unemployment Rate Women	Men	Women as % of Unemployed
1966	35.4	79.8	31.3	3.4	3.3	31.8
1967	36.5	79.3	32.1	3.7	3.9	31.4
1968	37.1	78.6	32.7	4.4	4.6	31.8
1969	38.0	78.3	33.2	4.7	4.3	35.1
1970	38.3	77.8	33.6	5.8	5.6	34.7
1971	39.4	77.3	34.2	6.6	6.0	36.8
1972	40.2	77.5	34.5	7.0	5.8	39.1
1973	41.8	78.2	35.1	6.7	4.9	42.6
1974	42.9	78.7	35.6	6.4	4.8	42.6
1975	44.2	78.4	36.3	8.1	6.2	43.2
1976	45.0	77.7	36.9	8.4	6.4	44.2
1977	45.9	77.7	37.3	9.5	7.3	44.1
1978	47.8	77.9	38.3	9.6	7.6	44.8
1979	48.9	78.4	38.3	8.8	6.6	46.1

SOURCES: For 1966-77, calculated from Statistics Canada, *Historical Labour Force Statistics: Actual Data, Seasonal Factors, Seasonally Adjusted Data* (Cat. No. 71-201) Ottawa, 1978. For 1978, calculated from Statistics Canada, *Labour Force Annual Averages, 1975-1978* (Cat. No. 71-529) Ottawa, 1979. For 1979, calculated from *The Labour Force, December 1979* (Cat. No. 71-001), Ottawa, 1980.

unemployment, the unemployment of "prime age" males.[6] The basic assumption of this argument may be challenged. Little evidence has been produced to show that men have a prior claim to any job. If the jobs are male by virtue of some unique male physical or mental capacity, women would not be able to take them away. If they are male by virtue of men's sole financial responsibility for spouses and/or children, then only one-quarter of male workers qualify[7] and many women would also meet this criterion.[8] Moreover, it is difficult to understand how single men would be more eligible by this criterion than single women.

Even if the logic of this argument is left unchallenged, however, it is clear that women, for the most part, are not taking jobs away from men. They are employed and unemployed in different industries and occupations. In our initial study of women's labor force participation and later in The Double Ghetto,[9] we show, using census data, that within the Labour Force Survey occupational categories, women and men tend to have and to seek different jobs. For example, the women in the teaching category are concentrated at the elementary level, while in medicine and health they are much more likely to be nurses, technicians, and therapists rather than doctors; and in the managerial and administrative category, they are much more likely to manage small boutiques and cosmetic counters than large department stores. Other researchers[10] report the same patterns of segregation. Almost two-thirds of the female labor force work in just 20 of the 500 occupational categories used in the census while just over one-third of male workers are employed in the 20 jobs with the most men. On the other hand, almost one-half of the men in the labor force work in occupations which together employ less than 5% of the women who work.

Unfortunately, Canada does a full census only once every 10 years and the results from the 1981 census will not be available for several years. The monthly Labour Force Survey does provide more current data, but the broad categories used for the sample survey techniques make detailed analysis impossible. A limited series of analyses, however, is possible on the basis of this more current sample data.[11]

In our previous work, we introduced two ways of measuring sex segregation. The degree of sex typing is measured by calculating the percentage of all workers in an industry or occupation who are female. The degree of female concentration is measured by calculating the percentage of all female workers who are employed in an industry or occupation. By making these calculations for the Labour Force Survey data on employment and unemployment by industry and occupation, it is possible to get some idea of the extent to which men and women compete with each other for work. Because the survey data for both industries and occupations yield similar results, we shall report only the occupation data here.

The occupational divisions given in Table 2 clearly reveal the limited competition between men and women. Just three different jobs— clerical, sales, and service—account for three-fifths (62.6%) of all female workers, but represent only 26.4% of all male workers. Unemployed women also are concentrated in these occupational groups. Well over one-half (57.4%) of unemployed women and less than one-quarter (23.9%) of unemployed men are seeking work in these occupations. Over one-third of male unemployment is found in construction, transportation, and materials handling and other crafts occupations, where very few women (about 3% in total) are either employed or unemployed. Once again, a closer examination of male and female unemployment rates reveals greater segregation. For instance, the male occupations (the natural sciences; fishing, hunting, and trapping; forestry and logging; and mining and quarrying) account for an additional 7% of male unemployment.

TABLE 2

PERCENTAGE DISTRIBUTION OF EMPLOYMENT AND UNEMPLOYMENT BY OCCUPATION AND SEX, CANADA, 1979

OCCUPATION	EMPLOYMENT			UNEMPLOYMENT		
	WOMEN AS % OF OCCUPATION	WOMEN AS % OF FEMALE WORKERS	MEN AS % OF MALE WORKERS	WOMEN AS % OF OCCUPATION	WOMEN AS % OF FEMALE WORKERS	MEN AS % OF MALE WORKERS
White collar						
Managerial, professional, technical	40.8	24.0	22.2	56.6	11.7	6.9
Clerical	77.2	34.0	6.4	81.6	26.4	5.1
Sales	39.9	10.7	10.3	51.7	8.0	6.4
Subtotal	**52.9**	**68.8**	**38.8**	**67.2**	**46.1**	**18.4**
Blue collar						
Service	53.9	17.9	9.7	61.4	23.1	12.4
Primary occupations	18.0	3.0	8.5	18.9	2.1	9.5
Processing	19.1	7.8	21.0	31.8	10.6	19.9
Construction	1.4	0.2	10.3	---	---	20.4
Transportation	5.3	0.6	6.4	---	---	7.1
Materials handling and other crafts	18.2	1.8	5.2	30.0	2.3	6.2
Subtotal	**24.5**	**31.3**	**61.2**	**31.4**	**38.1**	**75.4**
Unclassified				68.5	15.8	6.2
Total	**38.8**	**100.0**	**100.0**	**46.1**	**100.0**	**100.0**

SOURCE: Calculated from Statistics Canada, *The Labour Force, December 1979* (Cat. No. 71-001), Ottawa, 1980.

The data on part-time employment, although crude, give some indication of the sex segregation in full- and part-time work.[12] Recent data show that 2½ times as many women as men work part time, representing 71.9% of all part-time workers. Nearly one-quarter (23.3%) of the women as compared to 5.8% of the men work part time.[13]

The occupational data given in Table 3 indicate that there is also segregation in part-time work. Over two-thirds (67.1%) of all part-time jobs are in clerical, sales, and service occupations. While men working in sales and service are about as likely to work part time as are women in these occupations, it should be remembered that, as Table 2 shows, 62.6% of the women, but only 26.4% of the men work in clerical, sales, and service jobs. More than one in three of the women in sales, service, primary occupations, and transportation works part time. Although men and women usually work part time in different occupations and while sex segregation and concentration are more likely to encourage competition among women rather than between the sexes, the data do suggest that men and women may be seeking the same part-time work in sales and service jobs. But here too, more extensive information may reveal segregation within these broad categories which would mitigate against women taking part-time jobs from men. For example, in trade occupations, women are more likely to be cashiers, while men carry out the groceries, and to sell women's underwear rather than household appliances.

The Labour Force Survey data reaffirm the patterns evident in the census data. There are "men's jobs" and there are "women's jobs." Many more women are working but most continue to work in "women's jobs" and in jobs that are part time. These patterns of segregation suggest that the great influx of women into the labor force has not put many men out of work. They also suggest that the dramatic rise in female labor force participation may be related to the occupational structure, to the creation of more women's jobs.

Job Creation

Not only are more women entering the labor force but women are also getting more of the jobs. Since men and women are, for the most part, employed and unemployed in different industries and occupations, it seems likely then that more of the employment opportunities are in areas where women have traditionally been employed. Although an analysis of a long historical period would be useful, the data are difficult to obtain due to fundamental changes made to the Labour Force Survey in 1975. However, an examination of the jobs available in 1979, compared to those available in 1975, provides some indication of the growth in demand for workers and for specific types of workers.

Most of the 1,085,000 jobs created between 1975 and 1979 are in the occupations with the most women. Put another way, the occupations with the slowest growth tend to be those dominated by men.

TABLE 3

**PERCENTAGE DISTRIBUTION OF PART-TIME EMPLOYMENT
BY OCCUPATION AND SEX, CANADA, 1979**

OCCUPATION	PART-TIME % OF OCCUPATION			FEMALE % OF PART-TIME EMPLOYMENT	OCCUPATION'S % OF PART-TIME EMPLOYMENT		
	BOTH SEXES	WOMEN	MEN		BOTH SEXES	WOMEN	MEN
White collar							
Managerial, professional, technical	9.1	18.0	3.0	80.6	16.6	18.6	11.5
Clerical	16.4	18.8	8.0	88.6	22.3	27.5	8.8
Sales	20.4	37.3	9.0	72.9	17.0	17.2	16.2
Subtotal	**13.9**	**21.4**	**5.4**	**81.4**	**55.9**	**63.3**	**36.4**
Blue collar							
Service	27.1	36.5	16.0	72.4	27.8	28.0	27.1
Primary occupations	13.0	36.4	7.9	50.0	6.6	4.6	11.8
Processing	2.6	5.4	1.9	40.5	3.2	1.8	6.9
Construction	2.9	---	2.6	---	1.5	---	4.7
Transportation	4.9	34.8	3.2	38.1	1.6	0.9	3.6
Materials handling and other crafts	10.8	12.2	10.5	20.5	3.4	1.0	9.6
Subtotal	**11.2**	**27.0**	**6.0**	**59.1**	**44.1**	**36.3**	**63.6**
Total	**12.5**	**23.2**	**5.8**	**71.9**	**100.0**	**100.0**	**100.0**

SOURCE: Calculated from Statistics Canada, *The Labour Force, December 1979* (Cat. No. 71-001). Ottawa, 1980.

As Table 4 indicates, women are found mainly in white-collar occupations (68.8% of all female workers) and over one-half (51.8%) of all new jobs are in these occupations. Within the white-collar group, the least growth was in the sales area which has a high concentration of male workers. Within the blue-collar sector, the largest proportion of new jobs was created in the service occupations where women constitute more than one-half of the workers. The only area in which men predominate that experienced a large growth was processing, and men were hired for over three-quarters of these new jobs.

Women moved into almost all occupations. However, although the largest growth in the white-collar sector was in managerial, technical, and professional occupations, women were hired for a smaller percentage of jobs in these occupations than they were in other white-collar jobs. Women were hired for only a minority of new jobs in the managerial, technical, and professional category while they dominated the growth in clerical and sales as well as service jobs. On the other hand, 60% of the jobs created for women were in sales, service, and clerical occupations, which are among the lowest-paid and least-rewarding female jobs.

Although historical data on part-time job creation by industry and occupation and by sex are not directly available, it is possible, by comparing Table 3 with Table 4, to gain a sense of the relationship between part-time work and job creation. The significance of this comparison lies in the facts that not only has there been a sharp rise in the share of jobs that are part time, but that women have filled almost four-fifths of these new part-time jobs.

The jobs have been created in the industries and occupations where both part-time work and female workers are most prevalent. In the occupational structure, both part-time work and job creation have been relatively strong in the clerical and service occupations dominated by women. Aside from services, men make up a significant component of part-time workers (over one-quarter) only in occupations with little or no job creation for men between 1975 and 1979. In fact, in sales, the most important such occupation for male part-time jobs, the overall number of male workers actually declined during this period. Meanwhile, the most important occupation for male job creation, processing, is over 98% full time for men. More of the jobs are part time, more of the part-time jobs are in industries and occupations dominated by female workers, and more women take this part-time work. Given the evidence of segregation from the last census, there is every reason to believe that the next census in 1981 will confirm the argument presented here. It seems likely, on the basis of the data that are available, that more of the new jobs are secondary in terms of pay, working conditions, and job security. In short, more of the jobs are "women's jobs."

The data presented here also do not permit a detailed analysis of the effect of technology on skill levels. However, as Braverman and others[14] have suggested, the introduction of new technology often breaks down the

TABLE 4

**PERCENTAGE DISTRIBUTION OF JOB CREATION
BY OCCUPATION AND SEX, CANADA, 1975-1979**

OCCUPATION	TOTAL % OF JOBS CREATED MEN/WOMEN	TOTAL % OF JOBS CREATED WOMEN	TOTAL % OF EMPLOYED WOMEN	WOMEN'S % OF JOBS CREATED	WOMEN'S % OF EMPLOYMENT 1979
White collar					
Managerial, professional, technical	33.6	26.5	24.0	46.6	40.7
Clerical	13.3	22.8	34.0	101.4	77.1
Sales	4.9	12.6	10.7	152.8	39.9
Subtotal	**51.8**	**62.0**	**68.8**	**70.6**	**52.9**
Blue collar					
Service	18.8	24.6	17.9	77.5	53.9
Primary occupations	3.6	2.5	3.0	41.0	18.0
Processing	15.9	6.2	7.8	23.3	19.1
Construction	1.8	0.8	0.2	25.0	1.4
Transportation	4.2	1.7	0.6	23.9	5.3
Materials handling and other crafts	3.9	2.2	1.8	33.3	18.2
Subtotal	**48.2**	**38.1**	**31.3**	**46.7**	**24.5**
Total	**100.0**	**100.0**	**100.0**	**59.1**	**38.8**

SOURCES: For 1975, calculated from Statistics Canada, *Labour Force Annual Averages, 1975-1978* (Cat. No. 71-529), Ottawa, 1979. For 1979, calculated from Statistics Canada, *The Labour Force, December 1979* (Cat. No. 71-001), Ottawa, 1980.

work into repetitive tasks which require little skill, training, and strength, i.e., "women's work." At the beginning of the Industrial Revolution, machines encouraged employers to hire the inexpensive labor of women and children. It may be happening again. As Brecher points out in his study of the electrical products industry, "The degradation of jobs has been intimately connected with the allocation of male and female labor. Where possible, employers have been eager to break down jobs into lower-paid 'women's work'."[15] More women may be getting jobs in areas where technology has limited the skill, training, and time required. And these jobs themselves are insecure, since they too may be completely automated. The new "word processing equipment," so eloquently and frequently advertised as lowering employer costs and as providing relief from dull repetitive tasks, may contribute to even higher unemployment rates for women.

Female Unemployment

Although there was a large number of jobs to be filled in 1979 as compared with 1975, there was not enough work for the rapidly increasing number of women seeking employment. While the female share of unemployment growth was somewhat lower than their share of jobs created during this period (53.3% as against 59.1%), it was much higher than their 1979 shares of either jobs (38.8%) or classified unemployment (43.9%).

As Table 5 shows, when the data are broken down by occupation, the unemployment pattern is quite different from that revealed in job creation. While over one-half of the new jobs created are white-collar jobs, less than 40% of the increase in unemployment occurs within this group of occupations. However, over one-half of the female increase is for women seeking white-collar work. One-third of the growth in female unemployment is concentrated in clerical jobs. The increase in clerical unemployment consists entirely of women, and women also assume most of the unemployment burden in managerial, professional, and technical work. Over 70% of the increase in white-collar unemployment is experienced by women.

Blue-collar work, which accounts for three-fifths of the increase in unemployment, is usually considered a male domain, but even here over one-half (32.5% of 60.9%) of the unemployment rate increase is noted in service occupations, where women comprise the majority of workers and where they experience 65.5% of the increase in unemployment. In fact, 71.6% of the growth in overall unemployment, and 93.3% of the growth in female unemployment, are concentrated in the white-collar and service occupations. The rest of the occupations, while dominated by men in terms of the growth in both employment and unemployment, account for only 29.4% of the overall increase in employment and for only 28.4% of that in unemployment between 1975 and 1979.

TABLE 5

PERCENTAGE DISTRIBUTION OF UNEMPLOYMENT
BY OCCUPATION AND SEX, CANADA, 1975-1979

OCCUPATION	TOTAL CHANGE IN % MEN/WOMEN	TOTAL CHANGE IN % WOMEN	TOTAL WOMEN'S % 1979	WOMEN'S % OF CHANGE	WOMEN'S % OF TOTAL, 1979
White collar					
Managerial, professional, technical	13.0	13.3	13.8	54.5	56.6
Clerical	17.8	33.3	31.4	100.0	81.6
Sales	8.3	6.7	9.5	42.9	51.7
Subtotal	**39.1**	**53.3**	**54.8**	**72.7**	**67.2**
Blue collar					
Service	32.5	40.0	27.4	65.5	61.4
Primary occupations	6.5	4.4	2.5	36.4	18.9
Processing	5.9	1.1	12.6	10.0	31.8
Construction	9.5	---	---	---	---
Transportation	3.6	---	---	---	---
Materials handling and other crafts	3.0	1.1	2.8	20.0	30.0
Subtotal	**60.9**	**46.7**	**45.2**	**40.8**	**31.7**
Total classified	**100.0**	**100.0**	**100.0**	**53.3**	**43.9**

SOURCES:　For 1975, calculated from Statistics Canada, *Labour Force Annual Averages, 1975-1978* (Cat. No. 71-529), Ottawa, 1979.　For 1979, calculated from Statistics Canada, *The Labour Force December 1979* (Cat. No. 71-001), Ottawa, 1980.

CAUSES OF INCREASED FEMALE LABOR FORCE PARTICIPATION

An examination of the jobs available indicates the demand side of the equation; it does not explain supply or even surplus labor. To argue that more of the jobs are "women's jobs" does not explain why women take these jobs or why a growing proportion of women are unemployed. While personal characteristics such as age, marital status, education, and values are clearly important factors in women's labor force participation, economic need is the primary motivation for most women's job search. Most women are employed or unemployed because more of them need the income.

That most women, like most men, work because they require the income is not widely accepted in Canada. Indeed, some economists and politicians have argued that women, particularly married women, do not need to work and thus their employment, and especially their unemployment, may be dismissed as unimportant, as a matter of choice which has little consequence in terms of economic hardship. The argument assumes that most women have "prime age" males to support them and that these males are the people who need the work, who need the income. Although it is clear that jobs in Canada are not allocated on the basis of economic need (if economic need were the criterion, then many of these same politicians and economists would have difficulty justifying their right to work), there is ample evidence that many women rely heavily on their own earning capacity for financial support.

Economic Need: For Not-Married Women

Approximately 30% of the female labor force is single. It is difficult to determine what proportion of these women could rely on a "prime age" male for support but, in 1979, 60% of these women are observed to be 20 years of age or over.[16] Surely it can be assumed that the overwhelming majority of these women depend on their labor force incomes and not their fathers for support. The dramatic decline since World War II in the labor force participation of young, single women suggests that most of those who can rely on others for financial assistance do so and, therefore, stay out of the labor force.

Almost 10% of the female labor force is separated, widowed, or divorced. Some of these women may receive financial support from "real" workers, but Boyd's research[17] indicates that employment is the major source of income for nearly three-quarters of divorced women and one-half of the separated women. The National Council of Welfare argues that:

> When the labor force status of poor and non-poor, single-parent mothers is compared, it becomes clear that paid employment is the main factor preventing the better-off ones from succumbing to poverty.[18]

This study also shows that "Widows and other formerly married women living alone are most likely to be poor: 54% have incomes below the poverty line."[19] Economic need is a primary factor in the labor force participation of these women.

Economic Need: For Married Women and the Family

The other 60% of the female labor force is married. The movement of married women into the labor force accounts for 61.7% of the increase in female participation rates, and the financial requirements of these women who are assumed to have men to support them is more open to question. That married women's labor force participation is rising in spite of their poor job opportunities and low wages, in spite of the scarcity and quality of day care facilities,[20] and in spite of the "double burden" of two jobs, suggests that they need to work and they need the income. There is, however, more direct evidence of the growth in their income requirements. According to the National Council of Welfare,[21] there would be a 51% increase in the number of poor families in Canada if the wife did not have an income. "Women's jobs" often make the difference between low income and an adequate level of living.

As we argued in The Double Ghetto, and as MacLeod and Horner[22] have shown through their computer simulations, women's income is the primary way that low- and middle-income families have stabilized their living standard. In 1978, 57.4% of families with two income earners have more than $20,000 a year to live on. Less than half of that proportion (26.4%) of families with only one member working for pay earn more than $20,000 a year.[23] Furthermore, the increasing male unemployment rate and rising inflation may make the married woman's employment crucial for survival. The National Council of Welfare states that:

> While about three-quarters of non-poor husbands of all ages are employed on a full-time, full-year basis, less than a third of low-income husbands are. In addition to these working poor whose full-time jobs pay them less than a poverty line wage, another fifth of low-income husbands have occasional, part-time or seasonal employment that does not adequately support their families.[24]

Boyd's research[25] indicates that over 40% of married women rely on employment as their major source of income. Several researchers including Ostry, Spencer and Featherstone, Skoulas, and Nakamura et al.[26] have all found a strong negative relationship between the husband's income and the wife's employment. That is, the lower the husband's income, the more likely the wife's labor force participation.

Not only increasing unemployment but also falling wages may make it difficult to survive without more than one income in a family, especially given today's rising standards and sharply rising prices for the most necessary of commodities. While overall figures on wage rates often indicate an increase in real personal incomes, these figures may camouflage the inequality evident when the data are analyzed in greater detail. In their analysis of the occupational structure of earnings in Canada, Meltz and Stager found that "employees generally display a pattern of either no change or a decline in relative earnings from 1970 to 1975."[27] Moreover, when the percentage increase in wages is broken down by industry, it is clear that the smallest percentage increases in wages between 1975 and 1978 are in precisely those industries (service and trade) with the most job creation.[28] In other words, more of the jobs available are in the lower-paid (absolutely and relatively) sectors, thus increasing the need of many families for additional income.

Economic Need and Unemployment

Women's financial requirements also encourage them to stay in the labor force even when they cannot find work. They become part of the unemployed. A decade ago, Ostry argued that there were low female unemployment rates in Canada because women here were "less likely to remain in the market looking for work, but instead return to some non-labour force activity."[29] Now, Canadian women are staying in the labor force, in part at least, because their economic needs are even more pressing.

Approximately 45% of unemployed women are single, separated, widowed, or divorced. Because most of these women rely primarily on their own earnings for support, it can be assumed that most would suffer financially when unemployed. But what about the 55% of unemployed women who are married?[30] In their study of married women's employment and earnings, Nakamura et al. argue that:

> The 1970 average employment incomes of the husbands of unemployed wives were lower than the averages for the husbands both of wives who were not in the market labour force and of those who were working. Moreover, the average family asset incomes for wives at least 35 years of age, and the average per person family incomes (excluding the earnings of the wife) for wives at least 30 years of age, were also lowest for unemployed wives. Thus we find that, whatever the personal motivations for these wives for seeking work, there is an obvious need in their families for additional income. [31]

There also is evidence to show that unemployment insurance benefits are paid primarily to those who need the money. In 1973, 77% of total benefit expenditures went to individuals earning $6,000 a year or less.[32] Families whose combined yearly incomes totaled less than $10,000 received 58% of all benefits.[33] Futhermore, two-thirds of the families who received benefits and earned more than $10,000 a year had two or more income earners and were net contributors to the program.[34]

UNEMPLOYMENT INSURANCE

Because it is often assumed that women do not need to work, it is frequently argued that women illegitimately collect unemployment insurance benefits, and the translation of these arguments into policy then creates hardship for the majority of women who need the money. Most of the arguments ignore or downplay the fact that this is an insurance scheme that people contribute to on the basis of employment and have a right to collect, if they qualify, regardless of age or sex. But the attack on women as abusers appears to be primarily related to their right to collect unemployment insurance when they have another job at home, and, it is assumed, men to support them.

In People and Jobs, the Economic Council of Canada argues that "the increase in benefits had provided some disincentive to search for gainful employment or, more precisely, to remain idle voluntarily, particularly for women."[35] Green and Cousineau, in their study of unemployment in Canada, are suspicious that "where there is more than one earner in the family, some of what appears to be unemployment is really the enjoyment of leisure or the participation in non-labour market work activities."[36] It is clear from the preceding discussion that they primarily suspect married women. Little evidence has been produced, however, to prove that women who collect benefits do not qualify under the Act. Recent amendments to the legislation suggest just the opposite. In order to prevent women from qualifying, the regulations had to be changed.[37]

Under the amended version of the Unemployment Insurance Act, employers and employees contribute equal shares while the federal government pays only when unemployment rates are excessive—and more and more unemployment is considered normal. Benefits are related to weeks worked, to earnings, and, by a complicated formula, partially to regional unemployment. Benefits have been reduced from 66.7% of average earnings to 60% (a maximum of $159 per week) for everyone, regardless of dependents. As the combination of taxable benefits and deductible contributions still holds, a more regressive aspect of the plan has been retained, but some repayment is required from those with high net annual incomes. Claimants must now work longer, 10 to 14 weeks depending on regional unemployment rates, before they become eligible for benefits. In order to discourage people from entering the labor force simply to become eligible for unemployment insurance, an additional 14 weeks of insurable employment is required for "new entrants," or those working at the first job they have had in the last 52 weeks. In order to discourage voluntary quits, up to 6 weeks may be added to the normal 2-week waiting period if an employee resigns without what the Unemployment Insurance Commission officer considers "just cause." In order to discourage "repeaters," or people who claim benefits more than once, a claimant must now work up to 6 weeks longer than is necessary for the first claim. In order to eliminate people who are "not fully committed to the labour force," new regulations disqualify anyone who has either not worked at least 20 hours a week or not earned at least 40% of the maximum insurable earnings.[38] Since women are more likely to be short-term, part-time, and new workers, these new amendments are more likely to disqualify women (and young people) than "prime age" males.

Although it is difficult to obtain accurate historical data on payments to women, there is other evidence, in addition to the changes in the regulations, that indicates women legitimately receive unemployment insurance. Schwartzman,[39] a former employee of the Unemployment Insurance Commission, claims that benefit officers are expected to cut off between 40% and 60% of the people interviewed. Married women and young men are subject to particularly close scrutiny because they are thought to be the high-abuse category. The Comprehensive Review of the Unemployment Insurance Programme reports that "females have a significantly higher incidence of disqualification and disentitlement than males, regardless of insured weeks and dependency status."[40] Given that women face closer scrutiny and more rigid requirements than men, this higher disqualification rate is likely to be more a result of discrimination against women than proof of abuse. In any case, the high rate of disqualification clearly suggests that those women who do manage to collect benefits do so legitimately.

There is also evidence to suggest that women and young people do not receive their share of unemployment insurance. According to the Social Planning Council of Metropolitan Toronto:

> A comparison of the unemployment insurance claim-ant file with the official unemployment figures shows that young people and women significantly underuse the program, both in proportion to their numbers officially unemployed and in relation to older groups and women.[41]

Table 6 dramatically illustrates this underuse. Columns 1 and 2, compiled by Statistics Canada to indicate the use of the unemployment insurance program, suggest that young people and women receive more than their share of benefits. However, the addition of column 3 clearly shows that people under 25 and women over 25 are not getting their share of the insurance payments. While these groups constituted two-thirds of the unemployed in 1972, they received only half the benefits. By 1976, the gap had narrowed only slightly.

In addition, women receive less money if their claims are deemed legitimate. Although over one-half of the male unemployment insurance beneficiaries draw benefits which exceed the minimum wage in their province, this is the case for only one-tenth of the females.[42] Because benefits are based on wages and women's wages on the average are 60% of what men earn,[43] women would be expected to receive lower benefits than men. Another explanation for this difference, which is noted above, is that more women than men do not file an unemployment insurance claim.

TABLE 6

UNEMPLOYMENT INSURANCE BENEFITS FOR ALL WOMEN AND MEN
UNDER 25 YEARS, CANADA, 1972-1976

YEAR	% OF BENEFITS[1]	% OF LABOR FORCE	% OF UNEMPLOYED
1972	50.6	49.3	66.9
1973	52.4	50.6	69.4
1974	53.3	51.1	70.3
1975	54.4	51.8	70.4
1976	56.5	52.3	71.7

SOURCES: Columns 1 and 2 from Statistics Canada, *The Labour Force, September 1978* (Cat. No. 71-001),
p. 70. Column 3 calculated from Statistics Canada, *Historical Labour Force Statistics—Actual Data, Seasonal
Factors, Seasonally Adjusted Data, 1979c* (Cat. No. 71-201) pp. 54, 56, and 58.

[1] These data are from taxation statistics and thus are based on tax filers.

Women do constitute a relatively high proportion, 46%, of those who
voluntarily quit their jobs without just cause. But only "10% to 15% of
those who quit with or without just cause file a claim for Unemployment
Insurance."[44] Furthermore, voluntary quits without just cause account for
only 7.5% of the total insurance claim load.[45] Clearly, few women are
quitting their jobs in order to receive unemployment insurance benefits.

SECONDARY WORKERS

Part of the reason that women's unemployment has been dismissed as
insignificant, and part of the assumption that women do not need to work,
is the classification of women as secondary workers. Women are
repeatedly described, in Canada and elsewhere, as secondary workers, but
the meaning of the term is unclear. In a Statistics Canada publication,
Buckley comments:

> The usual definition is based on the regularity of
> attachment to the labour force so that the secondary
> worker is typically a student or a housewife who
> normally or regularly switches back and forth be-
> tween labour force and non-labour force activites. In
> popular usage, the reference to secondary workers
> often seems to have a broader connotation (possibly
> confused with secondary earner) taking in all married
> women who work and sometimes, young persons who
> have left school but still live at home.[46]

According to this frequently used, broader definition, all married women
and most young people are secondary workers and all men over the age of

25 are primary workers. Sex, age, and marital status seem to be the criteria for classification as primary or secondary workers, yet there appears to be little indication of why such a distinction is necessary, particularly when it is somewhat denigrating to those classified as secondary. It does not seem to refer to the importance of the job to the employer or to the economy as a whole. Nor does it identify those who bear sole financial responsibility for their families since, as we have seen already, only about a quarter of Canadian males would meet this criterion and many women would qualify.

The only way the distinction makes sense is in terms of jobs. Jobs are not primary and secondary because of their importance to the employer and the employee, but because of pay, prestige, working conditions, job security, and future opportunities.[47] And while there is not a precise match between women and these jobs, it is clear that women and young people are disproportionately slotted into jobs that are secondary with regard to these considerations. It is not the workers, but the jobs that are secondary.

As Buckley's definition indicates, however, the distinction implies that women are secondary workers because they move in and out of the labor force, they work sporadically and part time, they lack commitment, and they earn less than men. And it implies that these patterns reflect the choices of women. The available evidence, however, indicates that, to the extent that these patterns exist, they are at least as much a result of the nature of the jobs as they are a reflection of women's preferences and attitudes.

Female Employment Continuity

The information on the continuity of female employment is contradictory and limited. Sangster reports that "recent Canadian data on this subject suggest that the proportion of women in the labour force who usually work full-year (51 weeks or more) is only slightly lower than the corresponding male proportion (77.7% vs. 80.9%)."[48] He goes on to state that "while we do not report general turnover rates in Canada, the specific studies that have been done indicate little difference in male and female separation rates, especially in what are defined here as secondary jobs."[49] But other studies, for example Nakamura et al.,[50] suggest that women are increasingly working less than 40 weeks per year.

It is difficult, therefore, to tell if women are more irregular employees than men. There is evidence that women are as likely as men to stay in the same job for 10 years or less but the percentage of women, as compared to men, who stay in the same job for longer periods of time steadily declines.[51] The data for job tenure by occupation also indicate that, while the percentage of men who have been in the same occupation for over 5 years is consistently higher than the percentage of females, women stay as long in the same jobs as men. The lowest years of job tenure for both men and women are in service and sales occupations. The highest years of job tenure for both sexes are in managerial and professional jobs.[52] The Canadian Department of Labour has pointed out that "the proportion of continuous workers is greatest in occupations of the highest socio-economic class."[53] Women, however, are seldom found in these jobs. Furthermore, the data presented here support the Department's contention that women are less likely to change jobs "if the occupation is managerial, professional, or clerical than if it is commercial, factory, or service."[54] Like men, women are more likely to stay in the good jobs and to leave the ones that offer little.

Occupational Segregation into Secondary Jobs

Women are, however, more likely than men to have secondary jobs, and the jobs they do have are unlikely to encourage commitment. The sex segregation in the labor market, combined with women's later entry into the labor force, explain to a great extent the discontinuity that does exist in female work patterns. Much of the remaining explanation comes from the lack of alternatives to women's other job at home.[55] The jobs, both those in the market and those at home, produce interrupted work patterns.

Over 40% of all women who experience discontinuity in employment do so because they were laid off or their jobs were abolished.[56] Women are not only in jobs that are discontinued, they are often the first fired. As Gunderson points out, "Not having invested much in their female workers, firms are not concerned about losing them permanently should they be laid off in a recession."[57] It is also easier to fire women, because they are less likely than men to be unionized[58] and because they frequently have little seniority. Thus, discontinuity may become self-perpetuating for women: less job continuity producing even less job continuity.

It has also been suggested that women's higher turnover rate results from their lack of commitment to the labor force. In a study of work values sponsored by Manpower and Immigration, Burstein and Associates find:

> . . . no difference between the sexes in the degrees of commitment and loyalty the self-defined employed felt to their jobs and their employers. Both described themselves as equally conscientious, prompt, and attentive employees.[59]

Although the study also shows that women were less ready than men to make a long-term commitment to their job, they also report that, over time, women were less inclined both than men and women at an earlier period to agree that they fit their job.[60] Marchak's research on white-collar workers in the province of British Columbia and Archibald's study of public servants show that women plan to stay in or return to their jobs.[61] Available data indicate that women are less likely than men to leave their jobs because of dissatisfaction.[62] Furthermore, they are less likely than men to benefit from remaining at the same job.[63]

Part-time workers as secondary. That the overwhelming majority of part-time workers are women is also used as an indication of their secondary status. As we have already seen, more of the jobs available are part time. Many employers rely on part-time workers, save money by doing so, and would have difficulty replacing them with more expensive full-time workers. Women take these jobs because they are available and because their work at home may make it difficult for them to participate full time in the labor force. Given the paucity and expense of child care, and of afterschool, lunch hour, and summertime facilities for children, it is not surprising that many women take these part-time jobs. One-third (32.7% in 1979, up from 30.1% in 1975) of Canadian women selected "could only find part-time work" or "personal and family responsibilities" (out of five possible responses) as their reason for working less than 30 hours a week. Less than one-half (43.3% in 1979, down from 45.9% in 1975) indicated that they did not want full-time work and here the absence of a subsequent question on reasons for not wanting full-time work may conceal additional structural factors limiting choices.[64] The same is true for the one-fifth of women who selected the "going to school" response since many may be attending educational institutions because they can find only part-time work. When women do take these part-time jobs, they are likely to make a full commitment to the work. More than one-third (34.4%) part-time employees have been in the same job for 1 to 5 years and more than one-fifth (21.2%) have been in the same part-time job for 5 years or more. [65]

The wage differential. The final factor relegating women to secondary worker status is that of women's wages. Women are secondary workers if this means that they earn less than men. Women are paid less than men even when they perform similar tasks. According to the Economic Council of Canada, women are "overconcentrated in low-paying and underrepresented in high-paying industries. Similarly they are overrepresented in the least organized sectors and underrepresented in those that are organized."[66] But Ostry[67] argues that, even when this segregation is taken into account, there remains a sizable pay gap between male and female workers. While Nakamura et al. would quarrel with the argument based on female segregation, they do state that:

> The average woman earns less than the average for all wage earners (men and women taken together) in all industries. . . . Moreover, the average growth in income from wages and salaries is less for women than it is for all wage earners in all industries.[68]

And the gap between male and female earnings may be widening. Gunderson suggests that "females may be losing ground in occupations where the earning gap is small and gaining where the earning gap is large."[69] Furthermore, women are more likely to be offered and to take part-time work. Because women earn less than men they are classified as secondary workers and their work is viewed as less important. Women generally do not earn less than men by choice. Nor do their lower wages necessarily indicate that the work is less important to the employer or to the employee. Employers hire women and pay them less than men because there are many of them seeking work and because they lack the organization and resources to object.

CONCLUSION

Women have been flooding into the labor market in Canada. They have responded to their growing economic needs and to employer demands for female workers to fill jobs that have traditionally been "women's work." However, more of the jobs available are part time and many more of the women who seek employment fail to find work. Those who are unemployed find it increasingly difficult to return to the home. And, when technological developments and rising inflation are taken into account, there is every reason to believe that this trend will continue. These patterns cannot be dismissed as unimportant because women are secondary workers, as dangerous because women take jobs away from men, as devious because they allow claims on unemployment insurance, as a matter of choice because most women are married. More of the jobs created are secondary and part time, and women are not in competition with other men, but with each other for these jobs. Most women take them because they need the income. If these trends are to be altered, these structural factors must be dealt with directly, not redefined.

NOTES

1. This paper builds on work reported in P. Armstrong (1979) and H. Armstrong (1979). Financial support for this work has come in part from the Canadian Research Institute for the Advancement of Women, to which Pat Armstrong extends her thanks. Appreciation also is extended to Irene Christie for her excellent typing during the preparation of this paper. For the sake of convenience, "work" in this paper refers to labor market work. This is not to deny that almost all women work in the home.

2. U.S. Department of Labor, Bureau of Labor Statistics, 1979b:Table 2.

3. Calculated from Statistics Canada, 1979a, and Statistics Canada, 1979b. The figures are not precisely comparable since current data are based on the Revised Labour Force Survey.

4. Official statistics on unemployment usually underestimate actual unemployment. The hidden unemployed are not included and a significant number of the hidden unemployed are women. For discussion of hidden unemployment, see H. Armstrong, 1979; Gonick, 1978; People's Commission on Unemployment in Newfoundland and Labrador, 1978; and Stirling and Kovri, 1979.

5. Calculated from Statistics Canada, 1979a:Table 101. See also McIlveen and Sims, 1978:31-33.

6. In Statistics Canada's recent analysis (1980a:2) of Canada's female labor force, the following question is asked: "Is the high rate of unemployment in recent years a result of the rapid entry of females into the labour market or is high unemployment due to the fast growth in the total labour force plus changing economic conditions?" No answer to this question is offered in the Statistics Canada publication.

7. In 1978, 25.4% of Canadian families had one income recipient, according to Statistics Canada, 1980b:Table 55.

8. In 1978, women headed 9.6% of Canadian families, according to Statistics Canada, 1980b:Table 55. For a more complete discussion of the steady increase in female-headed households, see the National Council of Welfare, 1976.

9. See Armstrong and Armstrong, 1975, 1978.

10. Connelly, 1978; Gunderson, 1976; Lautard, 1976; Nakamura, Nakamura, and Cullen, 1979b.

11. Because of the significant changes undertaken in the Labour Force Survey in 1975, many of the data are available on a historically consistent basis from that time only. Statistics Canada, however, has recorded selected data back as far as 1966 on the basis of the Revised Labour Force Survey.

12. One aspect of the Labour Force Survey revisions of 1975 was to reduce the number of hours required for full-time work from 35 to 30, making historical comparisons difficult. The pre-1975, full-time data were recalculated by us to make them comparable. Some important historical comparisons, such as for full-time and part-time employment by industry or occupation and sex, are not even available from 1975.

13. Calculated from Statistics Canada, 1979a:Table 83.

14. Braverman, 1974; Baker, 1964; Zimbalist, 1979.

15. Brecher, 1979:226.

16. Calculated from Statistics Canada, 1979a:Table 59.

17. Boyd, 1977:56.

18. National Council of Welfare, 1979:12.

19. Ibid.:7.

20. For a discussion of the inadequacy of day care in Canada, see Canadian Council on Social Development, 1976:51-67.

21. National Council of Welfare, 1979:Table 3.

22. Armstrong and Armstrong, 1978:Chapter 6; MacLeod and Horner, 1980.

23. Calculated from Statistics Canada, 1980b:Table 14.

24. National Council of Welfare, 1979:10.

25. Boyd, 1977:55.

26. Ostry, 1968a; Spencer and Featherstone, 1970; Skoulas, 1974; Nakamura et al., 1979b.

27. Meltz and Stager, 1979:122.

28. Between 1975 and 1978, the following increases in current dollar wages were recorded:

Primary industries	31.5%
Manufacturing	33.9%

Construction	33.9%
Transportation, communication, and other utilities	33.9%
Trade	27.7%
Finance, insurance, and real estate	28.6%
Service	25.3%

Calculated from Canada, Department of Finance, Economic Review, 1980: Reference Table 49. These percentage increases do not reveal the larger differences in actual wages, as the trade and service industries started from much lower points in 1975.

29. Ostry, 1968b:7. The low official unemployment rate for women was in fact primarily the result of the indirect way of asking about labor force activity in the Former Labour Force survey. By changing to more direct questions in 1975, Statistics Canada discovered the presence of 223,000 more women in the labor force, of whom 79,000 were unemployed. As a result, their unemployment rate moved from 6.4 to 8.1, overtaking that of men.

30. Calculated from Statistics Canada, 1979c.

31. Nakamura et al., 1979b:17.

32. Unemployment Insurance Canada, 1977:H-14.

33. Ibid.:H-20.

34. Ibid.:H-22.

35. Economic Council of Canada, 1976:152.

36. Green and Cousineau, 1976:112.

37. See Armstrong, 1980a.

38. In response to the pressure from women's groups, the federal government has recently announced that it will reduce the minimum weekly hours of work from 20 to 15.

39. Schwartzman, 1978.

40. Unemployment Insurance Canada, 1977:C-14.

41. Social Planning Council of Metropolitan Toronto, 1978:12. Calculated from the Canada, Dominion Bureau of Statistics, 1963:various tables.

42. Economic Council of Canada, 1976:21.

43. Gunderson, 1976.

44. Employment and Immigration Canada, 1979:2.

45. Ibid.:Table 1.

46. Buckley, 1973:7.

47. Some jobs may require less skill and training and, thus, the employer may be less concerned about retaining any particular employee in the job. However, the job itself still has to be done and, therefore, is important. Garbage disposal and baby-sitting are useful examples here.

48. Sangster, 1973:30.

49. Ibid.:34.

50. Nakamura et al., 1979b:99.

51. By 1979, 36.4% of women and 29.0% of men had been in the same job 1 to 5 years, and 17.0% of women compared to 17.6% of men

had been in the same job for 6 to 10 years. Meanwhile, 11.5% of women compared to 16.8% of men had been in the same job for 11 to 20 years and 4.5% of women compared to 13.1% of men had been for 20 years or more. Calculated from Statistics Canada, 1979a:Table 78.

52. Calculated from Statistics Canada, 1979a:Table 78.
53. Canada, Department of Labour, 1960:32.
54. Ibid.:20.
55. Women, not men, leave the labor force because of family responsibilities. For a more complete discussion of the relationship between women's two roles, see Armstrong and Armstrong, 1978.
56. In 1979, 42.8% of all unemployed women in 1979 lost their job or were laid off. Calculated from Statistics Canada, 1979a:Table 99.
57. Gunderson, 1976:104.
58. As compared to 43% of male workers, 27% of female workers are unionized. For further information of Canadian women and unions, see White, 1980.
59. Burstein, Tienhaara, Hewson, and Warrander, 1975:54.
60. Ibid.:55.
61. Marchak, 1973:95; Archibald, 1970:95.
62. MacDonald, 1978.
63. Archibald, 1970:95.
64. Calculated for 1975 from Statistics Canada, 1979b:Table 25 and 1979a:Table 86.
65. Calculated from Statistics Canada, 1979a:Table 84.
66. Economic Council of Canada, 1976:106-107.
67. Ostry, 1968a:45.
68. Nakamura et al., 1979b:89.
69. Gunderson, 1976:122.

POVERTY VIEWED AS A WOMAN'S PROBLEM—THE U.S. CASE

Jane Roberts Chapman and Gordon R. Chapman

Center for Women Policy Studies
Washington, D.C.   U.S.A.

> For awhile the introduction of machinery which took away from the home so many industries deprived women of any importance as an economic factor; but presently she arose, and followed her lost wheel and loom to their new place, the mill. Today there is hardly an industry in the land in which some women are not found. Everywhere throughout America are women workers outside the unpaid labor of the home, the last census giving three million of them. This is so patent a fact, and makes itself felt in so many ways by so many persons, that it is frequently and widely discussed. Without here going into its immediate advantages or disadvantages from an industrial point of view, it is merely instanced as an undeniable proof of the radical change in the economic position of women that is advancing upon us. She is assuming new relations from year to year before our eyes: but we, seeing all social facts from a personal point of view, have failed to appreciate the nature of the change.
> --Charlotte Perkins Gilman, Women and Economics, 1898.

The economic status of women relative to their basic needs can be evaluated in key areas such as income level and income equity with men as well as occupational and career equity and mobility. The recent World Conference on the United Nations Decade for Women (July 14-30, 1980) summarizing progress toward these goals set by the first such conference in 1975, concluded that during this time women have lost ground in establishing equity with men and that the economic status of the majority of women in absolute terms has generally worsened.

This phenomenon is not new. As the above passage indicates, the economic situation of women has historical roots dating back at least to the Industrial Revolution. It is interesting to note how Gilman interpreted the impact of industrialization on women in England and America at the turn of the century—the movement of economic activity from the domestic to the specializing and differentiating spheres of business and industry—which first diminished the economic importance of women and then provided them with new, perhaps more meaningful and exciting, opportunities. In fact, as Gilman also foresaw, the conditions of the Industrial Revolution, which continued to place women in jobs in keeping with their roles as wife and mother, only served to increase the economic dependence of women.

Current statistics show that this inequity is the case today. Despite the emergence of the women's movement and growing publicity about equal rights, the economic status of American women has been deteriorating during the past two decades. Bearing in mind that the roots of women's dependence date back at least 100 years, the purpose of this chapter is to examine the reasons why women's economic status has continued to deteriorate and to suggest some possible solutions for reversing this trend in the future.

POVERTY AND THE ECONOMIC STATUS OF WOMEN IN THE CURRENT PERIOD

In a diversified economy, overall prosperity tends to grow in relation to the Gross National Product (GNP). The overall continued long-term increase in the GNP since the 1940s and the restructuring of the major economic sectors have resulted in an expansion of women's labor force participation. These changes, however, have not produced a general improvement in women's economic status. Wages paid to working women, like those paid to nonwhite men, have increased at slower rates than those of white men; the wage gaps between men and women have widened even during periods of general prosperity (Table 1). Men, on the other hand, have benefited not only from prosperity, but also from industrial diversification and the development of new, high technology industries.

These differences between men and women portray very clearly an overall picture regarding the status of women which is not hopeful: The more things change, the more they stay the same; or, more accurately, the more things seem to improve, the worse they actually get. In fact, what the forthcoming data will indicate is a long-term process of expansion and consolidation within the labor market. That is, while women apparently are moving into the labor force, and even in some instances into nontraditional occupations, in fact they are generally moving into marginal areas. At the same time, men have continued to become consolidated in the better, higher-paying jobs.

In order to understand the reasons for this expansion and consolidation, and the effects on women's economic status today, six major factors

TABLE 1

COMPARISON OF MEDIAN EARNINGS OF YEAR-ROUND, FULL-TIME WORKERS BY SEX, 1955-1978

YEAR	MEDIAN EARNINGS WOMEN (1)	MEDIAN EARNINGS MEN (2)	EARNINGS GAP IN DOLLARS (3)	WOMEN'S EARNINGS AS % OF MEN'S (4)	PERCENT MEN'S EARNINGS EXCEEDED WOMEN'S (5)	EARNINGS GAP IN 1967 DOLLARS (6)
1978	$9,350	$15,730	$6,380	59.4	68.2	$3,267
1977	8,618	14,626	6,008	58.9	69.7	3,310
1976	8,099	13,455	5,356	60.2	66.1	3,141
1975	7,504	12,758	5,254	58.8	70.0	3,259
1974	6,772	11,835	5,063	57.2	74.8	3,433
1973	6,335	11,186	4,851	56.6	76.6	3,649
1972	5,903	10,202	4,299	57.9	72.8	3,435
1971	5,593	9,399	3,806	59.5	68.0	3,136
1970	5,323	8,966	3,643	59.4	68.4	3,133
1969	4,977	8,227	3,250	60.5	65.3	2,961
1968	4,457	7,664	3,207	58.2	72.0	3,079
1967	4,150	7,182	3,032	57.8	73.1	3,032
1966	3,973	6,848	2,875	58.0	72.4	2,958
1965	3,823	6,375	2,552	60.0	66.8	2,700
1964	3,690	6,195	2,505	59.6	67.9	2,696
1963	3,561	5,978	2,417	59.6	67.9	2,637
1962	3,446	5,974	2,528	59.5	73.4	2,790
1961	3,351	5,644	2,293	59.4	68.4	2,559
1960	3,293	5,417	2,124	60.8	64.5	2,394
1959	3,193	5,209	2,016	61.3	63.1	2,308
1958	3,102	4,927	1,825	63.0	58.8	2,108
1957	3,008	4,713	1,705	63.8	56.7	2,023
1956	2,827	4,466	1,639	63.3	58.0	2,014
1955	2,719	4,252	1,533	63.9	56.4	1,911

SOURCE: Developments and Issues in the U.S., Table 12. Compiled from U.S. Department of Commerce, Bureau of the Census: "Money Income of Families and Persons in the United States," Current Population Reports, 1957 to 1977, and "Money Income and Poverty Status of Families and Persons in the United States: 1978" (Advance Report).

NOTES: For 1967-78, data include wage and salary income and earnings from self-employment; for 1955-66, data include wage and salary income only. Column 3 = column 2 minus column 1; column 4 = column 1 divided by column 2; column 5 = column 2 minus column 1 divided by column 1; column 6 = column 3 times the purchasing power of the consumer dollar (1967 = $1.00).

will be examined: general economic growth and development; wage and occupational status; education and training; labor force participation, unemployment, and unionization; changes in demography and family structure; and the war on poverty.

General Economic Growth and Development

The deterioration of women's economic status over the past two decades has accompanied unstable and failing rates of growth in the real GNP and GNP per capita. Similarly, manufacturing output has only been maintained at approximately 80% to 85% of capacity, with one precipitous drop during the 1974 to 1975 recession period. The sharp recovery from the 1975 recession was not sufficiently stable to generate much new employment through 1980.

Projections of current economic conditions indicate a basic lack of vitality, at least in major manufacturing sectors where modernization also has lagged. This sluggishness is expected to continue throughout the next few years; in fact, without major new efforts the situation may persist indefinitely as other countries, competing for world markets, incorporate increasingly higher rates of industrial research and development in product design and manufacturing.

Table 1 shows the income differential to be far greater in the relatively worse period of economic distress after 1974 than in the period of consistently greater growth between 1960 and 1966. These figures indicate that women's earnings, which were only 59% of men's in 1978, generally declined over the whole period from 1955 to 1978. It is interesting to note that women who have attempted to move into non-traditional areas have been able to do so only during periods when economic growth is sufficient to create temporary shortages of male employees. During subsequent downturns and readjustments, however, women are usually affected by the last-hired-first-fired rule. (This rule operates against them even during wartime, when they are encouraged to relieve male employees who, nevertheless, generally continue to accrue seniority in addition to receiving preferential re-employment rights on their return, at which time female employees are laid off.) Perhaps partly as a cushion against instability, the number of women with two jobs has increased, as more women attempt to break into nontraditional occupations. This increase also indicates the severity of hardship and the need for security due to recession and inflation.[1]

Since the wage gap is strongly affected by general economic conditions, it probably cannot be expected to narrow much in the future. Comparing trends in the GNP with trends in employment, it appears that it takes an annual growth in the real GNP of at least 3% per year to fuel the expansion and would probably take from 5% to 10% per year to produce any meaningful advance in terms of closing the male/female gap in occupational participation and income. Growth rates of 5% to 10% have not been experienced in the United States since the 1940s and are

TABLE 2

PERCENTAGE DISTRIBUTION OF EMPLOYED WOMEN BY MAJOR OCCUPATIONAL GROUP, 1960-1979[1]

OCCUPATIONAL GROUP	1979	1978	1977	1975	1970	1960
Total employed (in thousands)	40,446	38,881	36,685	33,553	29,667	21,874
Percent	100.0	100.0	100.0	100.0	100.0	100.0
Professional and technical workers	16.1	15.6	15.9	15.7	14.5	12.4
Managers and administrators (except farm)	6.4	6.1	5.9	5.2	4.5	5.0
Sales workers	6.9	6.9	6.8	6.9	7.0	7.7
Clerical workers	35.0	34.6	34.7	35.1	34.5	30.3
Craft and kindred workers	1.8	1.8	1.6	1.5	1.1	1.0
Operatives, except transport	10.8	11.1	11.2	11.0 }	14.5	15.2
Transport equipment operatives	.7	.7	.6	.5		
Nonfarm laborers	1.3	1.3	1.2	1.1	.5	.4
Private household workers	2.6	2.9	3.1	3.4	5.1	8.9
Other service workers	17.2	17.7	17.9	18.2	16.5	14.8
Farmers and farm managers	.3	.3	.3	.3	.3	.5
Farm laborers and supervisors	.9	1.0	1.0	1.1	1.5	3.2

SOURCE: Developments and Issues in the U. S., Table 11. Compiled from U. S. Department of Labor, Employment and Training Administration, "Employment and Training Report of the President," 1979, and "Employment and Earnings," January, 1980.

[1] Although this analysis and, probably, the underlying data are at some variance with Table 3, also prepared by BLS and published in *Women in the Labor Force: Some New Data Series*, p. 3, the general finding is further expanded and supported by it. In addition, distributions for the Professional-Technical category also increased only modestly from 1970-1978. See Table 3.

unlikely to occur at any time in the future except during the period of war. Growth rates in the range of 4% to 5%, however, are not impossible.

Women's Occupational and Wage Status

Tables 2 and 3 show that during the period 1960 to 1979, total women's employment in occupational groups in which they are most concentrated (professional, technical; sales and service; and clerical) rose from 65% to 75%. Table 4 shows that the difference between male and female earnings was substantial for these occupations, especially clerical work, which has the highest percentage of total female employment and which experienced the largest expansion of employment opportunities

TABLE 3

EMPLOYMENT IN SELECTED OCCUPATIONS, 1950, 1970, AND 1978[1]

OCCUPATION	TOTAL MEN/WOMEN			WOMEN AS PERCENT OF ALL WORKERS IN OCCUPATION		
	1950	1970	1978	1950	1970	1978
Professional, technical	4,858	11,452	14,245	40.1	40.0	42.7
Accountants	377	711	975	14.9	25.3	30.1
Engineers	518	1,233	1,265	1.2	1.6	2.8
Lawyers, judges	171	277	499	4.1	4.7	9.4
Physicians, osteopaths	184	280	424	6.5	8.9	11.3
Registered nurses	403	836	1,112	97.8	97.4	96.7
Teachers, except college and university	1,123	2,750	2,992	74.5	70.4	71.0
Teachers, college and university[2]	123	492	562	22.8	28.3	33.8
Writers, artists, entertainers	124	761	1,193	40.3	30.1	35.3
Managerial, administrative except farm	4,894	6,387	10,105	13.8	16.6	23.4
Bank officials, financial managers	111	313	573	11.7	17.6	30.4
Buyers, purchasing agents	64	361	370	9.4	20.8	30.5
Food service workers	343	323	589	27.1	33.7	33.8
Sales managers, department heads: retail trade	142	212	343	24.6	24.1	37.3
Clerical	6,865	13,783	16,904	62.2	73.6	79.6
Bank tellers	62	251	449	45.2	86.1	91.5
Bookkeepers	716	1,552	1,830	77.7	82.1	90.7
Cashiers	230	824	1,403	81.3	84.0	87.1
Office machine operators	143	563	827	81.1	73.5	74.2
Secretaries, typists	1,580	3,814	4,570	94.6	96.6	98.6
Shipping-receiving clerks	287	413	461	6.6	14.3	22.8

SOURCE: Bureau of Labor Statistics.
[1] Numbers are in thousands.
[2] Includes college and university presidents.

during the period (and where women's earnings were 63% of men's in 1979, as compared with 62% overall).

In the United States, although more women are working than ever before, the majority occupy low-level, low-paying jobs concentrated within a range of 20 occupations. As a result of stereotyping and segregation, women, like nonwhite men, have historically been confined to relatively few low-status occupations where competition tends to keep wages permanently depressed. This competition has increased in recent years because, while more women want to work, the number of occupa-

TABLE 4

MEDIAN USUAL WEEKLY EARNINGS
OF FULL-TIME WAGE AND SALARY WORKERS,[1]
BY SEX AND OCCUPATIONAL GROUP,
SECOND QUARTER 1979 (PRELIMINARY)

OCCUPATIONAL GROUP	WOMEN	MEN	WOMEN'S EARNINGS AS PERCENT OF MEN'S
Total	**$183**	**$295**	62
Professional, technical	261	375	70
Managerial, administrative, except farm	232	386	60
Sales	154	297	52
Clerical	180	287	63
Craft	189	305	62
Operatives, except transport	156	257	61
Transport equipment operatives	194	277	70
Nonfarm laborers	166	220	75
Service	138	203	68
Farm	125	153	82

SOURCE: Development and Issues in the U. S., Table 13. Compiled from U. S. Department of Labor, Bureau of Labor Statistics, "Women in the Labor Force: Some New Data Series," 1979.

[1] Workers 16 years of age and over.

tions open to them has remained nearly the same and economic growth has not been sufficient to create new jobs to meet the demand.

It is not likely that there will be much improvement in the quality of employment for the majority of women in the near future--in spite of the fact that there have been increases in female employment and that there have been modest gains in career opportunities for some women in the higher reaches of employment which can be expected to continue. One might assume that the wage gap would diminish if more of these high-level jobs were available to women. However, comparison of median income of men and women with their educational attainment shows that the income differential widens directly in relation to the number of years of education completed. For example, in 1977 the gap increased from $4,345 for fewer than 8 years of schooling to $7,603 for 5 or more years of college. This comparison suggests that there is an increasing disparity in wages between men and women for those women in the growing minority at the top as well as for the greater majority of women still working in traditional and marginal employment.

TABLE 5

WOMEN AS PERCENT OF TOTAL EMPLOYMENT IN SELECTED OCCUPATIONS, 1974—1979[1]

OCCUPATION	1979		1978		1977		1976		1975		1974		PERCENT CHANGE 1975-1979
	TOTAL EMPLOYMENT	WOMEN AS PERCENT OF TOTAL	TOTAL EMPLOYMENT	WOMEN AS PERCENT OF TOTAL	TOTAL EMPLOYMENT	WOMEN AS PERCENT OF TOTAL	TOTAL EMPLOYMENT	WOMEN AS PERCENT OF TOTAL	TOTAL EMPLOYMENT	WOMEN AS PERCENT OF TOTAL	TOTAL EMPLOYMENT	WOMEN AS PERCENT OF TOTAL	
Professional and technical	15,050	43.3	14,245	42.7	13,692	42.6	13,329	42.0	12,748	41.3	12,338	40.5	23.8
Accountants	1,045	32.9	975	30.1	868	27.5	866	26.9	782	24.6	803	23.7	79.2
Computer specialist	534	26.0	428	23.1	371	23.2	387	19.1	363	21.2	311	19.0	80.5
Industrial engineers	245	7.3	206	8.7	214	7.0	201	4.5	187	2.7	193	(1/)	260.0
Lawyers and judges	499	12.4	499	9.4	462	9.5	413	9.2	392	7.1	369	7.0	121.4
Librarians	188	80.9	187	84.5	193	83.4	183	82.4	180	81.1	N.A.	N.A.	4.1
Life and physical scientists	280	18.9	273	17.9	275	15.6	282	12.1	277	14.4	246	15.9	32.5
Physicians	431	10.7	424	11.3	403	11.2	368	12.8	354	13.0	346	9.8	*
Registered nurses	1,223	96.8	1,112	96.7	1,063	96.7	999	96.6	935	97.0	904	98.0	30.5
Elementary teachers	1,374	84.3	1,304	84.0	1,313	84.2	1,383	84.8	1,332	85.4	1,297	84.3	1.8
Secondary teachers	1,213	50.7	1,154	51.6	1,157	51.2	1,188	50.5	1,184	49.2	1,186	48.3	5.5
Surveyors	85	3.5	82	2.4	68	1.5	69	1.4	70	*	N.A.	N.A.	*
Airplane pilots	72	*	69	1.4	64	*	64	*	60	*	N.A.	N.A.	*
Managers and administrators	10,516	24.6	10,105	23.4	9,662	22.3	9,315	20.8	8,891	19.4	8,941	18.5	49.9
Bank officials/financial managers	620	31.6	573	30.4	543	27.3	546	24.7	518	23.6	510	21.4	60.7
School administrators (elementary and secondary)	299	37.5	275	35.6	265	36.2	282	32.6	263	28.1	352	27.8	51.4
Clerical workers	17,613	80.3	16,904	79.6	16,106	78.9	15,558	78.7	15,128	77.8	15,043	77.6	20.2
Typists	1,020	96.7	1,044	96.6	1,006	96.3	983	96.7	1,025	96.6	1,038	96.2	.4

	1		2		3		4		5		6		% change
Craft and kindred workers	12,880	5.7	12,386	5.6	11,881	5.0	11,278	4.8	10,972	4.6	11,477	4.5	46.1
Carpenters	1,276	1.3	1,253	1.0	1,171	.9	1,021	.7	988	.6	1,073	(1/)	183.3
Painters, construction and maintenance	483	5.0	484	5.2	461	3.3	413	2.9	420	3.8	456	(1/)	50.0
Machinists and job setters	642	3.3	591	3.0	576	2.6	570	2.9	557	2.5	558	(1/)	50.0
Auto mechanics	1,272	.6	1,209	.6	1,161	.9	1,124	.6	1,102	.5	1,041	(1/)	33.3
Printing craft workers	455	22.2	417	21.8	389	22.4	380	19.2	375	17.6	386	18.1	53.0
Telephone installers/repairers	302	9.9	297	6.7	279	5.0	282	5.0	314	4.8	349	4.9	100.0
Operatives, including transport	14,521	32.0	14,416	31.7	13,830	31.4	13,356	31.2	12,856	30.2	13,919	31.1	19.6
Meat cutters and butchers, manufacturing	89	31.5	114	28.9	88	35.2	87	29.9	100	27.0	N.A.	N.A.	3.7
Punch and stamping press operatives	158	29.1	156	30.1	152	36.2	155	32.9	130	27.7	170	30.6	27.8
Sewers and stitchers	810	95.3	814	94.8	820	95.2	812	95.9	803	95.8	858	95.8	.4
Bus drivers	358	45.5	337	45.1	339	42.2	332	39.5	310	37.7	265	37.4	39.3
Truck drivers	1,965	2.1	1,923	1.9	1,898	1.3	1,741	1.2	1,694	1.1	1,752	(1/)	121.1
Service workers	12,834	62.4	12,839	62.6	12,392	62.0	12,005	61.5	11,657	62.3	11,373	62.9	10.3
Cleaners and servants	485	97.3	530	97.0	574	96.5	553	97.1	599	97.3	588	97.6	-23.5
Waiters	1,363	89.4	1,383	90.5	1,310	90.4	1,259	90.7	1,183	91.1	1,182	91.8	13.1
Nursing aides, orderlies	1,024	87.5	1,037	87.0	1,008	86.3	1,002	86.8	1,001	85.8	959	86.9	4.3
Hairdressers and cosmetologists	575	89.2	542	89.1	526	88.2	534	88.0	504	90.5	498	92.4	12.5
Protective service	1,406	8.8	1,358	8.5	1,324	7.9	1,302	6.4	1,290	6.3	1,254	6.4	51.2

SOURCE: Developments and Issues in the U. S., Table 12, compiled from U. S. Department of Commerce, Bureau of the Census: "Money Income of Families and Persons in the United States," Current Population Reports, 1957 to 1977, and "Money Income in Poverty Status of Families and Persons in the United States: 1978" (Advance Report).

¹ Numbers are in thousands.

*Percent not shown where employment estimate is less than 35,000.

TABLE 6

EARNED DEGREES CONFERRED BY FIELD OF STUDY, 1972 TO 1977, AND BY LEVEL OF DEGREE AND SEX, 1977

MAJOR FIELD OF STUDY	BACHELORS[1] (1,000)							MASTERS, 1977		DOCTORATES, 1977	
	1972	1973	1974	1975	1976	1977 TOTAL	1977 MALE	TOTAL	MALE	TOTAL	MALE
All fields	894.1	930.3	954.4	931.7	934.4	919.5	495.5	317,164	167,783	33,232	25,142
Agriculture	13.6	14.8	16.3	17.6	19.5	21.5	16.7	3,724	3,177	893	831
Animal science	2.7	2.9	3.1	3.4	3.9	4.1	2.8	429	370	122	113
Forestry	2.0	2.0	2.3	2.6	2.7	2.7	2.4	445	402	84	79
All other	8.9	9.9	10.9	11.5	12.9	14.7	11.5	2,850	2,405	687	639
Architecture	6.5	7.0	7.8	8.2	9.2	9.2	7.2	3,213	2,489	73	62
Area studies	2.8	3.1	3.2	3.1	3.1	3.0	1.3	989	525	153	104
Biological sciences	37.6	42.7	48.9	52.2	54.9	53.6	34.2	7,114	4,718	3,397	2,671
Biology, general	27.8	31.6	36.6	39.2	40.7	39.5	25.1	3,322	2,183	608	454
Zoology, general	5.2	5.4	5.8	5.7	5.6	5.0	3.5	521	364	284	224
All other	4.6	5.7	6.4	7.4	8.6	9.1	5.6	3,271	2,171	2,505	1,993
Business & management	123.3	128.2	133.9	135.5	145.0	152.1	116.5	46,545	39,881	869	814
Business management & administration	32.1	33.3	37.1	39.3	44.5	47.4	37.4	23,938	20,789	472	441
Business, commerce, general	31.7	30.3	33.0	31.5	30.4	30.2	23.3	9,845	8,442	99	95
Accounting	25.1	28.3	29.8	31.6	36.4	39.2	29.2	3,278	2,642	53	49
All other	34.4	36.4	34.1	33.0	33.7	35.4	26.5	9,484	8,008	245	229
Communications	12.3	14.3	17.1	19.2	21.3	23.2	12.9	3,091	1,719	171	130
Computer & information sciences	3.4	4.3	4.8	5.0	5.7	6.4	4.9	2,798	2,332	216	197
Education	192.4	195.6	186.6	168.7	156.5	143.7	39.9	126,375	43,174	7,955	5,186
Elementary, general	93.7	90.1	81.0	69.6	60.9	53.0	6.6	21,795	2,915	199	72
Physical	26.3	27.0	27.8	24.7	24.3	23.3	12.8	4,716	2,824	247	169
Music	7.1	7.5	7.8	8.1	7.9	7.6	3.2	1,437	692	74	54
Special education, general	4.1	5.5	6.7	7.8	8.2	8.5	1.0	8,249	1,479	225	139
Other education fields	61.2	65.6	63.3	58.6	55.2	51.2	16.4	90,178	35,264	7,210	4,752
Engineering	51.5	51.6	50.7	47.3	46.7	49.3	47.1	16,245	15,525	2,586	2,513

Fine & applied arts	33.9	36.1	40.0	41.1	42.4	41.8	16.2	8,636	4,211	662	447
Art[2]	13.5	15.3	16.4	16.2	16.6	16.3	5.3	2,636	1,199	95	41
Music[3]	6.3	6.8	7.9	8.7	9.1	9.2	4.3	3,147	1,637	402	300
Dramatic arts	4.5	4.8	5.4	5.5	5.7	5.2	2.2	1,315	664	98	71
All other	9.6	9.3	10.3	10.7	10.9	11.1	4.4	1,538	711	67	35
Foreign languages	19.4	19.5	19.5	18.2	15.6	13.9	3.4	3,147	965	752	365
Health professions	28.9	33.9	41.9	49.5	54.3	57.3	11.9	12,951	4,163	538	366
Nursing	13.2	15.5	19.4	23.8	26.8	28.4	1.5	3,257	102	24	2
Pharmacy	4.7	5.1	5.8	6.5	7.0	7.5	5.1	274	202	75	65
Medical lab. technology	3.4	3.8	4.8	5.1	5.4	5.3	1.0	333	149	3	1
All other	7.6	9.5	11.8	14.1	15.0	16.1	4.3	9,087	3,710	436	298
Home economics	12.2	13.6	15.4	16.9	17.5	17.4	.7	2,334	207	160	37
Law (excl. first professional)	.5	.5	.5	.4	.5	.6	.4	1,574	1,366	60	52
Letters	73.6	71.3	65.3	57.9	52.3	47.1	20.5	10,451	4,237	2,199	1,358
Library sciences	1.0	1.2	1.2	1.1	.8	.8	.1	7,572	1,546	75	35
Mathematical subjects	23.8	23.2	21.8	18.3	16.1	14.2	8.3	3,695	2,396	823	714
Military sciences	.4	.3	.3	.4	1.2	.9	.9	43	42	--	--
Physical sciences	20.9	20.8	21.3	20.9	21.6	22.5	18.0	5,331	4,450	3,341	3,022
Chemistry, general	10.7	10.2	10.4	10.5	11.1	11.1	8.5	1,669	1,249	1,441	1,270
Physics	4.6	4.1	3.9	3.7	3.5	3.4	3.0	1,290	1,170	920	869
Geology	2.5	2.8	3.2	3.2	3.3	3.7	2.9	920	811	257	237
All other	3.1	3.6	3.8	3.5	3.8	4.3	3.5	1,452	1,220	723	646
Psychology	43.4	48.1	52.3	51.4	50.4	47.4	20.6	8,301	4,313	2,761	1,770
Public affairs & services	12.7	18.0	24.3	28.6	33.6	36.3	20.1	19,454	10,663	335	225
Social sciences	159.6	157.7	152.2	136.8	127.9	117.4	71.2	15,458	10,369	3,784	2,949
History	44.0	41.2	37.4	31.8	28.6	25.4	16.5	3,393	2,199	921	715
Sociology	35.6	36.0	35.9	31.8	28.0	24.7	9.7	1,830	1,018	714	480
Political science or government	28.3	30.2	30.9	29.3	28.5	26.4	19.0	2,222	1,718	641	541
Social sciences, general	21.4	18.3	17.1	14.0	12.5	10.8	6.0	2,019	1,234	72	48
Economics	15.3	14.9	14.4	14.1	14.9	15.3	11.8	2,158	1,775	758	672
All other	14.9	17.1	16.5	15.8	15.5	14.7	8.3	3,836	2,425	678	493
Theology	3.9	3.5	4.2	4.8	5.5	6.1	4.5	3,625	2,488	1,125	1,083
Theological professions, general	2.0	1.7	2.5	2.8	3.5	3.9	3.3	1,831	1,459	1,048	1,015
Religious education	1.1	1.0	1.3	1.4	1.5	1.6	.8	1,325	688	41	36
All other	.7	.8	.4	.7	.5	.7	.4	469	341	36	32
Interdisciplinary studies[4]	16.7	20.8	24.9	28.5	32.8	33.9	18.0	4,498	2,827	304	211

SOURCE: Statistical Abstract of the U. S. for 1979, Table No. 283, p. 169. U. S. National Center for Education Statistics, *Earned Degrees Conferred*, annual.

NOTE: Includes Puerto Rico and outlying areas.

— Represents zero. [1] Requiring 4 or 5 years. [2] Includes history and appreciation. [3] Includes performing and liberal arts, history, and appreciation.

[4] Comprises general liberal arts and sciences, biological and physical sciences, humanities and social sciences, engineering, and other disciplines.

A major government objective for improving the employment status of women has been increased occupational mobility.   Table 5 shows phenomenal rates of growth for some occupations during the period 1974 and 1975 to 1979. In evaluating these data, however, it is important to remember that 1974 to 1975 was the worst recessionary period since the 1930s. Female unemployment fell from 9.3% in 1975 to 6.8% in 1979 while the female labor force participation rate increased from 45.9% to 50.7% during the 1975 to 1979 period.   Even so, while the number of women employed as industrial engineers (the fastest-growing occupation) rose 260%, the number employed in relative terms rose from about 3% of total employment to about 7%. (Table 3 shows an increase from 1.2% in 1950 to 2.8% in 1978.) By and large, women continued to be employed in significant numbers in 1979 in their traditional fields with small numbers of them affected by fluctuations in industries most influenced by economic growth and recession and where last-hired-first-fired rules apply. Only 10 of the occupations listed in Table 5 had 50% or more women employees, essentially the same as observed in the recession years of 1974 to 1975.

What is suggested by these data on occupations and earnings is the kind of expansion and consolidation discovered for employment as a whole. Where women are expanding into nontraditional occupations, their levels of pay seem to suggest that these may often be transitional and redefined positions (para-legal, etc.) or have entry-level or lower-level job content.

Education and Training

During the past two decades, women increased their participation more than men at all levels in education and training.   In higher education, the percentage of female participation in terms of degrees earned rose steadily from 34.2% in 1960 to 44.6% in 1977. The greatest increase in degrees conferred was for the doctorate which rose from 1,000 to 8,100, while the number of master's degrees increased from 24,000 to 150,000 and baccalaureates rose from 139,000 to 441,000. But in 1977, women's participation in higher education continued to cluster around traditional areas: mainly education, with health, sociology, language, home economics, letters, psychology, and library science the only other areas of significant participation (Table 6).   While these fields reflect women's career choices, they do not necessarily reflect demand for women employees.

Professional education and training also saw substantial increases in numbers of women graduates during the period from 1960 to 1977.[2]   In medicine, degrees conferred on women rose from 5.5% to 19.1% of the total, while in law the number rose from 2.5% to 22.4%. Although these percentages appear great, only 2% of all women college graduates received a law or medical degree in 1977, as compared with 8% of men graduates.   (This percentage for women does represent a substantial increase, however, since the value in 1960 was only about 0.5%.)

Labor Force Participation, Unemployment, and Unionization

The trend in the wage differential between men and women has followed related gaps in employment and unemployment. Female labor force participation rose steadily from about 34.8% in 1960 to about 51.7% in 1980. Significantly, the gap between participation rates for women and the much higher rates for men has narrowed at a somewhat faster pace. While female participation rates have increased substantially, male participation rates declined slightly over the period. Despite this, a U.S. Department of Labor projection predicts that the difference between male and female participation rates will decrease less drastically in the future.

Although male labor force participation rates declined in the period from 1960 to 1979, total employment of males rose from 49 million to 61 million. However, the unemployment rate was the same or slightly lower for men in 1979 as in 1960 (5.4 to 5.1). The opposite was true for women. That is, employment of women rose by almost 50% while their participation rate rose almost as much. At the same time, the unemployment rate for women, always somewhat higher than for men, rose steadily from 6% to 7% to become significantly higher than men's rates. Unemployment, although affected by definitional biases, has always been higher for women than for men during economic recession; and hidden unemployment (those who have dropped out of the market entirely because of the lack of opportunities) has no doubt always been higher for women than men, as indicated by differences in labor force participation rates for times of prosperity and decline and the overwhelming influx of women into the market during times of economic expansion.

Although female membership in organized labor has experienced rapid growth, 28% in the period from 1970 to 1976 as compared with a slight decline in male membership,[3] almost all major union administrative positions are occupied by men, and the union structure tends to emulate the male/female orientation of business and industry itself. Therefore, regardless of claims by union officials, women employees inevitably lag behind male employees in job and occupational mobility in organized employment.

Changes in Demography and Family Structure

Demographic changes, including continued lower population increases due to falling birth rates added to the effects of increased longevity and immigration, have resulted in changing population and related consumption patterns that are also affecting employment patterns. Regional movements of population and industry have created employment opportunities in new areas while leaving growing pockets of economic distress and unemployment in "abandoned" industrial areas.

The number of working women who are also heads of household rose dramatically from 1960 to 1979, as shown in Table 7. Despite continued

TABLE 7

PERCENTAGE DISTRIBUTION OF
WOMEN'S LABOR FORCE PARTICIPATION
BY MARITAL STATUS, MARCH 1950 TO MARCH 1979

MARITAL STATUS	1979	1975	1970	1960	1950
Total	50.7	45.9	42.6	34.8	31.4
Single	62.7	56.7	53.0	44.1	50.5
Married:					
Husband present	49.4	44.4	40.8	30.5	23.8
Husband absent	58.8	54.8	52.1	51.8	47.4
Widowed	22.6	24.3	26.4	29.8 ⎫	36.0
Divorced	74.0	72.1	71.5	71.6 ⎭	

SOURCE: Developments and Issues in the U. S., Table 5. Compiled from U. S. Department of Commerce, Bureau of the Census, Current Population Report P-50, No. 29, and U. S. Department of Labor, Bureau of Labor Statistics, Special Labor Force Reports 13, 130, and 183, and unpublished data.

NOTE: Data for 1950 and 1960 are for persons 14 years of age and over; data for 1970, 1975, and 1979 are for persons 16 years of age and over.

declines in the birthrate (as well as changes in marriage, divorce, and child support practices), this trend can be expected to continue through the 1980s. The standard statistical approach has counted as heads of household only those women who are either divorced, separated, or widowed, but the percentages shown in Table 7 suggest that a high incidence of women contribute to support for married couples and families where both spouses are present. Therefore, given the socio-economic changes brought on by the women's movement, it can be expected that more women will be heads of household in the future. Bureau of Labor Statistics data in 1980 showed that 52% of all married women or 24.4 million were in the labor force. The median income for married working women amounted to $6,375. This figure was considerably less than $9,784 for working women heads of household and $18,915 for working male heads of household.

The 1980 labor statistics data also show the continued decline in marriage rates. Although married persons living with their spouses still predominate, their share fell from 69% in 1970 to 61% in 1980, when never-married and divorced persons made up nearly one-third of the work force. Single-mother heads of household accounted for 67% of the labor force participation rates compared with 54% for women in two-parent families.

Although there has been growth in the number of single mothers and unmarried couples living together, the increasing divorce rate has been the major cause of the increase in the number of female-headed families with children.   Significantly, the number of male-headed families has grown only modestly and, not surprisingly, the number of husband-wife families with children has declined.   The trend in divorce and female-headed families has accompanied the increased movement of women into the work force.   There also is some evidence that "separation rates are correlated with the amount of family income contributed by the wife, ... which suggests that new economic opportunities for women are one explanation for rising divorce rates."[4]   On the other hand, declines in alimony and child support settlements and increased default rates suggest the push of necessity, rather than the pull of new-found wealth, as an increasing cause.

Although the number of female-headed families with children has been increasing in all income groups in the population, most of these families are poor.   In 1977, the average income of households headed by women was less than one-half that for male-headed households, and 41% of families headed by women with husbands absent were below the poverty level.[5]   In 1978, families maintained by women with no husband present had a poverty rate of 31.4%, far higher than those of husband-wife families or families with a male householder and no wife present (5.4% and 9.2%, respectively).[6]

While some correlation analyses indicate the possibility that welfare payments provide a similar influence on families as does income from employment by the wife, this association does not bear out completely. On this point, Ross and Sawhill conclude that:

> ... welfare reform by itself is not likely to reduce dramatically the present trend toward female-headed families. As relative income from sources other than welfare, and cultural patterns of thinking and behavior continue to shift, female-headed families will almost certainly continue to grow.[7]

Another group that suffers the consequences of women's low economic status is older women.   Women over 65 who are not employed constitute a substantial segment of the poor in general and represent the single poorest group in the population.[8]   Especially since 1970, inflation has diminished the buying power of pensions and Social Security benefits for a population that is growing rapidly, a growth that is attributed to demographic changes and increased longevity.   Because of their lower status as workers or their dependency as housewives on a husband's income, pension and Social Security benefits for older and widowed women are predominantly at the lowest levels.

The War on Poverty

This major government effort, which began in the 1960s and continued with modifications during the 1970s, was dedicated to the eradication of poverty that was primarily related to denial of civil rights. This vast program expanded opportunities in education, training, and employment. It was perhaps most effective, however, in its analysis of civil-rights-related poverty and the establishment of criteria and levels of poverty. In general, while gains in employment and income were made by men, the number of poor women actually increased somewhat. In addition, because inequities in employment and income were reflected in the related government effort, it tended to foster poverty among women and a permanent class of female welfare recipients. For example, between 1975 and 1977, the number of male family heads below the poverty level who were unable to find work dropped 35% while the number of their female counterparts rose 24%.[9]

The acuteness of the problem of women and poverty was reflected in a recent report issued by the National Advisory Council on Economic Opportunity. The Council reported that: "The feminization of poverty has become one of the most compelling social facts of the decade. . . . The poverty population will be composed solely of women and their children by the year 2000."[10]

GENERAL CHARACTERISTICS OF FEMALE POVERTY

During the 1960 to 1980 period, the general decline in the economic status of women has been accompanied by some structural changes. In keeping with the expansion-consolidation of female/male employment described above, more poor women are in paid employment than ever before; more of them are married; and more of them are single parents and heads of households. These changes indicate that the movement of women into paid employment, based as it is on economic necessity, constitutes a permanent institutional change in American society.

In specific terms, poverty is defined in the United States in relation to a poverty line or poverty index. There are actually three major poverty lines and one poverty index developed by separate agencies of the federal government. Poverty is defined by the Department of Labor, Department of Commerce, and Office of Management and Budget in connection with the administration of social welfare programs. The poverty index is a measure derived by the Social Security Administration as an attempt to specify in dollar terms a minimum level of income that is adequate for families of different types and consistent with American consumption patterns. It is linked to the Consumer Price Index and, therefore, is a real or deflated indicator of poverty.

It can be argued that the established hourly minimum wage provides a more realistic poverty base line. It, too, is linked generally to an index of inflation, but for purposes of assessing the situation of women, it provides an annual income below the poverty lines. In 1980, for example, the federal minimum wage level has been set at $3.10 per hour. On an annual basis, this income is below the poverty level set by the Department of Labor and the income of most women on an hourly basis is at or below the minimum wage. In 1978, the median income for women working full time was $9,032 (compared with $16,062 for men), which is especially significant because of the large number of women who are employed as domestic servants and in other situations not covered by the minimum wage. (In 1978, about 64% of all household workers were covered, according to the Department of Commerce.) [11]

The poverty rate cited earlier is calculated on the basis of the poverty index (derived by the Bureau of the Census), and although it is statistically sound, it does not appear entirely reliable. The rate is determined on the basis of extremely low money incomes and is intended to describe general or median incomes based on various populations. A more realistic level of poverty is provided by the Department of Labor, based on the cost of goods and services at a prescribed minimum level of consumption. In using this index, many more women would be found poor and the difference in the number of men and women below poverty level is larger.

Poverty, therefore, is both a relative and a specific phenomenon. Especially in the case of women, it may be less precise and more insidious due to their basically unsound and frequently dependent financial status. It is clear, therefore, that poverty has become a predominantly female condition in recent years. In 1960, there were 40 million poor people in the United States. By 1972, after concerted efforts by federal anti-poverty programs, this number had fallen to 24.5 million, but virtually the entire decline was accounted for by improvements in the economic status of male-headed families. In 1976, nearly one-third of all families headed by women were below the poverty level, more than five times the rate for men. Female-headed families had an average income of only $4,600 a year in 1977, substantially below the poverty level, and at present most poor families with children are headed by women.

With wholesale prices for goods and services being led by the highest increase among intermediate purchases (goods and services used in the production of these items) and retail prices tending to rise in keeping with median income levels, the substandard economic status of most women tends to be chronic. The effect becomes crucial if earnings are at or below the minimum wage level (the poverty level), which is the case for most employed women and an even larger proportion of women who are dependent upon welfare or spouses.

Alternatives for Women in Poverty

Few services are available to aid poor women in improving job skills or attaining economic independence. For a woman with children, public assistance and/or criminal activities may provide a better living than being employed in a mimimum-wage job.

Women are increasingly participating in economic crime. As has been established above, women who are single parents and heads of household with children are increasingly among the poor. It is no co-incidence that these characteristics also are found among women offenders. In Economic Realities and the Female Offender, Chapman[12] found that between 1960 and 1978 economic crime (as defined in the study) increased from 25% to 56%. The study also found a strong correlation between female arrests and labor force participation.

Table 8 compares long-term trends in female population, labor force participation, and persons arrested for the period from 1930 to 1970. To examine the relationship between the incidence of arrests and labor force participation by women, percentages were calculated for the female labor force as a percentage of the female population. Finally, 10-year rates of change were calculated for the female labor force and arrests.

TABLE 8

TOTAL NUMBER AND PERCENTAGE INCREASES OF FEMALE POPULATION, LABOR FORCE, AND PERSONS ARRESTED, 1930 TO 1970

	FEMALE POPULATION[1]	FEMALE LABOR FORCE[2]	WOMEN ARRESTED[3]	FEMALE LABOR FORCE AS PERCENT OF FEMALE POPULATION[4]	FEMALE ARRESTS AS PERCENT OF FEMALE POPULATION[5]
1970	104,309,000	30,756,000	947,000	29.5	00.90
Percent increase	15	37	133	20	100
1960	90,992,000	22,410,000	406,000	24.6	00.45
Percent increase	20	35	427	13	350
1950	75,864,000	16,553,000	77,000	21.8	00.10
Percent increase	16	27	48	10	43
1940	65,608,000	13,007,000	52,000	19.8	00.07
Percent increase	8	22	174	13	133
1930	60,638,000	10,632,000	19,000[6]	17.5	00.03

SOURCE: Economic Realities and the Female Offender.

[1] Census data for total population table 24, Historical Statistics.
[2] Census data for persons fifteen years old and older, 1930, fourteen years old and older, 1940-1966; sixteen years and older, thereafter (Series 049-62 Historical Statistics).
[3] FBI Data: Series H999-1011 Historical Statistics.
[4] Column 2 divided by Column 1.
[5] Column 3 divided by Column 1.
[6] Estimated by simple extrapolation of estimates for 1932-1934.

The 10-year rate increases were consistently higher for females arrested than for the other rates of change. The female labor force also grew faster than the female population as a whole. That the increases in persons arrested were so consistent and so much greater than the increases in the female labor force, suggests that criminal activity results from demand for employment greater than the number of jobs available.

This latter suggestion, criminal activity as an alternative to employment, receives some additional support when variations in 10-year increases are examined. The largest increases in persons arrested came during periods of substantial economic decline, 1930 to 1940 and 1950 to 1960. These also were periods of slower labor force growth. On the other hand, the smallest increments in persons arrested occurred during the periods of increased economic activity as well as greater labor force growth, 1940 to 1950 and 1960 to 1970.

As indicated above, women who are heads of household are in the lowest wage category. For this group, which is faced with low and declining wages in the presence of welfare alternatives, the opportunity costs of working tend to diminish (thus, bringing them to an even lower level of poverty). During the same period that female employment has increased, unemployment rates and the number of women on welfare also have increased. During the period that the number of poor people was being reduced from 40 to 24.5 million, the federal government invested several billion dollars in antipoverty programs. Now, when the population of poor people is predominantly female, most of the aid for these programs is gone. Although the female offender population closely coincides with the poor family head, which was the primary target of the "war on poverty," correctional administrators are now operating without the same networks of social programs that were available to this group in the 1960s and early 1970s.

Correlations between Female Poverty and Traditional versus Nontraditional Occupations

Female poverty is a concomitant of the growth in female employment. As The Employment of Women made clear, the form and nature of women's labor force participation follows deep-seated traditional patterns even under conditions of seemingly rapid socioeconomic growth and innovation. The following is a quote from The Employment of Women:

> For the most part, women did not retain the non-traditional jobs they had held during World War II. Furthermore, they failed to obtain nontraditional jobs in the two major growth sectors of the economy—housing and consumer goods production.
>
> The most conservative projections through the 1980s forecast large increases in the labor force participation of women, especially those between the

ages of 20 and 54. Even when assuming a low economic growth rate and a significant rebound in the birthrate (traditional barriers to women's employment), the Bureau of Labor Statistics expects participation and the high estimate is over 70 percent during the 1980s. All indications are that women in the United States will continue to play an important and permanent role in the labor force.

In recent years, there has been some increase in women's employment in nontraditional jobs, such as skilled trades in construction, truck driving, airplane piloting, telephone repair work, law and medicine. Although this trend is expected to continue to the 1980s, the overwhelming majority of women will continue to hold traditionally female occupations which are low paid in comparison to men's occupations. This fact can, at times, hide progress that has been and will continue to be made.

Furthermore, in certain professional areas requiring advanced degrees, demographic and industrial changes have resulted in mismatching between levels of education and work opportunities. In some cases, these mismatches may have serious consequences for traditionally educated women in the 1980s. For example, the number of available teachers exceeds the number of job opportunities in several fields. At the elementary and secondary school levels, this will continue to be an especially serious problem for women. Also, serious questions remain about the relatively high level of training required for low-paying jobs that are traditionally held by women compared with those traditionally held by men.

As noted earlier, women have supplied workers for the growing service sector. Projections to 1985 indicate that this will continue, particularly in the fields of health care, maintenance and repair, advertising, and commercial cleaning. While previous projections for transportation and public utility employment indicated only moderate growth into the 1980s, recent reallocation of fuel resources may increase the demand for mass transportation employees. The finance, insurance, and real estate industries are projected to grow substantially. Women are likely to hold increasing numbers of jobs in all these areas. However, as in banking and credit, most are likely to remain primarily in those jobs with lower status and lower pay.

[During the 1970s], the largest numerical growth of women employees occurred in retail trade, services, and State and local government, all of which were overall growth industries. These facts indicate that, although women are making progress in finding employment in nontraditional industries, the bulk of women are now, and are likely to continue to be, employed in industries that have traditionally employed them.[13]

CURRENT POSSIBILITIES FOR IMPROVING WOMEN'S ECONOMIC STATUS

One possible avenue for changing the economic status of women and helping women in poverty would seem to be the women's movement. Women in poverty, however, not only seem unaided by any advances the women's movement may have achieved (and many argue that earning differentials show that women have made no economic gains in the past decade), but their situation has worsened because of other changes in economic responsibilities. A large proportion of women with children under age 18 now work; support payments from absent fathers are seldom made, even in middle-class families; and there is a more widespread expectation that women should be financially self-sufficient. The women's movement must become considerably stronger and more highly directed to underlying sources of improvement in the economic status of women if women are to reverse the long-term trends which have been examined here.

Present government action in the United States does not appear to be adequate in scope, sufficiently realistic, or specifically directed enough to counter the two major trends which we have found to be endemic and chronic in nature. That is, we have observed the consolidation of men into a range of better-paying jobs and the expansion of women into lower-level service and support jobs, with a minority of women moving into higher-level and better-paying nontraditional occupations (though at lower pay than their male counterparts). A General Accounting Office Staff Study of laws and executive orders for nondiscrimination and equal opportunity programs, conducted in 1978, found some 43 such laws and executive orders based on sex. Most of these laws have been found to nibble at the edges of the problem and are difficult to enforce. For example, Title VII of the 1964 Civil Rights Act prohibits discrimination based on pregnancy. The Fair Labor Standards Act mandates payment of a minimum wage which, at $3.10 an hour in 1980, produces a gross income well below the poverty level. Although some modest occupational change has taken place, it is probably due mainly to job redefinition by federal contractors for the sake of contract compliance.

In addition to the 43 laws and orders relating to sex discrimination, approximately the same number is found to be associated with race, color, religion, and national origin. This similarity is based on the fact that all of these aspects of discrimination were mentioned in the same order. Discrimination against women is a civil rights problem, but too general a treatment of it has actually impeded progress—as data on male/female poverty and income have shown.

The laws and orders relating to discrimination have been ineffective because they have not been specific enough in their direction and because compliance with them is not easily enforced. In fact, the whole question of compliance can often be avoided by employers by the segregation of jobs through redefinition. This approach has produced a degree of failure which has not yet been assessed.

> Federal law has prohibited some forms of employ-
> ment discrimination with respect to wages since the
> passage of the Equal Pay Act (EPA) of 1963. How-
> ever, neither the EPA, nor the more general prohibi-
> tion on employment discrimination because of race,
> sex, religion, or national origin contained in Title VII
> of the Civil Rights Act of 1964, has been applied to
> the question of wage rates paid for jobs into which
> minorities and women have been traditionally seg-
> regated.

> Most of the jobs to which women have been
> allowed access are different from the jobs which men
> have traditionally done. Therefore, the EPA, which
> has been interpreted to apply only where men and
> women work on similar jobs, has had little impact. It
> has, however, been used by courts to restrict the
> scope of the broad prohibition on discrimination con-
> tained in Title VII when the question of wage rate
> discrimination has been raised. [14]

Another issue that affects women's labor force participation and their subsequent economic status is that of sexual harassment. All women who work must do so with the expectation that they are likely to experience harassment at some time in their working lives. Many women experience it on a daily basis as part of the standard work environment. However, definitive knowledge of the extent of sexual harassment in employment has been difficult to attain. A recent survey of sexual harassment in the federal workplace was conducted by the Merit Systems Protection Board of the U.S. Government. On the basis of a random sample of 23,000 employees (men and women), the survey results showed that 42% of the women respondents had experienced some form of overt sexual harassment in the 24-month period covered by the survey. In terms of the extent of the problem, 42% would place sexual harassment in employment in the category of other major sex-related social problems affecting the status of women: family violence, rape, and incest.

A study of sexual harassment and sex discrimination of women in employment, which was based in part on an analysis of the Merit Systems Study, was conducted by the Center for Women Policy Studies and concluded:

> Sexual harassment, whether narrowly or comprehensively defined, has been found to be one of the most serious employment problems facing women. In the restrictions which it imposes on women, it is the means as well as the message, paralleling in the workplace methods found in society as a whole for subduing and directing the aspirations of women. That this is so is indicated by the results of studies and court cases which show the heavy involvement of supervisory personnel in the most serious forms of overt sexual harassment as well as the most subtle sex discrimination. It is also shown by the indifference, unwillingness, inability, and even opposition which is often shown by management to providing relief for victims of even extreme forms of harassment, not to mention to providing a harassment-free and egalitarian work environment.[15]

CONCLUSION

The data cited in this report indicate that despite greater participation in employment, women have become increasingly concentrated in poverty-level incomes with very modest changes for them in upper-income levels. The opposite has been the case for men, however. The wage gaps between men and women have widened even during periods of general prosperity. Men, moreover, have benefited not only from prosperity, but also from industrial diversification and the development of new, high technology industries.

The differences between women and men in employment and earnings are reinforced by discriminatory employment practices which tend to keep women employees confined to a few traditional occupations. Even when women find employment in higher management and professional levels and nontraditional occupations, wages paid to them are usually far lower than those for their male counterparts. In addition, they almost always find upward progress to be blocked by discrimination and very frequently they are subjected to sexual harassment which, whether intended or not, usually has serious adverse effects on their job and career development.

The trends in female employment and income since the 1960s combined with their increasing economic responsibilities have resulted in a squeezing action: greater participation by women in economic activities with less-than-satisfactory results. This situation is reflected in their increasing participation in welfare programs. It is also reflected in their increasing participation in economic crime.

Attempts to diminish poverty among women should include, if not begin with, equal opportunities for employment and earnings. Before this step can be meaningful, there must be equal opportunities and access to educational and vocational training. To make such opportunities truly effective, however, there must also be social and cultural change so that women, like men, are able to view educational training and job and career opportunities in terms of the labor market without the impediments of stereotypes and traditional categorizations. Viewed from this perspective, welfare payments provide a stopgap measure at best and at worst nothing less than a means of perpetuating poverty in its traditional form.

Female poverty has been found to be a concomitant of the growth in female employment. The basis of this phenomenon is competition for a relatively limited number of jobs in a job market influenced by discrimination. More women are entering the job market, but they do so by accepting the low-paying jobs that are increasingly being abandoned by men who continue to improve their overall economic position. On the basis primarily of continued discrimination in employment, the growth in female poverty is expected to continue in the future. As noted above, the National Advisory Council on Economic Opportunity concluded in its 1980 report that at the present rate, the poverty population of the United States by the year 2000 will be composed solely of women and their children. It was noted earlier that it takes an annual growth in real GNP of at least 5% to 10% per year to produce any meaningful advance in terms of closing the male/female gap in occupational participation and income. Although such high rates of growth are hardly likely for the foreseeable future, some improvements in the status of women and female poverty would be possible under conditions of moderate growth given adequate, enforced, legal provisions prohibiting sex discrimination in employment.

These analyses have indicated a chronic situation regarding the status of women over the past 20 years that can be expected to continue indefinitely in the absence of counteracting forces from strong and sustained economic growth, meaningful and highly directed government policies and programs, and widespread litigation and enforcement relating to equal employment opportunity and equal pay (or passage of the Equal Rights Amendment and its enforcement). Although these three conditions appear essential, they have only been partially met despite the pressures for change brought about by the women's movement. Without strong countermeasures, we may be seeing the creation (or expansion) of a social institution, perhaps best characterized as the modification of the role of housewife or domestic woman into that of the "work woman."

With the expansion of female employment, female poverty in both relative and absolute terms can be expected to continue to increase as it has steadily done since 1950. In terms of the male/female difference in median income, without attempting to compute ratios (for which the data probably are not adequate), the gap has generally widened with the expansion of employment. Factors such as growth in the GNP have also been found to have a strong bearing on wage levels. In view of the predictions of low growth and high inflation for the U.S. economy through the 1980s, it can be expected that poverty among women will continue to grow at a rate slightly greater than the rate of expansion of female employment.

The poverty-related policies and programs in the United States improved the conditions for some men at the expense of women. That is, the "war on poverty" diminished male poverty somewhat but increased the extent of poverty among women. While the women's movement has accompanied the increased participation of women in the work force, it also has accompanied the deterioration in their economic status. In this regard, one cannot help but agree with the findings of the second U.N. Conference on the Decade for Women, that the "economic crises of the 1970s have settled on women's shoulders."

NOTES

1. Sekscenski, 1980.
2. Based on Tables 281 and 282 of Statistical Abstract of the United States, U.S. Department of Commerce, 1979:168-169.
3. Ibid.:427, Table 704.
4. Ross and Sawhill, 1975.
5. U.S. Department of Commerce, 1979:459, Table 752; and 1979:467, Table 767.
6. U.S. Bureau of the Census, 1979a.
7. Ross and Sawhill, 1975:126.
8. In 1979, the average income of women 65 and older who were below the poverty line was $3,000 for whites and $2,000 for blacks.
9. U.S. Department of Commerce, 1979:467, Table 768.
10. Twelfth Annual Report of the National Advisory Council on Economic Opportunity, August, 1980:147-8.
11. U.S. Department of Commerce, 1979:123, Table 697.
12. Chapman, 1980:54, Table 4-3.
13. U.S. Department of Labor, Women's Bureau, 1980b.
14. Blumrosen, 1979:399-400.
15. Candea and Chapman, 1981.

POVERTY, WORK, AND MENTAL HEALTH:

THE EXPERIENCE OF LOW-INCOME MOTHERS[1]

Deborah E. Belle

Harvard Graduate School of Education
Cambridge, Massachusetts   U.S.A.

Ruth F. Tebbets

University of California
Berkeley, California   U.S.A.

This paper examines the relationship between work and well-being for low-income mothers with young children.   For these women, paid work can be a means of raising income and reducing the severity of poverty, yet it is likely to offer low wages, few benefits, and few opportunities for advancement while reducing considerably the time and energy needed to perform the tasks associated with motherhood. Because of the stresses of poverty and the responsibility for young children, low-income mothers are a high-risk group for such mental health problems as depression.[2] An important consideration of this study is to determine whether work is a protective factor for these women or simply an additional stress.

Research that has examined the impact of work on women has produced inconsistent results. There is evidence that work and well-being are highly related for women with exceptional IQs and for women with a college to professional-level education.[3] Ferree also found this relationship to hold for a working-class sample whereas Baruch and Barnett did not find this association in their study of middle- to upper-class women.[4] Two recent large-scale surveys found no relationship between work status and depression among women.[5] One review article of the research literature in this area confidently states that there is evidence for an overall positive relationship between work and well-being.[6] However, given the bias of studying well-educated, professionally oriented women in much of the literature reviewed, the generality of this finding is questionable.

Most employed women do not have advanced degrees and work at jobs that offer low pay, inadequate benefits, and few opportunities for advancement.[7] Occupational segregation in the United States is widespread with women predominating in poorly paid jobs. Fully 80% of all working women are to be found in only four occupational categories: clerical work, sales, manual labor, and service jobs.[8] Sawhill[9] calculated the percentage of predominately female occupations in which a female high school graduate, aged 25 to 34, would earn less than $3,000 per year working full time as well as the percentage of predominately male jobs with a similarly dismal earning potential. She found that while 20% of the male occupations had such poverty-level wages, over one-half of the predominately female occupations had a comparable low-level wage.

Women's earnings, relative to those of men, have actually declined in recent years. In the mid-1950s, a full-time employed woman earned, on the average, 63 cents for every dollar earned by her male counterpart. By the mid-1960s and 1970s, this figure had declined to below 60 cents.[10] Further, the gap between men's and women's earnings is considerably wider in jobs that are unskilled and already poorly paid. Among sales workers, for instance, women earn only 39% of male wages.

These facts are particularly alarming when we realize that for the vast majority of working women, financial need is a major factor in their decision to work. Among married women, the rate of labor force participation generally increases as their husbands' income declines. The spiraling divorce rate and the financial need which so often accompanies divorce have propelled many women into the labor force. Few families receive regular child support payments from the absent father.[11] Almost one-half of all female-headed families are poor, and the majority of poor families with children are now headed by women.[12] Government welfare benefits do not eliminate the need to work, because these payments rarely raise recipients' incomes to above the poverty line. Three-fourths of the families receiving Aid to Families with Dependent Children (AFDC) benefits still have poverty-level incomes.[13] Thus, the unequal earning power of women, the economics of divorce, and the low level of government welfare benefits lead to the fact that many American women and their families live in poverty.

THE STRESS AND FAMILIES PROJECT

This paper reports on a group of women for whom these problems are acutely relevant: low-income mothers with young children. The data to be discussed were obtained from the Stress and Families Project, an intensive field study of 43 low-income mothers in the Boston, Massachusetts area.[14] The paper describes the work experiences and work ambitions of these women, their experiences in combining work with family life, their attitudes towards work, and the barriers that stand in the way of their employment. The paper also explores the relationships between aspects of women's work experience and their emotional well-being.

The Stress and Families Project was initiated to investigate the relationship between life situation and mental health among low-income mothers, the group at greatest risk for the mental health problem of depression. The low-income mothers who participated in this research were recruited without regard to their current mental health status. They ranged in age from 21 to 44 and represented every legal marital status category (never married, married, informally separated, formally separated, widowed, and divorced). Each woman had at least one child between 3 and 7 years of age. Twenty of these women were single-parent heads of household, 12 lived with their husbands, and 11 lived with boyfriends. Per capita household income for the families ranged from $500 to $4,000, which meant that families in this study included those far below federal poverty lines as well as some living just above poverty lines. Thirty-three of the 43 respondents received government welfare payments. The sample was almost evenly divided on race; there were 21 black women and 22 white women. Educational attainments of the respondents ranged from the fourth grade to graduate school with a mean educational level of 11.8 grades.

Each of the 43 women was interviewed by a woman of her own race on a variety of topics, including work experiences, daily routine, life stress, social networks, coping strategies, family nutrition, experiences with social service institutions, and early family history.

A measure of depressive symptomatology was chosen as the major index of well-being because of its relevance on several counts. Most generally, nationwide concern has been expressed over the growing, epidemic-like prevalence of depression.[15] Women are much more likely than men to experience depressive symptomatology, and poverty and responsibility for young children are additional risk factors associated with depression.[16] These results suggest that the mothers interviewed in this study were part of the demographic group most at risk for depression. In addition to these associations, depressive symptomatology has demonstrated sensitivity to other variables of importance to this research. That is, stressful life events and conditions can precipitate depression, and depression has been isolated as a psychological condition which can profoundly affect family interactions.[17]

Depressive symptomatology was measured by a brief self-report questionnaire, the CES-Depression Scale, developed at the National Institute of Mental Health. The scale includes 20 items which represent the major components of the depressive syndrome: depressed mood, feelings of guilt and worthlessness, helplessness, despair, retarded activity, and change in eating and sleeping patterns. Scores on this measure do not necessarily determine whether or not a person is suffering from the clinical syndrome of depression, but they do indicate the extent of depressive symptomatology and unhappiness. The measure was administered twice, at the beginning and the end of the research period. Respondents' scores on the two administrations were highly correlated (r = .51, p < .001) and the average of a respondent's two scores was used for the analysis reported here.

The distribution of respondents' scores approximates a normal distribution and encompasses a range of 43 points out of a possible 60. Compared to the general population, the respondents as a group reported a very high level of depressive symptoms. One-half of these women had scores which the authors of the instrument consider to be indicative of "depression." The mean score for this group (17 points) was similar to that of people who have recently lost a spouse to death or marital separation.[18] Thus, the respondents as a group exhibited an alarming level of depressive symptomatology, yet among the respondents there was a great variation in symptom level, and this variability enables us to search further for the specific features associated with depression among low-income mothers.

Work Status and Experiences

At the time of the study, one-half of the respondents were engaged in a form of paid employment. For some women, the paid employment was a full-time job, while for others it was a part-time position which only brought in a few dollars a week. During the course of the study, several respondents changed their work status by quitting jobs, taking new jobs, or changing their work hours. When we compared the depressive symptom scores of women who had worked during the study with those of women who were nonemployed throughout the designated time period, we found no significant difference between the two groups. However, this finding is not particularly informative since the women we characterized as "working" varied greatly in their degree of involvement with work.

A composite picture was created to illustrate a typical respondent's work history. This typical respondent began work as a youngster, picking fruit during school vacations and contributing the proceeds to the family budget. On her own at 17 years of age, she held a succession of jobs at franchise restaurants which never paid more than $2.00 an hour and often required working late shifts. She married, and stopped working when she was 5 months pregnant with her first child. When the baby could walk, she started working in part-time cleaning jobs to supplement her husband's salary, leaving the baby with her sister when possible and taking him with her when necessary. She had two more children in the next 4 years. No longer able to get away during the day, she sometimes worked evenings at the fast food restaurant where she had worked before marriage. Now separated from her husband, she works part time in a nursing home to supplement her government welfare payments.

For Stress and Families respondents, paid work was a familiar part of life. Only two women had not worked over their lifetimes—one woman had a work history of 15 jobs—and the respondents as a group had an average of 4 jobs during their lifetimes. These patterns correspond to nationwide statistics which show that over 90% of welfare mothers have worked at some time and three-fourths have worked full time.[19] The nature of the respondents' jobs again parallels nationwide patterns. Waitressing and kitchen work alone accounted for over one-sixth of the jobs named, and factory work rivaled these jobs in frequency. Nursing, child care, and housekeeping also accounted for large numbers of jobs. Almost all of the jobs listed by respondents could be categorized as service jobs, manual labor, sales positions, and clerical positions. Thus, Stress and Families respondents have held exactly the kinds of jobs that national studies have shown to be typical for women. These jobs are generally low in required education, pay, and prestige, thereby demanding hard work for minimal rewards.

When asked about the bad features of each job, the most common themes were low pay, bad hours, boring work, high pressure, and environmental hazards. Robbery was an omnipresent threat for one woman who worked behind a counter in a record store. Another woman would walk up and down six flights of stairs to a card factory rather than trust a broken elevator. Transportation to and from work frequently posed problems of both safety and convenience.

On the positive side, good pay, good hours, and the intrinsic interest of the work were frequently mentioned as favorable aspects of jobs, as were pleasant co-workers and supervisors. Other responses were unique to the job or the respondent: pleasant customers at a doughnut shop, the freedom of setting one's own hours as a housekeeper, the belief that one was needed at a nursing home, and the acquisition of skills in a bookkeeping job.

Even within the restricted range of these jobs, we found that pay and prestige did make a difference to the woman's emotional well-being. That is, the higher the salary earned in the 3 years prior to the study, the lower was the level of depressive symptoms. Similarly, when jobs were scored for occupational prestige according to the Blau-Duncan Index of Occupational Status,[20] a strong association between occupational prestige and well-being emerged. With higher levels of occupational prestige, the respondents tended to have lower depressive scores.

Work Aspirations

The respondents' work ambitions were generally realistic in light of their past experiences and current situations. When respondents were asked to imagine the ideal job for them, only a few mentioned glamorous careers (e.g., a model or a clothes designer) for which they had no relevant work experience. Most of the ideal jobs mentioned by respondents were those which would build upon their earlier work. For example, one woman selected nursing as her ideal job and she had trained and worked as a health attendant and nursing aide.

Pay level frequently was mentioned as an important attribute of a job. Many women mentioned that their ideal jobs would provide a decent salary and allow them enough time with their children. For one respondent, an ideal job was one in which she could work 20 hours a week and earn enough money to support herself and her daughter.

The respondents' quest for a decent salary takes on additional meaning against the backdrop of poverty. It is poverty that places these women at such great risk for depression, and a well-paying job would enable them to escape from poverty. Among the respondents, economic problems were at the root of many difficulties. One-third of the respondents said that they did not have enough money to buy adequate food for their families. Low income often meant poor housing, discrimination, and exposure to violent crime. Money difficulties were related to worries in many areas of life, such as the living environment, parenting, and mental health. Even within the restricted income range represented by the respondents, women with the highest per capita household incomes reported significantly fewer depressive symptoms.

Coordinating Work and Family Responsibilities

Not surprisingly, securing work compatible with motherhood was seen as a problem for many women. The most common reasons given for stopping work were family events such as a pregnancy, birth of a child, or problems with child care arrangements. Three-fourths of the respondents agreed with the statement, "Some women find it very hard to get the kinds of jobs they can do while being a mother."

The cost of child care can also destroy the economic argument for work. One respondent told the interviewer, "More than half of my pay check would be gone before I see it. It's hopeless. Why bother?" Another woman stated, "The problem with most jobs is that you can't afford to take them."

Respondents often turned to part-time work, work during evening or nighttime hours, and work which could be done at home or on a flexible schedule in order to minimize the conflicts between work and family. In the 3 years prior to the study, when all respondents had child care responsibilities, only about 20% of all jobs were full-time, nine-to-five jobs. The respondents' home-based jobs included telephone sales, knitting and crocheting, child care, and janitorial work. These jobs generally were very poorly paid.

While many respondents had at least occasional child care help from paid sitters, day care centers, and the public schools, relatives were the most common source of child care assistance. Although such help is often characterized as a simple gift, it is not without cost, as many relatives often expect assistance in return. For example, in return for the care of her two children in the evenings, one respondent felt obligated to provide baby-sitting help every morning for her sister's two children. Her child care responsibilities, therefore, were not substantially reduced by this help, but merely shifted to different children and to different parts of the day.

Attitudes Towards Work

Many respondents mentioned the kinds of jobs they had already held so often—waitressing, kitchen work, and housekeeping—as work they would now take only as a last resort. Sometimes a job could be so draining and unrewarding that the end of the job marked a change for the better.

On the other hand, the work ethic is alive, well, and flourishing among the Stress and Families respondents. These women tended to associate work with confidence, self-esteem, accomplishment, dignity, and independence. As one woman said, "I want to feel I can say I took care of myself." Some women spoke of the boredom that sets in when there is no work to structure a day. Of the nonemployed women, most expressed a strong desire to work. Two-thirds of the women reported that they felt lonely "every day" or at least a few times a week because they were not working. Few nonemployed women felt happy more than occasionally that they were not working.

For some of the women, the frustrated desire for work was very troubling. One respondent described herself as being in a tense and irritable state before she finally found a job. The nonemployed respondents who wanted to work had a higher level of depressive symptoms than working respondents and nonworking respondents who did not want to work.

Barriers to Employment

Child care responsibilities place many obstacles in the way of women's employment, as has been discussed. For many women, another major barrier is the fear of losing medical insurance should their income rise above limits set by the welfare department. Few of the respondents' jobs provided medical coverage for them and their children, and many families experienced serious health problems requiring expensive treatment. One woman in the study was paying installments on a large medical bill accrued when she was hosptialized while she and her husband had no medical insurance. Her weekly payments of $2.00 toward this bill placed additional strain on an already tight budget, and at this rate of repayment it will take at least 100 years to cancel the debt. When asked whether they would give up AFDC benefits and start work if they could keep their Medicaid and food stamp benefits, most respondents said that they would do so, but most of the other women said that they would not do so because of continuing child care responsibilities and their inability to earn enough to support their families without welfare benefits.

Some respondents saw their lack of education as the most important barrier in the way of decent employment. Like many other women in the study, one respondent expressed the hope that things would be different for her own children.

> For all of my kids, my daughter included, I would push for education. . . . The more education you have, the more freedom you have because you have more money. You don't owe anybody anything. The more money you have, the more power you have. . . . I would push for the education, a good job, and somewhat of a career for all my kids. . . . And I would hold myself up as an example. I married young; I didn't complete my education, we ended up in the projects. If I'd had an education, we'd be able to get a job and everything else and wouldn't end up in the projects. I have no skills in anything.

Even though child care responsibilities and poverty made the enterprise a particularly difficult one, several women in the study had returned to school to complete their high school educations and to go to college.

CONCLUSION

Work is important to the well-being of low-income mothers, as these findings show. Women in the study who had worked at relatively well-paid and prestigious jobs enjoyed mental health advantages over women who had been confined to the most poorly paid and least prestigious jobs. Women who wanted to be working, but were unable to do so, experienced more symptoms of depression than employed women or those who did not want to work.

The low-income mothers emerge as a hard-working group of women who juggle paid employment and child care responsibilities in an attempt to maximize benefits for their families. They place a high value on work and self-reliance, and they know that additional income could alleviate many of their most pressing problems. However, most of the jobs open to them do not provide a decent standard of living for their families, and drain away time and energy which could be used to care for their children. Furthermore, even minor increments in income can mean the loss of government-provided medical insurance without enabling women to purchase their own medical coverage. Child care assistance is often extremely costly, either financially or in terms of obligations to those who provide assistance without pay. In order to work at all and increase family income, women often have to settle for poorly paid part-time work with inadequate benefits and no opportunities for advancement. Their work ambitions are reasonable and even modest, and yet the training or education needed to achieve important credentials often requires time and money which is simply not available.

The respondents, as a group, cannot be described as unskilled workers. Many had trained and worked for years in highly skilled (although underpaid) professions such as teaching and nursing. They did not lack the skills, merely the credentials which would certify their skills and help them to command a reasonable salary. Other women did suffer from a lack of formal education and a lack of job experience in the field they wished to pursue in the future.

Some mothers of young children were fully occupied with child care responsibilities. While they often wanted to work, they recognized that they were needed at home for a certain number of years. For such women, government prodding to enter the labor force is unsound. A higher level of assistance payments would do much to promote the emotional well-being of these mothers and, in fact, the stability and well-being of their families.

For mothers without such pressing child care responsibilities and with work experience in the field they wish to pursue, credential programs which are affordable in terms of time and money could promote a raise in the standard of living and improve the well-being of many women and their families. Current government training programs often offer women several preselected career options without regard to their prior experience and interests. Would it not be reasonable to meet women where they are and reap the benefits of their work experience and enthusiasm?

Still other women will need further education or training in fields that are new to them. Since most traditional women's jobs are underpaid, efforts should be made to inform women about nontraditional career options that will provide decent pay and working conditions. The upgrading of pay scales for part-time work and traditional women's work is also urgently needed.

While much can be done to help low-income mothers with young children, an ounce of prevention is worth a pound of cure. No young woman should graduate or drop out of high school without a skill that could support a family at a decent standard of living. With a climbing divorce rate, the financial support from a husband cannot be counted on to last a lifetime. More and more families also find that a husband's income is not enough to support the family at a reasonable standard of living. We should not be turning young women out into the world without the skills and the credentials they will need to escape poverty and the depression that so often accompanies it.

NOTES

1. The authors express their gratitude to Stephen Piper and to members of the Stress and Families Project staff, including Diana Dill, Ellen Feld, Elizabeth Greywolf, Cynthia Longfellow, Nancy Marshall, Maureen Reese, Linda Tsang, and Susan Zur, who reviewed earlier drafts of this paper. Linda Tsang also organized much of the case history material for the paper.
2. Pearlin and Johnson, 1977; Brown, Bhrolchain, and Harris, 1975; Radloff, 1975.
3. Sears and Barbee, 1977; Manis and Markus, 1978.
4. Ferree, 1976:76-80; Baruch and Barnett, 1978.

5. Pearlin, 1978; Radloff, 1975.
6. Kanter, 1977b.
7. National Commission on Working Women, 1979.
8. Ibid.
9. Sawhill, 1976.
10. Barrett, 1979b.
11. Ross and Sawhill, 1975.
12. Ibid.
13. Pearce, 1979.
14. This research was supported by the National Institute of Mental Health, Mental Health Service Branch, Grant Number MH28830 (Susan Salasin, Project Officer).
15. Klerman, 1979; Secunda, 1973.
16. Brown et al., 1975; Radloff, 1975; Weissman and Klerman, 1977; Pearlin and Johnson, 1977.
17. Brown et al., 1975; Markush and Favero, 1974; Weissman and Paykel, 1974.
18. Radloff, 1977.
19. Pearce, 1979.
20. Blau and Duncan, 1967.

PART III

WOMEN AND THE WORLD OF WORK:

SOCIOECONOMIC FACTORS AFFECTING WOMEN'S INCREASED

LABOR FORCE PARTICIPATION

INTRODUCTION

Camille Kim Cook

Naval Health Research Center
San Diego, California   U.S.A.

During the past two decades, women in all NATO-allied and many other countries have begun to enter the labor force in ever-increasing numbers. This widespread movement has raised the question of what effects various economic factors, public and private policies, family considerations, and women's own self-perceptions have upon their increased labor force participation. The papers presented in this section examine these factors, and comparisons among presentations raise several interesting questions concerning how these factors differ from country to country and to what extent they are influenced by the cultural and socioeconomic situation in each country. For example, in most of the modern and progressive countries, there continues to be evidence of sexual inequality in the workplace similar to that found in developing countries. Some of the variables that are examined in an attempt to explain and understand the existing inequality in the workplace and other aspects of women's labor force participation include: (1) membership in labor unions and women's influence within them, (2) educational level and vocational training, (3) legislation affecting the employment of women, (4) the differential between women's and men's wages, (5) part-time work, and (6) economic growth. Additionally, the papers delineate a range of personal factors that influence women's labor force decisions, such as age, marital status, number and age of children, husband's income, and distance to work.

What purpose can be served by examining these variables? Pintasilgo has stated that women have a profound and positive effect upon the economy both by their numbers in the labor force and because of the types of jobs they perform, which often are crucial in meeting basic human needs. However, she also points out that the effects upon women themselves are not equally positive; for example, in many countries women are becoming a new form of slave labor. The research of

191

Armstrong and Armstrong, Belle and Tebbets, Chapman and Chapman, and others indicate that, all too often, women are employed in jobs that are tedious, underpaid, undervalued, and sometimes even dangerous to their health. In order to improve these workplace conditions, the factors affecting women's participation in the labor force need to be examined with a view toward how the impact of these variables can be altered, eliminated, or enhanced in the future in order to promote a healthier economy and a more egalitarian society.

WOMEN IN TRADE UNIONS

In free market economies, labor unions are to a great extent the pacesetters for increasing wages and for improving working conditions. They are, in fact, the major representatives of workers in the labor market. Consequently, the policies, goals, and accomplishments of labor unions have a tremendous impact on the economic position of women. In the presentation, "The Most Difficult Revolution: Women in Trade Unions in Four European Countries,"[1] Alice Cook begins with the assumption that unions increase wages and enhance working conditions but questions their efforts to advance women's equality in the labor market, the workplace, and the unions themselves. In her analyses, Cook identifies six key areas as being those in which the unions could do the most to advance equality: (1) women's place in union life, (2) trade union education, (3) vocational training by employers, (4) collective bargaining and equal pay, (5) part-time work, and (6) health and safety.

Women's Role in Union Life

The first and most important aspect of women's role in union life is that women are not members of unions in an equal proportion to their numbers in the labor force. This fact is a reflection of both union history and women's socialization into certain narrowly defined roles in society and the world of work. Cook concludes that there have been scattered, primarily experimental, efforts to integrate women into unions but, thus far, such programs have not been a major priority, nor have they been adequately discussed or evaluated.

Trade Union Education and Vocational Training

In the area of trade union education, several salient issues are brought to the fore in Cook's paper. Education, she contends, is directly related to job mobility; therefore, it is essential to provide education, at least for in-shop responsibilities (including health and safety), during regular working hours. Also, special introductory courses for women, in which they can simultaneously build their self-confidence and fill in any gaps in their educational preparation, need to be continued and expanded. According to Cook, in order for such courses to be truly effective in reaching women and integrating them into unions, their planners also must be cognizant of women's dual responsibilities: Courses should be conducted in fairly short blocks of time at locations close to women's homes, and child care should be provided. Additionally, material on women's issues should be introduced into existing curricula.

One of the main reasons for women's low wages is their segregation into so-called pink-collar jobs. Cook maintains that vocational education of young people is crucial in eliminating these occupational ghettos. Although unions can exert influence and participate in this area, she contends that vocational education of the young is primarily the responsibility of school boards. It should be noted, however, that the military also plays a significant role in providing vocational training and education, particularly in the United States.[2] Perhaps trade unions, school boards, and the military could work more closely together in designing and implementing training programs that would more effectively integrate women into nontraditional occupations. Given the scarcity of funds available for education, this type of cooperation could decrease overlap and duplications in material, thereby conserving the limited financial resources of both government and industry.

Collective Bargaining and Equal Pay

It is a well-established fact that in most countries, women's salaries are 25% to 40% below men's. Cook presents the wage differentials for 16 European countries and the United States, as reported by the European Commission on Economics.[3] Although these figures are derived from a variety of sources at different times, a comparison across countries reveals that, on the average, women earn approximately 67% of what men do. The country with the smallest wage differential between men and women is Sweden. But, even in this progressive country, which, as Rosén points out, has long prided itself on sexual equality, the figure is approximately 80%. By contrast, the widest wage gap is found in the United Kingdom where women's wages are about 55% of men's. The United States follows closely with 60% and recent U.S. Department of Labor statistics indicate that the figure has since dropped to 57%.

How can these wage differentials persist in apparent defiance of widespread legislation against wage and sex discrimination? This question is raised by several of the contributors to this book, including Rosén, Chapman and Chapman, and Connelly. In the case of labor unions, Cook maintains that unions can adopt one of two policies in dealing with this problem through collective bargaining, which is the major process for establishing wages in free market economies. Since the initiative for such bargaining is with the unions, their policies affect the amount of increases in wage scales and salaries. The results of union bargaining also are beneficial to unorganized workers whose wages tend to follow those negotiated by the unions.

The first approach that unions can adopt is one of narrowing the wage gap between low- and high-income workers. This goal would have to be a major tenet agreed upon by union members prior to entering into collective negotiations with employers. Because of their concentration in low-paying occupations, women would benefit greatly from such a program. The second approach to wage equality is job evaluation. This method involves measuring the value of a job according to such criteria as skill, responsibility, physical exertion, etc., and then establishing an appropriate pay schedule. The former policy has proved much more effective in raising women's wages than the latter.

Women in Part-time Work

In her discussion of the problems that unions encounter in dealing with part-time work, Cook points out that the vast majority of part-time workers are women. (Armstrong and Armstrong indicate that in Canada 72% of the part-time workers are women.) Few countries have adequate legislation protecting part-time workers and unions have traditionally either opposed the hiring of part-time workers or ignored their problems. This adverse treatment is evidenced by the fact that part-time employees generally do not receive benefits comparable to those provided for full-time workers. Thus, part-time employees are often more cost-effective for employers--a fact that sometimes causes resentment among full-time union workers. Recently, there have been efforts in Europe to organize part-time workers and to provide equitable employee benefits for them, but, overall, the problem is still not being adequately addressed.

Health and Safety

In the area of health and safety, Cook lists five problems that pertain to women: (1) hazards that women share with men because they are employed in the same work environment, (2) hazards that are limited almost exclusively to women because of their segregation into a small number of predominantly female occupations, (3) stress-related problems attributable in part to the double burden of paid labor in the workplace and unpaid labor in the home, (4) risks involved when pregnant women are employed in certain jobs, and (5) risks to women in jobs beyond their physical capabilities. Traditionally, unions have tended to bargain for hazard pay rather than demand control or removal of unsafe conditions. Even when changes in such conditions have been effected, little attention has been paid to the special health hazards of women.

Integrating Women into Unions

Cook concludes her paper by suggesting several policies for integrating women into unions in an effort to more successfully identify and solve their problems. For example, the following issues should become an integral part of union objectives and be placed on every union agenda: support for equal employment opportunities in both general and vocational education, support for and implementation of maternity leave, and expansion of child care services. The paper ends on an optimistic note by stating that to be even more effective, current programs and policies require expansion, a clarification of focus, and greater continuity.

Union Women in Lower-Level Management

Dafna Izraeli's paper on self-confidence of men and women shop stewards in Israel ties in closely with many of the ideas and solutions expressed by Cook. In Israel, women have been fairly successful in attaining positions as shop stewards and, therefore, as committee members in lower-level management. Thus, they have at least a limited power base which, according to Cook, is an essential starting point in resolving the particular problems of women in the work force. Izraeli investigates women's perceptions of their own competence at the committee deliberation stage and how these perceptions affect their influence within the committee and within the union. The author studied 111 women and 148 men in 65 Israeli firms, all of which are in the food, textiles, or electronics industries, and have at least 100 employees and one or more woman committee members. The results confirm the following six hypotheses. First, women's assessment of their overall performance in committee work and their specific skills relevant to the committee is lower than men's assessment of themselves on the same criteria. Second, women's evaluation of their overall performance is more strongly associated with their self-ratings on committee relevant skills than is the case for men. Third, women who constitute majorities on committees perceive themselves as more influential than do women who serve on otherwise all-male committees. Men's perceptions, on the other hand, are not strongly affected by the proportion of men on the committee. Fourth, the findings indicate that men's perceptions of their individual influence over the committee are higher than those obtained for women. Fifth, men believe that within the committee, they are more influential than women. Finally, when there is only one woman on a committee, there is no difference between men's or women's perceptions of her influence; however, men's self-ratings are lower than women's self-ratings on committees with more than one woman.

These conclusions point out the inherent difficulties in being a skewer--the term used by Kanter [4] and Stiehm to refer to women who are tokens or minorities within a decision-making group. On the basis of Kanter's theory, Stiehm explains that women who are skewers will automatically be perceived as less competent and less influential and will have a different power base within a group. This evaluation supports Izraeli's findings that as the number of women on a committee increases, women's influence as perceived by both women and men rises proportionately. Izraeli concludes, just as Stiehm does, that for women to be successful in improving their status in the workplace, they need proportional representation at all levels of power within business, industry, and government.[5] This conclusion also is substantiated by the findings of Lorna Marsden.

WOMEN IN THE NAVY WORK FORCE

The U.S. Department of the Navy work force comprises over 1,000,000 individuals, which includes approximately 500,000 Navy military, 200,000 Marine Corps military, and 300,000 federal employees. During the past 14 years, the percentage of women in the naval labor force has increased from 1% to 10% and approximately 65% of the total number are in civil service occupations. In his paper, "Women in the U.S. Department of the Navy Work Force," Richard Niehaus provides this information and states that the vast majority of women in the Navy work force are concentrated in civil service jobs which do not have a high technological content. Thus, Niehaus identifies one of the major problems involved in integrating women into all aspects of the military: The Navy is highly technologically oriented. For example, over 75% of the top-level, higher-paying, civil service positions require a degree in engineering. According to Hornig, in 1978 only 6.7% of the baccalaureate degrees in engineering were awarded to women; consequently, there is an obvious shortage of qualified women to fill these jobs. Similar to women in trade unions, women in the U.S. military seem to be confronted with the problems of socialization and segregation into certain "women's jobs." Again, we see the need for increased occupational and technological education of women.

Integrating Women into the Navy

With the problem of a scarcity of technologically qualified women as his major premise, Niehaus contends that the Navy is making a concerted effort to expand the role and increase the number of women serving on active duty. Many of the changes implemented were brought about by a 1978 statutory amendment which provides for: (1) permanent assignment of women to noncombat ships, (2) temporary assignment of women for up to 6 months on any ship for which a combat mission is not anticipated, and (3) assignment of women to any aircraft not engaged in a combat mission. An outgrowth of this amendment is the Women in Ships program which is a large-scale commitment of the U.S. Navy to bring women into mainline jobs. This program has been effective in placing women in almost all shipboard jobs including those in Deck, Electronics, and Engineering/Hull—of course, women are still excluded from all combat-related specialties. The fact that this amendment and the resultant program have contributed greatly to expanding and increasing the role of women in the U.S. Navy emphasizes the extent to which legislation can be an important factor in women's increased labor force utilization.

Also contributing to women's participation in all branches of the military is the research that has been and is being conducted on the physical requirements for all occupational specialties. This research will serve as the basis for defining physical standards for these jobs and also may lead to the development of new methods and technologies for performing physically strenuous jobs, innovations that will be beneficial to both men and women.[6]

ECONOMIC ANALYSES: A CROSS-CULTURAL STUDY
OF WOMEN'S INCREASED LABOR FORCE PARTICIPATION

Whereas the preceding papers discuss variables related to women's participation in different sectors of the economy, each of the following papers analyzes economic and other factors affecting women in a specific country. Additionally, the following questions are raised: What effect does women's increased participation in labor market work have upon the economies of different countries? How has women's role in the economy evolved historically and what trends can be predicted in the future?

Women in Paid Employment: Reapportioning the Division of Labor

Barbara Reagan's chapter, "Microeconomic Effects of Women Entering the Labor Force," provides an informative and in-depth comparison between full-time homemaker families and those in which the wife works outside the home. By comparing such factors as income, savings behavior, spending behavior, hours worked in the home, and travel time to work, the author develops a comprehensive picture of the impact employed women have on family structure. Reagan's chapter contains detailed documentation of the double burden; for example, when a woman begins working full time in the labor force, the work she has performed at home is not, as a rule, divided equally between her and her husband or among family members. Instead, the wife continues to spend more than triple the time on household chores as her spouse. Despite this situation, Reagan appears hopeful that many families are adapting to their shared economic contributions and working toward an equal division of labor in the home as well.

Reagan is not equally optimistic about the outlook for the rate of economic growth, which, she predicts, will continue to increase slowly (this trend also is noted by Chapman and Chapman). The consequences of this minimal economic growth are that the advancement of women within the economy will remain slow and tenuous. Reagan's conclusion concerning the growth of the economy also is supported by Patricia Connelly in her chapter on women's role in the Canadian economy. A major thesis in Reagan's paper is the economic importance and necessity of two earners in most families; Connelly, in turn, investigates the historical evolution of the two-earner family in Canada.

In identifying the long-term effects of capitalism, industrialization, trade unionism, and the world wars upon the movement of women into the labor force, Connelly educes why women's work in Canada continues to have such a low monetary value. She maintains that there is still a basic assumption on the part of employers that women are working for "extras," although, historically, this has never been the case. This assumption, which also is refuted by Armstrong and Armstrong, allows employers to pay women less and to restrict women's opportunities for occupational advancement. Connelly emphasizes the importance of women's participation in market work by citing a national study which states that 51%

more husband-wife families would have incomes below the poverty line if the wives were not employed. Connelly's historical overview of women in trade unions also provides relevant evidence in support of Cook's statements about the inequitable, and often adverse treatment of women in unions. For example, Connelly explains that in the early trade union movement women were often excluded from membership because they were perceived as a threat to men's wages and even their jobs.

A somewhat similar historical perspective was presented by Florence Wilhelm-Rezende on "Women's Work and the Evolution of Family Structures." In her paper, she traces the movement of women into the French labor force from 1830 to the present and, similar to Connelly, she notes the impact of the Industrial Revolution and World Wars I and II on women's increased participation in the work force. After establishing this historical background, Wilhelm-Rezende focuses on the chronic and well-known difficulties that women have encountered in working outside the home while still residing in traditional families. A comparison of her main points with those presented by Connelly and Reagan shows the pervasiveness and similarity of the issues faced by employed women in different parts of the world. Not surprisingly, women in France also are confronting the daily realities of the double burden. Wilhelm-Rezende gives figures on the amount and type of household work performed by French men that correlate closely with figures provided by Reagan on the household division of labor in American families. For example, in two-earner French families, men perform slightly less than one-third of the total household chores and the types of jobs that they perform are still largely determined on the basis of sex.

Factors Affecting Women's Labor Force Participation in Greece

In his presentation, Stylianos Athanassiou examines several variables that influence women's labor force participation in Greece and analyzes the effect of this participation on the population and socioeconomic development of that country. Women's labor force participation in Greece is fairly low (20.9% in 1975) as compared to other NATO-allied countries. In fact, participation rates actually decreased from 1961 to 1971. Athanassiou attributes this decline in women's employment to three factors: the migration of workers to other countries, urbanization, and a relatively low population growth.

Although overall participation rates for women are low, the author identifies several variables that have a positive effect on increasing women's labor force participation. For example, Athanassiou's data indicate that Greek women with higher levels of education have an increased tendency to be employed outside the home, particularly in skilled occupations. However, the percentage distribution of educational levels compiled by Athanassiou indicates that higher education for women is not the norm in Greece. Athanassiou does not elaborate on this association between education and paid employment, but it seems probable that as more Greek women achieve higher levels of education, their participation rates will likewise increase. In his conclusion, the author states that women in paid employment have had a positive influence on the overall economy; therefore, an increase in the level of women's educational attainment could contribute, in part, to the continued economic growth of Greece.

Factors Influencing Women's Labor Force Decisions

The chapters of Michelle Riboud and Wolfgang Franz are economic analyses of labor supply and earnings, although the research pertains to the two contrasting environments of Andalusia and the Federal Republic of Germany. Comparison between the two is interesting because the Federal Republic of Germany is highly industrialized whereas Andalusia has a largely agrarian population.

One of the most interesting findings of Riboud's paper concerns the strong continuity of attachment to the labor force on the part of Andalusian married women: only 16.8% report having interrupted their working life once. However, only 19% of all Andalusian women participate in the labor force and 36% of married women have never even entered the labor market. This pattern is in direct contrast to that followed by the large majority of American women, most of whom are employed at some point in their lives but who have periods of nonparticipation occasioned by marriage, childbearing, child care, and migration. Wilhelm-Rezende indicates that these familial considerations also affect women's work patterns in France. The factors listed above apparently have little effect upon the labor force decisions of Andalusian women who work outside the home. However, in Andalusia, as in Greece, a woman's level of educational attainment has a marked effect on her probability of being employed outside the home, again emphasizing the importance of education as a correlate of women's increased labor force participation. According to Jacob Mincer,[7] the discussant for these symposium papers, Riboud's findings parallel similar studies in the United States, France, Great Britain, Sweden, and Canada.

The research of Franz indicates that, as in the United States and France, factors related to marriage and children are very important in the labor force decisions of West German women. Franz's theory explains women's labor force decisions in terms of a life cycle model. According to this theory, such variables as a woman's age, the number of people in her household, the types of jobs available to her, and the family's total income, all affect her decisions about how much and how often she will participate in the work force over her entire life cycle. From Franz's perspective, when women do work outside the home, it is mainly to increase family income and, consequently, consumption and leisure. This finding is consistent with the conclusion reached by both Reagan and Connelly: It has become a necessity for most families to have two earners in order to maintain an adequate level of living.

CONCLUSION

The factors that have the greatest impact on women's increased labor force participation are: public policies concerning women's education and legal rights, familial and life cycle factors affecting women's employment decisions, and aspects of women's self-perception resulting from their socialization.

From the papers discussed in this section, it can be concluded that the participation of women in paid employment has a positive impact in several different areas. In general, their participation increases a country's level of living, has a positive effect on the Gross National Product, and raises family income. Additionally, many of the jobs that women do are invaluable to society. As pointed out by Pintasilgo, women perform jobs related to our basic needs—such as value givers (teachers and child care attendants), health care dispensers (nurses and other health care professionals and aides), and food providers (agricultural, processing, and food service workers). Yet women's work continues to be undervalued and underpaid. Moreover, the participation of women in paid employment often has negative effects on the woman's leisure time, the personal care of her family, and even on her economic circumstances.

What factors can be changed in order to eliminate these negative effects? First, in order to lighten the double burden of paid employment and household tasks, child care services need to be greatly improved, more part-time work should be made available for both sexes, and there needs to be a general restructuring of the traditional division of houshold duties. These changes would help to release women from many of the personal and family-related concerns which currently constrain their labor force participation.

The second goal is to narrow the difference between the earnings of women and men. If men can be integrated into female-dominated occupations, it may become evident to larger segments of the population that many "women's jobs" are not unrewarding in terms of personal fulfillment, merely undervalued. However, at this stage, given the higher pay scale of "men's jobs," we can begin by increasing the number of women in the higher-paying skilled labor and professional occupations. Education, of course, is crucial in accomplishing this goal—the level and type of education that women receive now and in the future is directly linked to their remunerative value in the labor market.

But how optimistic can we allow ourselves to be regarding the educational, occupational, and economic advancement of women? The conclusions of both Reagan and Chapman and Chapman, for example, are equally pessimistic: The economy will continue to grow slowly which will result in fewer job opportunities and women will again bear the brunt of this slow rate of economic growth in the form of unemployment, under-employment, and poverty. When women face these economic disadvantages, they are, of course, not the only ones affected; if women can find only low-paying jobs or no jobs at all, their children and families are likewise forced to subsist at a lower level of living (a level that is often below the poverty line). Yet, as Athanassiou and Pintasilgo point out, women's labor force participation has a generally positive effect on a nation's economy.

The economic position of women, therefore, is not a special interest issue but a universal one. How then can this pattern of ignoring the economic problems and realities of women continue without resolution? Education of the general public about the facts of this issue coupled with the formulation of specific strategies and policies, such as those brought out in several of the papers are two key steps toward bringing about the changes in the workplace, government, and family structure that are so essential, not only for women and their families, but also for the growth and well-being of the world's economy.

NOTES

1. This paper is a shortened version of a recently published book by the same title.
2. Hoiberg, 1980a.
3. United Nations, 1979:13-15.
4. Kanter, 1977a.
5. For an additional discussion of the specifics of power expansion within middle management, see Izraeli, 1975.
6. For an informative survey of much of this research, see Hoiberg, 1980b.
7. See, for example, Mincer and Ofek, 1979.

MICROECONOMIC EFFECTS OF

WOMEN ENTERING THE LABOR FORCE

Barbara B. Reagan

Southern Methodist University
Dallas, Texas   U.S.A.

In the United States, increasing numbers of women, especially those with children, now are choosing to contribute financially to their families by earning money in paid employment.[1] The basis for this decision is that they can take employment hours and contribute more from net cash earnings than by staying at home and producing additional products and services with those hours.   Their income helps with family bills and enhances the family's level of living.   It also adds to the real national income by increasing the total quantity of goods and services produced in our economy.   In the 1970s, real per capita income after taxes rose more than 25%.[2] As the size of the average family declines, even constant income would mean more per capita.   If women had not moved into the labor force, our national level of living might have fallen sharply, instead of increasing.

Not all women see jobs away from home as increasing their contributions to their families.   If the kinds of jobs they can qualify for pay less than the value of what they can produce at home with those hours, it is rational for them to stay at home.   However, the significant point here is that women should be free to choose between working in the market economy, without sexist barriers, or staying at home.   Both choices—to seek paid employment or to work in the nonmarket economy of home--are valid.

The transition period as women in large numbers move into the labor market is often disruptive to the family with a working wife.   The wife and her family have to deal with both workplace institutions and work roles of men that developed when mothers usually did not work outside the home.   It is also disruptive to the once-idealized family with a full-time homemaker.   The family with this pattern may feel under attack and find its income and assets falling in relation to two-earner families of comparable age and education.[3]

The effects of the wife's labor force participation on the family's economic situation are discussed in this paper. I shall begin, however, with the need for economists to consider the changing nature of the household and the family.

THE FAMILY AND ECONOMIC MODELS

In traditional neoclassical models in economics, the decision-making unit is assumed to be the household (economic family). Until 1974 or so, most work on labor supply was based on the conventional static family without consideration as to how decisions were made within the unit or how changing family composition affected the results. Changes in consumption, labor supplied to the market, or leisure occurring in response to small changes in one's own wages, other family members' wages, prices, or property income were observed.

Recently, efforts have been made to extend the conventional model explicitly to cover costs of working, such as commuting costs, which are seen to be more important in women's labor supply; joint labor supply models have modeled the husband's and the wife's utility functions separately and jointly estimated the parameters of the husband's and the wife's decisions; or dynamic aspects have been incorporated in life cycle models of labor supply of husbands and wives. Uncertainty is not yet incorporated in most models, although such work is on the horizon. Even the dynamic models, however, tend to be tested with data limited to husband-wife families each with the same two spouses over the extended period.

The sharp expansion since 1974 in studies by economists on labor supply work is largely the result of a growing awareness of the inadequacies in the former simplistic models as women have become more visible as workers and the increasing need to improve estimates of the labor supply of women. Many issues are unresolved in the theory of, and econometric measurement of, labor supply particularly related to women. A major unresolved issue is how the family unit is to be defined so as to reflect changes in composition.

For years, consumption and market analysis assumed that the household unit was the traditional family of an employed husband, a full-time homemaking wife, and children under 18. This picture is unrealistic today given actual living patterns. Such a traditional family type represents less than 16% of the economic decision-making units. This low percentage does not mean that women's working per se increases divorce. In the past 50 years, divorce rates and living standards have increased together.[4] The divorce rate is now between 30% and 40%. It is predicted that about 50% of the marriages of the 1970s will eventually end in divorce.[5] When women have economic alternatives for jobs in paid labor markets, divorce from unsatisfactory marriages is facilitated.

The volatility in the American family composition does not mean that the family is of less importance today as an institution, but it does mean an increase in family re-formation over time. The changing composition of families, along with the increase in two-earner families and the increase in families with women as the sole earner, make inappropriate the former assumptions underlying analytical paradigms in research relating to American women.[6]

Watts and Skidmore[7] are among the economists who have called for an end to the use of the stereotypical family and recommended a new tabulation of data on individuals already being collected by the U.S. Bureau of the Census and other agencies. The proposed tabulation would determine the diverse composition of family relationships and the way the functional organization of the families changes from one time period to another. Watts and Skidmore's classification includes seven household units: single persons, couples, parent-child units subdivided into one-parent and two-parent units, other household units that include children[8] or other dependents, related adult units, nonfamilial adult groups, and institutions. This scheme will cover virtually every person. There is no presumption that the units will be permanent; a person can change household status numerous times during a lifetime.[9]

Labor supply and marital status need to be treated as jointly determined choice variables. The marital status variable might be most useful if it follows a typology that allows for family re-formation over time.

WOMEN'S WILLINGNESS TO DO MARKET WORK

A typical woman enters the labor force by working slightly more than half time, and higher wages induce her to work somewhat more hours, an estimated wage elasticity of .79.[10] The relatively high wage elasticities of married women's estimated labor supply functions (an example is Heckman's 1.23 from his 1974 paper on shadow prices)[11] are caused by a variation in work schedules of women entering the labor force, not from the variability in hours worked among already employed women.[12]

Cogan's work is based on a model that allows for an initial discontinuity in the married woman's labor supply function caused by costs of work.[13] The estimated costs of work are large, about $1,000 a year or 16% of her average earnings.[14] The costs of work mean that the woman either does not work or works at least enough to cover both the costs of work and the value of her lost time at home (i.e., the reservation wage). The reservation hours of work estimated by Cogan are 1,150 or slightly more than half-time work.

Cogan found that preschool children are the most expensive source of work costs, with each preschooler adding nearly $400. Children also increase the time costs of work, particularly younger children. Such time costs could be search costs for adequate day care. High-school-age children decrease the costs of the wife's work by about $50 each, suggesting that they contribute to household production such as the care of younger children.[15]

Hanock[16] found that married white women who have more than 8 years of education tend to reduce the time they work in the market according to the number of their children whereas those with less education increase their labor supplied with each additional child. He concludes that this negative interaction indicates that education has a positive effect on home productivity in child rearing and that educated women place a high value on the quality of their children's upbringing.

In general, while researchers agree that women's labor supply function is upward rising, they do not concur on women's labor market behavior and the magnitude of their wage elasticity.[17] They agree that children, particularly very young ones, deter a woman's labor force participation; a wife decreases her amount of work as the husband's earnings increase, as does her husband to a lesser extent when her wages go up; and the wife decreases her labor market work as her age and education increase. The wife (or the husband) decreases the amount of work supplied to the market as the family's property income (nonwage income) increases.

For the two areas of little agreement, it does seem clearer now that the labor supply curve for women is not as responsive to wage changes as previously thought, but it is still thought to be less wage inelastic than men's. It now seems likely that some women are on the backward bending portion of the supply curve: a small proportion of wives with higher wages and no children at home and a larger proportion of wives with children at home. According to earlier work, most husbands are on the backward bending portion of the supply curve, which in recent work has been evidenced especially among husbands with children as contrasted with childless husbands.[18]

On the basis of Mincer's work,[19] it has been widely believed that women tend to work primarily to smooth out dips in the family's income, such as during a husband's period of unemployment. Recent work, however, disproves this. Heckman and MaCurdy,[20] for instance, report no evidence that transitory variability in income affects women's supply of labor to the market at any point in time.

One indication of the permanence of women's labor force participation has been the dramatic decline in the probability of exit from the labor force for both full- and part-time female workers. Thus, the growth in the female labor force has been associated with a drop in the exit rate.[21] Quit rates for men and women are very similar for the same job. Among all ages, the proportion of quits to exit from the labor force are larger for women whereas those for quits to move are larger for men. Younger and older women workers, especially the 20- to 24-year-olds, have higher quit rates than the middle aged.[22]

Not too surprising, but new in quantified results, are the econometric results which indicate that the wife's health and the husband's attitudes affect her willingness to do market work.[23]

The fact that estimates of women's labor supply elasticities are decreasing, as models are not only improved but also tested with data from the 1970s instead of the 1960s, may suggest that the attitudes of women and men have shifted toward a greater acceptance at home of employed wives and their permanent attachment to the labor force. If so, this is capturing a mid-transition change, and we can expect more to come.

OCCUPATIONAL CROWDING AND EARNING EFFECTS

Occupational segregation by sex characterizes the American labor market.   Women are segregated and crowded into certain occupations, leading to lower wages in these female jobs.   More than 80% of the female work force is employed in clerical, nursing, teaching (not college or university), finance, insurance, real estate, and service jobs, each of which is more than 50% female.   Clerical occupations alone employ 40% of the female work force.[24]   These occupations are, for the most part, low paying and provide few opportunities for advancement.[25]

The media's emphasis on a few upwardly mobile women executives or women entering male-dominated fields gives the false impression that women recently have made considerable progress.   In fact, institutional barriers to equal opportunity for women remain in effect.   One barrier is the common practice of using different job titles or classifications for similar work performed by men and women in the same firm.   Barriers are often subtle, invoking the invisibility of women, inequitable on-the-job training programs, a denial of access to informal networks that can play a major role in career building, and an unwillingness of male superiors to serve in a mentor role as readily for women as for men.

Many jobs traditionally considered appropriate for women have little or no training opportunities and no career ladders (e.g., health service workers, elementary school teachers, secretaries).   This structuring of jobs was based on the now out-moded idea that women would be in and out of the labor force so much that investment in on-the-job training would not pay employers.

Women working full time, year round in 1976 on the average earned only from 57% to 61% of the wage of comparably educated men.[26] This wage differential is attributable to: a disproportionately high number of women at starting levels, occupational segregation by sex, and unequal pay for work of equal value. On the other hand, the decline in exit rates has meant increases in the average years of experience for women, which should be reflected in a narrowing of the wage gap between men and women. In 1976, however, women with 4 or more years of college who worked full time, year round had a median income of $10,520, which was only a little higher than the median income for male high school dropouts, $10,040, and was below that of the average male high school graduate, $12,260.[27]

Full-time women workers do not experience the same career paths as their male counterparts. Women gain very little in earnings throughout the life cycle.[28] The pattern of male earnings, on the other hand, is to rise sharply when men are in their mid 20s and 30s to a level whereby men aged 25 to 45 years are earning twice the median income of the 18- to 24-year group.

INCOME AND ASSETS

In 1975, those wives employed full time with husbands present contributed 60% of the household income for families who earned less than $10,000 and 45% for families with incomes from $10,000 to $15,000.[29] Clearly, the husband-wife family with a wife who works has a larger total income than if she does not work. However, that family type averages less income from the husband's earnings or from property income than the family with a full-time homemaker. Strober,[30] using 1968 data from the Michigan Survey Center, concludes that wives' earnings tend, on the average, to raise the income of the two-earner family up to the level of those families in the same life cycle group where the wife is not employed. In a similar study, Vickery[31] reports that in 1972 families with working wives had $2,600 more income than families with full-time homemakers. For a given income group of two-earner families, however, estimated savings and market value of financial assets as well as the net change in assets minus liabilities for the year are generally not as high as for families with full-time homemakers. The values are presented in Table 1. Two-earner families also had lower rates of home ownership and lower house values. In part, this finding may be a function of the fact that credit institutions in 1971 to 1972 still were not recognizing women's earnings in computing mortgage loan capability.

It can be argued that two-earner families have a lower propensity to save than traditional families because employed wives have recent job experience and because employed wives' attachment to the labor force is now relatively permanent and families feel less compelled to hedge against unemployment. Strober[32] also has shown that two-earner families save less and have fewer assets than similarly situated families with full-time homemakers.[33] Strober's ratio of current consumption to income is higher (and thus savings to income is lower) for families with working wives than for husband-homemaker families in each age group, holding income constant. Overall, the ratio of savings to income after income taxes is .26 for two-earner families and .46 for husband-homemaker families, after controlling for family income and composition.[34]

Strober[35] also reanalyzed the 1972-1973 Consumer Expenditure Survey data, which was used by Vickery, to ascertain whether the lower financial assets previously found for the two-earner family would still be

TABLE 1

INCOME AND ASSETS OF WORKING-WIFE AND HOMEMAKER FAMILIES BY INCOME GROUP, 1972

Four-Person, Husband-Wife Families with Husband Age 25-64	Income Group		
	Lower Middle	Middle	Upper Middle
Working-Wife Families			
Income after income taxes	$7,908	$12,164	$18,194
Financial assets	1,271	2,512	5,852
Net Δ in assets-liabilities	882	231	985
% of after-tax income	11	2	5
Homemaker Families			
Income after income taxes	7,932	11,560	17,555
Financial assets	2,139	3,557	8,854
Net Δ in assets-liabilities	890	780	2,925
% of after-tax income	11	7	17

SOURCE: Vickery, Table 24, pp. 180-81, 1979. Based on tabulations of 1972-73 Consumer Expenditure Survey.

NOTE: Vickery concludes that these figures are representative of averages for husband-wife families of other sizes as well as four-person families. Wife was classified as "employed" if she worked at least 14 weeks during the year or earned at least $900.

lower than that of the husband-homemaker family when life cycle stages as well as income are controlled. Two-earner family assets again are lower, and the greater the wife's earnings in comparison with the husband's earnings, the lower are both the average value of financial assets and the estimated market values of their houses.

Lazear and Michael[36] question whether the approximately 20% higher cash income of a two-earner family than a husband-homemaker family means a higher level of living. The family's level of living is the market value of the bundle of service flows consumed by the family, i.e., the market goods and services plus an allowance for family technology that transfers market goods into service flows.[37] The rough estimates in their study are limited to renters aged 35 years or less who have no children. Lazear and Michael report that the average two-earner family has to earn 30% more money income with the wife working than before to achieve the same level of living as the one-earner family. As they note, this is a research topic that needs further study.

Asset and saving behavior needs to be viewed in terms of multi-periods over the life cycle, rather than just a single point in time, before it can be related to hours of work. Smith[38] has developed such a model and extended it to cover the two-earner family. Wealth at any point is a function of past savings behavior. Savings are affected by the number and age of children, not through their effect on consumption expenditures, but through their effect on labor force behavior. Children of any age increase the husbands' hours of work, younger children decrease the wives' hours of work, and older children increase the wives' hours of work. Women's wages had no apparent effect on the saving or the consumption functions.

EXPENDITURES AND RESIDENTIAL LOCATION

Current Expenditures

Holding total income constant will reveal whether there is a taste or preference effect that accounts for differences in expenditures between the two-earner family and the husband-homemaker family. Strober[39] proposes four types of theoretical arguments for a taste effect of wives' employment.

Time-pressure theory. If the two-earner family places a higher marginal utility on time-saving goods and services, then they would spend more on such items than the husband-wife family with a full-time homemaker.

Transitory-income theory. If families consider that wives' working is temporary, wives' earnings have a large transitory component. We might expect that the two-earner family would save more and spend more on durable goods, which are considered savings in the permanent income theory.[40]

Higher work-related expenses theory. The work-related expenditures of the two-earner family will be higher than that of the single-earner family; these include both the direct expenses such as transportation, Social Security taxes, and suitable clothing as well as indirect ones such as buying more meals away from home, more child care, and more household and clothing care. The indirect costs of working result from substituting market services for those provided at home by the full-time homemaker.

Lower propensity to save. Two-earner families do not feel as great a need as other families to hedge against job loss and, thus, as a specific part of their life plan, save less than families with full-time homemakers at the same income level and life cycle stage. Alternatively, if families view savings as a residual item in their plan, it can be argued that after deducting the work-related expenses when the wife works, the two-earner family has less discretionary income available as a base for either savings or non-work-related consumption expenditures even though the two types of families have the same inherent propensity to save.

Empirical findings support the latter two theories, but not the first two.[41] Apparently, both types of families discussed here enjoy time-saving household services and spend about the same amount on them. The transitory income effect may not be applicable any more because women are now more strongly attached to the labor force, as discussed above. However, the data are consistent with the higher work-related expense theory and the lower propensity to save theory. We have already seen that the savings of two-earner families are lower. The empirical findings related to current expenditures are summarized below.

Higher work-related expenses. Holding family composition and income constant in regression analysis,[42] husband-wife families with wives who worked full time (here called a two-earner family) spent 21% or about $690 more a year (1972 dollars) than those families with full-time homemakers. Of this increase, a total of about $280 was for transportation, $210 for Social Security, and $100 for clothes.[43] Differences in child care expenses between the two family types were only about $5 more a year for the two-earner family.[44] The increases in these work-related expenses were partially offset by a lower amount spent on housing, approximately $70 less. The lower housing expenditures probably are associated with differences in residential location preferences.[45] In the average two-earner family, the wife's income taxes were estimated to be $950 on her earnings of $4,800. Thus, her income taxes and work expenses were a third of her before-tax pay.[46]

If the full-time homemaker family has an increase of $5,000 income from a pay raise for the husband to a total family income of about $14,500 (1972 dollars), then the expenditures would be increased primarily for the categories of shelter and personal care plus smaller increases for recreation, foods, gifts, and contributions.[47] In contrast, when an employed wife adds $5,000 to the family income (total income of about $14,500), the above categories are allocated only small increases. Instead, Social Security and retirement payments are at least four times higher than increases from the husband's $5,000 increment in pay. The two-earner family also increases expenditures by 70% to 85% more than the husband-homemaker family for transportation, dry cleaning, and clothing repair as well as 50% more for clothing. Again, these figures show that the major changes in family expenditures are work-related when the wife is employed. In general, increased expenditures account for nearly 60% of the incremental pay gained by the wife working; the rest is used for income taxes, savings, or debt retirement. The overall expenditures from a $5,000 increase in pay for the traditional family are only 30%, leaving a much larger amount for savings or debt retirement. Clearly, the increase in the Social Security payments (on benefits that average over $440 per month for this analysis) is taxed at a far higher rate for Social Security than an additional $5,000 paid to a husband because more of his additional $5,000 is beyond the taxable base.[48]

Durable goods and time-saving services. Families do not treat the wife's earnings differently from the husband's in spending disposable income after income taxes and work-related expenses are covered.[49] Adjustments to save the employed wife's time are small for food expenditures (e.g., eating at home versus eating out) and expenditures for care of clothing. Spending for furniture, house furnishings, appliances, and health care are not increased when the wife works.

Using 1968 urban data from the Michigan Survey of Consumer Finances and the 1977 Market Facts Consumer Mail Panel made available by Needham, Harper, and Steers, it has been shown by Strober[50] that, when income and other variables are held constant, there are no significant differences in the decision to buy or the amount of family income expended for durable goods whether the wife is a labor force employee or a full-time homemaker. The purchase of durable goods such as refrigerators, stoves, televisions, and furniture is affected by income, family life cycle stage, and whether the family moved recently, not whether the wife is employed.[51]

Residential Location

Employed women who are single parents and unmarried women who live alone and who tend to have low earnings will live closer to the city center and commute shorter distances than traditional families or two-earner, husband-wife families.[52] Higher-income families, on the other hand, buy more housing and find the lower-cost land in the suburbs more attractive than do lower-income families; their commuting costs are also higher because their travel time is more valuable. Lower-income families outbid higher-income families for centrally located land.[53] Differences in residential location are associated with income disparities rather than with family type.[54]

Madden[55] concludes that there is no difference in housing location (or size or quality) between two-earner, husband-wife families and husband-homemaker families after the effects of fertility, income, and job characteristics are explained. Married women who are employed in the market and who have children tend to live and work in more suburban areas than other earners.

SEXUAL DIVISION OF HOUSEHOLD LABOR

Examination of the expenditure patterns reveals that the working-wife family does not purchase market substitutes for services normally produced at home. According to the Walker study of time use in 1967 to 1968, sharing of household work by husbands and children is slight whether or not the wife is employed outside the home. Husbands who work less than 40 hours a week, work 15 hours a week at home; those who work more than 40 hours a week in the market, work 8 hours a week at home. Similarly, 6- to 11-year-old children work 4 hours a week at home; teenagers average 8 hours a week at home.[56]

When the wife is employed in the work force, there is a reduction in the total amount of time she spends on housework. For example, a working wife with a husband and two children works 30% less time (-17 hours) than her full-time homemaker counterpart. If the youngest of the two children is between 2 and 11 years old, then the working wife cuts her housework time by only 12 hours a week or by 25%.[57] The housework activities most reduced are physical and nonphysical care of family members. Some of the physical care may have been assumed by the children themselves since the Walker study covered the 1960s when most of the children of employed mothers were older than they would be if the study were repeated today.

Marketing and management time is the same whether or not the wife is employed. Wage-earning wives reduce somewhat the time needed to perform other household tasks, but overall they lengthen their total work time in both market and home. The employed wife's total work time averages 71 hours a week (37 hours away and 34 hours at home) or about 30% more than her husband's 55 hours (44 hours away and 11 at home). The full-time homemaker works about 57 hours a week while her husband averages 55 hours on the job (including travel time) plus another 11 hours in work around the house and yard, or about 16% more than his full-time homemaker wife.

The division of time spent in housework by two-earner urban families in 1971 is estimated by Wales and Woodland[58] with a model in which housework time is a function of a change in the husband's wage, a change in the wife's wage, and income. Husband-wife families with and without children at home are considered separately. Employed wives invest $3\frac{1}{4}$ times as much time in home production and maintenance as their husbands. In families without children, wives average about 19 hours a week; with children, wives average 24 hours a week doing housework. Children at home add about one-fourth to the time input of husbands as well as that of wives.[59]

From the correlation of residuals, Wales and Woodland conclude that unanticipated increases in expenditures will be met by both the husband and wife increasing their work hours with little change in housework hours, but with a cutback in their leisure hours. Unusually long hours of work by the husbands are associated very closely with their reduced leisure ($r = -.97$) and not closely with their reduced housework ($r = -.26$). For wives, unusually long hours of work are associated more weakly with reduced leisure ($r = -.74$) and more strongly with reduced housework ($r = -.32$) than is true for their husbands.[60] These findings suggest that men have their hours of home production and maintenance work trimmed to what they see as an irreducible minimum.

Results of a study of families in 1976[61] show that the average employed wife still spends twice as much time in housework as the average husband. The division of labor between husbands and wives appears to be less specialized in 1976 than was noted in a similar survey in 1965. In 1976, married men were employed about 4 fewer hours a week, and worked another hour or more a week around the house than reported in 1965.[62] The information available suggests that the employed wife has cut her housework time by about 7 hours a week since 1965.

Clearly, the uneven time in total work between husbands and wives in the two-earner family will continue to cause strain until both men's and women's work roles adjust. Some see the rise of the two-earner family as an early stage of the breakdown of the sexual division of labor, a process from which there is no turning back.[63] So far, the entry of women into the labor force has increased family income, the wife's economic independence, and her security, but at the expense of the wife's leisure and her personal care of family members.

IMPLICATIONS FOR LENGTH OF WORK TRIP

It is well established that women have shorter work trips than men, either measured in time or distance. The shorter commuter trip for women probably is related to workplace factors—wages, tenure in job, length of work week, and the location of jobs typically considered women's jobs—rather than to the residential land market.[64] However, it is difficult to disentangle the interactions because a comprehensive model to simultaneously clear spatial labor markets and urban land markets has not been developed.

In competitive markets where the wage differentials just compensate for commuting cost differentials, short-distance commuters, who are primarily women, are not disadvantaged. Wages net of commuting costs are equalized. On the other hand, if the local employers have monopsony power, then short-distance commuters have wages net of commuting costs that are lower by the monopsonistic exploitation differential.[65] That differential will increase as the spatial labor supply is more inelastic.

Employed women have a larger housework responsibility than employed men, as we have seen. Madden[66] concludes that in spite of this, the family is more likely to select a residential location conditional on the husband's job location rather than the wife's. There are several considerations that may explain why women look for work closer to home than men. Sex differences in household roles may push women to seek work near the home. The value of travel time is higher for women than for men because household responsibilities increase the cost of commuting. Women also choose jobs closer to their residences because their lower wages and shorter hours reduce the earnings' return to commuting. Further, women have different job opportunities than men (e.g., more clerical and service jobs), and their employers may have been better able to locate near the women's residences than the men's employers.

If housing is predetermined, then a woman would choose the length of work trip in order to ensure that the marginal change in length of trip to work would equal the marginal change in wages. A worker with a more isolated residence (which is correlated with lower housing price and greater quality of housing) may find few work opportunities nearby and greater wage returns to commuting farther. The higher wage compensates for the increased travel cost.

If the job location is predetermined, the housing location and length of work trip would be chosen so that the marginal change in the length of the work trip would equal the marginal change in housing expenditures.

Wages

Madden[67] found that of the factors considered in 1976 the most important determinant of length of work trip is wages. Both length of work trip and wages are measured in logarithms and the relationship is positive as would be expected whether the labor market is competitive or monopsonistic.

Women heading families with children adjust the length of their work trip in response to higher wages ($\eta = 2.24$) about as much as the most wage responsive group which is composed of men in two-earner households with children ($\eta = 2.33$). Women are as responsive to wages in length of trip as their husbands in two-earner couples without children ($\eta = 1.62$ and 1.64).

Women earners in other family types, such as two-earner households with children and single-person households, are less responsive in adjusting length of work trip to wages than are the men in those same family types (two earners with children: 1.85 and 2.33; singles: 1.55 and 1.90, respectively). Women earners also are less responsive in adjusting length of work trip to wages than are men in traditional families with non-employed wives where the wage coefficients are 2.15 for traditional families with children and 1.70 for traditional couples without children.[68]

Tenure

All workers have a tendency to remain longer on jobs located closer to their homes, but this effect is more pronounced for women.[69]

Hours in the Work Week

The duration of the work week is also positively related to length of the work trip for women; that is, women will travel somewhat farther for more work. When children are present in two-earner families, an increase in the husband's hours of employment per week decreases the work trip of both spouses, but an increment in the wife's work week increases the work trip of both. When children are not present, there are little or no such effects. The results suggest that two-earner households with children select residences with reference to the husband's job and husbands who work longer hours select residences closer to their work, while wives reduce their own commuting time and hours of work to make more time for family responsibilities. On the other hand, those wives in this group who take jobs with longer hours select jobs farther from home because the wage returns per trip are greater, and such wives have more impact on the residence decision, thus increasing the travel trips for both husband and wife. Men who average the longest trip to work and women who make the shortest are in two-earner families with children.[70]

CONCLUSION

Two-earner families in all stages of the family life cycle are finding new economic security in two pay checks. The effects of a wife working in the labor market on the family economic status are summarized as follows, comparing the two-earner, husband-wife family with the traditional one-earner, husband-wife family.

A wife's employment increases:

1. Family income by two-thirds of the wife's gross pay (income after income taxes is increased by 85% or less of her pay after income taxes because of work-related expenses);

2. Economic independence and future security of the wife who works;

3. Family spending on insurance and retirement (including Social Security), transportation, and clothing, which are primarily expenses related to the wife's working and are greater than would occur with a comparable increase in the husband's income;

4. Family net change in assets and liabilities, although not as much as would be expected given the higher income; and

5. The family's dependence on the market for goods and services thereby continuing the long-term trend of increased industrialization. This dependency makes the two-earner family somewhat more susceptible to business cycle swings, and makes it rely more on government actions such as fiscal and monetary policies and/or employment compensation or job programs.[71] On the other hand, two earners are less likely to be unemployed at the same time and, thus, are less sensitive to business cycle swings.

A wife's employment decreases:

1. Her leisure time (even though her length of commuter time to work is shorter than other workers) because she continues to do the major part of the housework while increasing her time in the marketplace;

2. Her personalized care of family members which means less care of elderly dependents at home as well as less care of members of the nuclear family; and

3. Housing expenditures (rent, utilities, etc.), the proportion owning houses, and value of the house, when compared with the average for

others in the new higher income group.

A wife's employment has little effect on:

1. Location of housing; and

2. Expenditures of the family for current consumption other than those changes brought about by the higher income. It is especially noteworthy that consumer durable purchases are not increased.

Unanticipated increases in expenditures will be met in two-earner families by both the husband and wife increasing their hours in market work and reducing their leisure time as well as the household work hours of some wives, but not those for men.

Although inflation has accelerated movement of women into the labor force, it seems likely that a slowdown in the inflation rate will not send hordes of women back to their kitchens like lemmings to the sea. Greater attachment to the labor force and career commitment are new and lasting patterns for most women.

Current predictions for the 1980s in the United States are that:

1. Two out of three of the new entrants into the paid labor force will be women;

2. Educational levels of women applicants will remain high;

3. Occupational segregation by sex may break down somewhat under this influx of women;

4. Those in the age cohort of the last baby boom, which peaked in 1955, will have to face the most competition in job opportunities;

5. Inflation, although not rampant, will continue; and

6. Economic growth will remain at a low rate of about 1% making advancement harder for women than it would be at a higher level of economic growth.

From this picture and from reviewing the recent past, it can be predicted that barriers to the advancement of women in top management jobs undoubtedly will continue to be subtle and, unfortunately, often effective. Equality of opportunity and of returns to investment in human capital will increasingly become the goals of both employed women and their husbands. Thus, two-earner families will be among those who are at the beginning of their careers and their family life cycles are forging the new patterns of family life and shared parenting to go with the new patterns of shared economic responsibilities.

NOTES

The author wishes to acknowledge with gratitude Patricia Kirby Cantrell for research assistance.

1. Between 1965 and 1976 the number of women working or looking for work increased by more than 12.2 million while the number of men in the labor force increased by 8.1 million. Both spouses were employed in 1976 in nearly half of the husband-wife families (U.S. Department of Labor, Bureau of Labor Statistics, 1977, pp. 5 and 37).
2. As more women enter the labor force and buy more services in the market instead of producing them at home, an unreal increase is recorded in the national accounts.
3. Vickery, 1979:159-160.
4. Hofferth and Moore (1979) find no evidence of this, but do find evidence that employment leads to an increase in the age of first marriage, for many women to over age 30.
5. Barrett, 1979a:9.
6. Reagan, 1979a:133.
7. Watts and Skidmore, 1979:63.
8. Persons under 18 years old are classified as children unless they are parents, members of couples, living alone, or living in an institution.
9. There is no inherent need in this classification to designate a "head of household."
10. Cogan, 1980:350.
11. This is the maximum likelihood (Tobit) wage elasticity, corrected for censoring bias when data cover only women who are employed instead of all women, which has been computed from data in Tables 1 and 3 by Heckman, Killingsworth, and MaCurdy, 1979:IV-Table 1.
12. Supply of work ideally would include not only labor force participation and hours of work, but also intensity and quality of work.
13. The model is tested using data from the 1976 Michigan Panel of Income Dynamics for white, nonfarm, husband-wife families in which the wife is not retired, in school, or disabled. About two-thirds of the wives covered reported some market work in 1975 and one-third reported no market work.
14. Cogan, 1980:354.
15. Ibid.:Table 7.5.
16. Hanock, 1980:297.
17. Variations in the coefficients of wage elasticity of women's labor supply (using annual hours of work) for joint labor supply models are reviewed by McElroy in Heckman, Killingsworth, and MaCurdy, 1979.
18. Wales and Woodland, 1977:125. Data are from 1960 to 1961.
19. Mincer, 1962.
20. Heckman and MaCurdy, 1980:16.
21. Barrett, 1979a:Table 2, quoting Len and Bednarzik, Monthly Labor Review, October, 1978.

22. Barnes and Jones, 1974:445.

23. Brown and Manser, 1978:27.

24. U.S. Department of Labor, Bureau of Labor Statistics, 1977:8-9.

25. For summaries of occupational segregation by sex, see Barrett, 1979b and Reagan, 1979b. For more detailed discussion, see Blaxall and Reagan, 1976.

26. U.S. Department of Labor, Department of Labor Statistics, 1977:16.

27. Ibid.:36.

28. U.S. Bureau of the Census, 1976b:47.

29. U.S. Department of Labor, Bureau of Labor Statistics, 1977:38.

30. Strober, 1977:417.

31. Vickery, 1979.

32. Strober, 1977:Table 1.

33. Less than one-fourth of the income group, and less than 5% of all the families studied.

34. Data from Michigan Survey Center for 1968, husbands aged 25 to 64.

35. Strober, 1979:14-15.

36. Lazear and Michael, 1980.

37. Lazear and Michael (Ibid.) call this "standard of living" which more appropriately means desired goal; level of living connotes the level actually achieved.

38. Smith, 1980.

39. Strober, 1979:2-4.

40. Mincer, 1960a, 1960b.

41. Strober, 1979:5-15.

42. Husband under 65, data from 1972-1973 Consumer Expenditure Survey.

43. Strober (1977:413) found that two-earner families have more cars and spend more on them. Also, in most cases under current legislation, the wife's contribution to Social Security and that of her employer on her behalf have little or no effect on her or her family's benefits. The Social Security system was designed to protect the traditional single-earner, husband-wife family with children. Women who work often have little increase in retirement benefits from working and paying taxes than they are due as dependent spouses. Thus, the two-earner family is likely to have paid more Social Security taxes and the survivor actually receives lower benefits than if one spouse had remained at home, out of the labor force.

44. Average cost of child care for a preschool child would, of course, be higher per family with a child that age. Full-time homemakers and employed women both send their children to nursery school and have baby-sitters.

45. Madden and White, 1980.

46. Vickery, 1979:Table 25, p. 184.

47. Ibid.:Table 26, p. 186.

48. More recently, of course, the taxable base has been increased by Congress.

49. Because many differences in expenditures are related to family age and sex composition, this finding is based on regressions that

took into account number and age group of children, life cycle stage, family assets, and after-tax income (Vickery, 1979:179-181.)

50. Strober, 1977, 1979.

51. In Strober's view (1979:4), families purchase durable goods for their consumption with the saving attributes being considered only incidentally by consumers. The only exceptions are that wife's employment is related (positively) to purchase of black and white TVs, and wife's entry into the labor force in the past year has a positive effect on the purchase of furniture using the 1977 Market Facts data (Strober, 1979:7). In these comparisons, level of net assets, income, life cycle stage, and whether the family had changed residence in the last 2 years are held constant.

52. Madden and White, 1980.

53. Mills, 1972, as quoted in Madden and White, 1980:13.

54. The only exception is that in the lowest income group in 1977 (family income < $12,000 per year), two-earner families used frozen TV dinners more frequently than husband-wife families with full-time homemakers.

55. Madden, 1980.

56. Vickery, 1979:188.

57. Ibid.:189-190.

58. Wales and Woodland, 1977.

59. Ibid.

60. Ibid.:129.

61. Vickery, 1979:197.

62. Ibid.:Table 27.

63. Matthaei, 1980:202.

64. Madden, forthcoming.

65. Madden and White, 1980:10. The same can be said of length of work trip. Madden and White say sex of family head instead of "family type."

66. Madden, forthcoming.

67. Ibid. Also, Madden and White (1980:10) note that A. Rees (1968) and others found evidence of systematic spatial wage gradients in the Chicago labor market in 1963. For blue-collar workers, they found "a regional wage gradient which peaks in the southeast and falls off toward the northwest section of the city. Labor markets for the two primarily female clerical occupations studied—keypunch operators and typists—were found to be of smaller geographic size. In fact, they were immediate neighborhood markets. Employers actually expressed preferences that female employees live nearby and some even had rigid hiring standards related to distance for females. [This was prior to the Civil Rights Act of 1964 which made such procedures illegal.] As a result, women experienced very little return from increasing commuting. . . . These results are consistent with a spatial monopsony labor market for female workers and a competitive market for male workers."

68. Madden, forthcoming: Tables 3 and 4.

69. Ibid.:12.
70. Ibid.:7.
71. If women select jobs from a smaller geographic area due either to
     their commuting preference and/or residential immobility, fewer
     employers compete in hiring them.   Madden has noted the
     possibility of monopsonistic or oligopolistic exploitation of
     women employees contributing to the fact that, overall, women's
     wages are lower than those for men. (It would not contribute to
     differences within a firm.)

WOMEN WORKERS AND THE FAMILY WAGE IN CANADA

M. Patricia Connelly

St. Mary's University
Halifax, Nova Scotia   Canada

Women's work in Canada is either unpaid work in the home, low-paid wage labor, or both. It is a fairly well-known fact that women in the labor force earn on the average only 60% of what men earn and that the dollar gap between their average earnings is increasing.[1] This disparity is not a new phenomenon; women in the Canadian labor force have always received lower wages than men for comparable work.

Roberts quotes the Globe of June 1897, which states that "almost universally . . . women receive for performing the same labor from one-third to one-half less wages than men."[2] MacDonald provides some historical data on women's and men's average annual earnings for each census year from 1901 to 1971. Women earned 52.8% of what men earned in 1911 and the overall picture provided "is of an increasing gap in absolute terms, roughly stable in long-run percentage terms."[3]

When we turn our attention to women's work in the home, we find that, in economic terms, the consequences of marriage are a small amount of security and a great deal of dependency. In return for their unpaid domestic labor, wives receive legal assurance that their husbands are responsible for ensuring that they are housed, fed, and clothed. With the exception of Manitoba (where wives have a legal right to periodically receive and spend, at their discretion, a reasonable amount of money for clothing and other personal expenses), husbands are not obliged to give any money to their wives.[4]

Married women who work full time in the home are not eligible for unemployment insurance or other benefits offered by employers such as pensions and disability insurance. For many women who are divorced, deserted, or widowed, this lack of financial protection can result in a poverty level existence. For example, a recent report by the National

Council of Welfare shows that widows are seldom adequately protected by their husbands' pension.

> All told, less than one widow in four can expect to get any regular benefits from her deceased husband's employer. Even then, what she receives will in most cases amount to only 50% of her husband's pension entitlement.

The council concludes by pointing out that "after fifty years of unpaid, faithful service a woman's only reward is likely to be poverty."[5]

Because most women work throughout their lives, they will have to confront these types of problems. What is the solution? Why do women receive little or no monetary reward whether they work in the home or in the labor force? The answers, I believe, must be sought in an examination of the conditions created by the development of capitalism, the dominant mode of production in Canadian society. In this paper, using census data and secondary sources, I will outline changes in the economic system and discuss how the conditions created by these changes have affected the family and, most importantly, women's domestic and wage labor in Canada.

CAPITALISM AND THE FAMILY

Before turning to the Canadian case, let us look more generally at the changes that occurred within the family with the rise of industrial capitalism. In precapitalist society, the family was a self-sufficient economic unit. There was a division of labor by sex and age, but the work of all was necessary for survival. Family members working together within the unit produced the goods needed to both maintain themselves and to raise the next generation.

With the rise of capitalism, there came a major change in the family. Since the family no longer owned the means by which its livelihood could be produced, it was unable to remain self-sufficient. It became necessary for some members to go outside the family to sell their ability to work, that is, their labor power for a wage. Work and family life were separated into public and private spheres of activity.

The expectation in a capitalist economy is that each person entering the labor force will sell his or her labor power for an individual wage, yet the assumption remains that the family as a unit is responsible for maintaining and reproducing the new labor supply. Moreover, this assumption underlies the unpaid domestic work and the low-paid, labor force activity of women.

Women and men are not paid as individuals while both must be concerned with maintaining and reproducing themselves. Rather, they are paid as family members who must necessarily combine some amount of wages and domestic labor to meet the cost of raising their children. Always paying women less than men ensures that women will remain primarily responsible for the necessary but unpaid work that must be done in the home. In this way, capitalism maintains a flexibility with regard to the cost and use of the labor supply that would not otherwise exist.

During the very early industrialization period in England, for example, all members of the working-class family were drawn into the capitalist production process. In some cases, they were hired as a family unit and the entire wage went to the male head of the household. Later, as capitalist production expanded and the division of labor increased, family members were separated and paid an individual wage which they pooled for their survival. The amount of domestic labor in which goods and services were produced by the family and only for the family's use became minimal. Later still, as more sophisticated machinery was introduced and the productivity of labor increased, the demand for labor declined. At the same time, the trade union movement was growing in strength and men demanded that they receive a family wage, that is, a wage high enough to support themselves and their families. Under these conditions married women began to move out of the labor force and into the home to provide the necessary but unpaid domestic labor on a full-time basis.[6]

It seems clear that with the family responsible for the new labor supply, the cost can be spread over several members in a number of ways according to the changing conditions of the capitalist system.

THE CANADIAN CASE: A HISTORICAL PERSPECTIVE

Preindustrial Society

Let us now look at some of the conditions created by capitalism as it developed in Canada and the effect of these conditions on the family in general and women's work in particular. Canada in the nineteenth century was a rural society. Most people lived on farms and the family was the central productive unit. Farm families usually built their homes, constructed their furnishings, made their clothes, and produced their food. Very little was bought or sold. Since most families were able to meet their daily needs, there was no market for surplus products. At most, local tradespeople such as blacksmiths and itinerant workers such as tailors or cobblers were hired on occasion and were paid in kind rather than in money.[7] The early nineteenth-century farm family was essentially a self-sufficient unit and the labor of all family members (men, women, and children) was necessary to maintain it as such.[8]

At the time of Confederation in 1867, 85% of Canada's population still lived on farms or in villages.[9] However, industrial capitalism and urbanization were clearly on the horizon. The construction of railroads was opening up the country and pulling it together into one economic unit. Railroads were fostering the centralization of production, the growth of cities, and the development of urban and rural consumer markets. The conditions were being created for the growth of production based on machinery and unskilled labor rather than production by skilled crafts people.[10]

The Advent of Industrial Capitalism

By 1875, industrial capitalism was on its way and industrialists were beginning to dominate Canadian affairs. City and provincial markets had expanded; cities such as Toronto, Montreal, and Hamilton almost doubled in population between 1851 and 1871.[11] Both farms and factories were increasing their capacity to produce surpluses and in turn were consuming these surpluses.[12] The trade union movement was developing.[13]

Once it had begun, industrial capitalism progressed relatively rapidly in Canada and so did the problems associated with it. In 1889, the Report of the Royal Commission on the Relations of Labour and Capital stated:

> With us the factory system has not grown slowly; it sprang into existence almost at one bound. . . . But it also has to be pointed out that in acquiring the industries at one bound we have also become possessed, just as quickly, of the evils which accompany the factory system, and which, in other lands, were creatures of a gradual growth.[14]

The major evil referred to by the Royal Commission was the desire of the capitalist to obtain the largest possible profit in the smallest amount of time with no regard for the good of either the worker in particular or the society in general. "To obtain a very large percentage of work with the smallest possible outlay of wages appears to be the one fixed and dominant idea" of the employer.[15] In other words, the capitalist was looking for the cheapest labor available.

As the British had before them, Canadian capitalists in need of a labor supply drew on families whose small farms had gone bankrupt and on tradespeople and artisans who had been forced out of their independent work by competition from large-scale production. Machinery and capital equipment introduced into agricultural production "freed" many farm workers who had no choice but to become wage laborers.[16]

More and more people were being drawn into wage labor, and self-sufficiency was no longer a possibility for an increasing number of Canadian families. Dependency on the wage was fast becoming a way of life. Women and children were among those drawn into wage labor; since they received lower wages than men, they were an important source of cheap labor for the employer.

Once again the Royal Commission on the Relations of Labour and Capital states:

> To arrive at the greatest results for the smallest
> expenditures, the mills and factories are filled with
> women and children, to the practical exclusion of
> adult males. The reason for this is obvious. Females
> and children may be counted upon to work for small
> wages, to submit to petty and exasperating exactions,
> and to work uncomplainingly for long hours.[17]

Unlike Britain, however, few married women were in the Canadian
work force in the early stages of industrialization. Jean Scott, writing in
1892, describes how uncommon it was to find married women working
outside the home:

> The employment of married women in factories and
> stores in Ontario is not general. In a large number of
> factories and stores there are no married women at
> all; at most only one or two widows. Married women
> in Canada do not seem to go out to work as long as
> their husbands are at all able to support them.[18]

It was the expectation at this time that men should be able to
support their families, that is, earn a family wage. Working-class men's
wages, however, were very low and at best they could support their
families at subsistence level and many could not do that. For example,
when asked if his wages could support his family, a laborer answered that
he could barely maintain it. He outlined his daily cost for food and
showed that after costs were deducted, he had six cents a day left. He
said:

> The rent has to be paid and fuel has to be bought, and
> what will be done for wear and tear of house--will 6
> cents meet all these demands? I just put that to
> show how a man can maintain a family on $1.10 a
> day; that is a low figure; he has wear and tear of
> house and fuel to meet and everything else to find in
> the house; I am not allowing him any meat, only fish,
> and if he gets meat on Sunday he has to keep it from
> the grocery, which is not fair to the grocer.[19]

Copp looked at the standard of living of Montreal wage earners from
1900 to 1930 by relating changes in their incomes to changes in the prices
of a minimum standardized budget for a family of five. He found that
"for the two-thirds of the adult male labor force employed as hourly wage
earners there was little chance of earning sufficient income, even at
maturity, to provide an average family with the minimum standard of
living."[20] The vast majority of Montreal families at this time needed at
least two wage earners per family to earn the minimum income level.

In most cases throughout the country, it was children and young
married women who contributed support to these families that had more

than one income earner. Some married women did find jobs outside the home, usually as domestic servants or daycleaners.[21] Others did work in the home to earn money. Those in the more rural areas did market gardening and raised chickens to sell eggs. Urban women took in laundry or did embroidery or dressmaking in their homes.[22] A 1916 (approximate) report on the tailoring trade in Toronto indicates the position of these women:

> The work suits women because they can carry on household work at the same time; several women are mothers of families who eke out their husband's wages in this way; others are obliged to be at home to care for aged or sick relatives.[23]

One of the most prevalent ways urban women earned income was to take in boarders. Many working-class households had at least one boarder, if not more.[24]

Despite the difficult conditions under which domestic labor was performed (i.e., primitive birth control methods and a low level of household technology), it is likely that many working-class wives in the late nineteenth and early twentieth century would have entered wage labor had the opportunity existed. However, the opportunity did not exist because the supply of women apparently was greater than the demand in every area of women's work, except that of agriculture and domestic service.[25] Since there was no labor market demand for married women, they remained in the home stretching their husbands' wage by intensifying their labor and by bringing in extra money whenever possible. With regard to labor force participation, married women were held in reserve until World War II. Available evidence indicates that fewer than 3% or 4% of all married women in Canada worked outside their homes before that time.[26]

Effects of Canada's immigration policy on the labor force. There was no labor market demand for married women during the early industrialization period in Canada because there was a more desirable reserve labor source in the form of immigrants. Between 1900 and 1920, the Canadian government, in response to employers' needs, initiated an open-door immigration policy. These were years of unprecedented immigration and both the labor force and the population grew more rapidly than ever before in the history of Canada.

Immigrants were more important than women for several reasons. First, like women they were cheap labor and they were, initially at least, a docile labor supply since many of them did not know Canadian customs or the English language. According to Ostry and Zaidi, "the bulk of these immigrants were from southern, central, and eastern Europe and were, on the whole, less skilled and less educated than those from the British Isles and northwestern Europe, the previous major sources of immigrant inflow."[27]

Second, unlike Canadian women, immigrants would add to the small Canadian population, thereby increasing the consumer market. In 1901, the population of Canada was less than 5.5 million. Between 1901 and 1914, 2.9 million immigrants were added to the population.[28]

And finally, occupational growth took place in areas where there was a demand for unskilled male laborers, for example, in the building trades, manufacturing, transportation, forestry, and farming.[29] It was primarily male immigrants who entered the country to perform these jobs. The increase in the male immigrant population was reflected by the abrupt increase in the ratio of males to females in the population. In 1901, there were 105 males per 100 females and by 1911 there were 112.9 males per 100 females.[30]

Women in the labor force. Young single women, however, remained a source of cheap labor for employers. As employment in agriculture declined, many farmers' daughters recognized that there was no future for them on the farm and moved to the cities to find work.[31] They were immediately drawn to those jobs which were related to work they had previously done in the home, such as domestic service, garment, textile, and food-processing jobs. Seventy-two percent of the female labor force in 1901 was employed in these jobs.[32]

Canadian women preferred factory to domestic jobs since their hours of work were more clearly defined and they had more time for themselves. As a result, there was always a demand for domestic workers and immigrant women were often brought to Canada for the express purpose of filling these jobs. "In 1911, for example, immigrant women formed 35% of the female work force in domestic and personal service, even though they made up only 24% of the female work force . . . ."[33]

At the turn of the twentieth century, wages in general were very low but women's wages barely allowed them enough to live on. They certainly could not support dependents. Using data for the year 1889, Rotenberg shows that women without dependents were left with a surplus of $2.43 once the total cost of living had been deducted from their total yearly earnings, while women with dependents incurred a deficit of $14.23.[34] "In the Labour Gazette of June, 1913, Professor C.M. Derick of McGill University stated that the average wages of female factory workers in Canada was $216/year or $5/week. The living wage at the time was considered to be $390/year or $7.50/week."[35]

Because of their low wages, most working women had no choice but to live at home with their families or in inexpensive boarding houses. Their wages were often an essential addition to the family income. In fact, the assumption made by the employer was that these young women would supplement their family's income and in turn their family would look after them.[36] When they married, they would leave the labor force because they would then have husbands to support them.

Women and trade unionism. Although women made up only a small percentage of the total labor force, 13.3% in 1901 and 15% in 1931, the male-dominated trade unions viewed their labor force activity as a matter of great concern. Klein and Roberts have summarized the position taken by trade unionists.

> The view of the trade union movement towards the entrance of women into the work force was obviously one of apprehension. They feared this process because of its effects on the male work force and on the traditional role of women. They were anxious to define very clearly the limits to which women's sphere could be extended and still be compatible with their notion of femininity. Yet often the logic of their own interests pressed them to take a progressive stance on key issues of civil and industrial rights, thus causing considerable internal tension and ambivalence in their overall view of women workers.[37]

Given the ambivalent feelings of trade unionists, it is not surprising that "at least three different, and contradictory, positions were taken by the union movement: exclusion of women from the labor force, protective legislation, and unionization."[38] As men, they accepted the prevailing traditional moralistic definition of women's place. They worried that women's health and morals would be damaged by the appalling conditions existing in most work settings. Consequently, they joined with reformers in advocating protective legislation for women.[39]

As workers, women were perceived by men as dangerous competitors who threatened men's wages and even their jobs. The early unions were organized along craft lines and unskilled women and men workers were a major threat. In the case of women, however, the unions not only excluded them from membership, but openly advocated their exclusion from the labor force. In this way, they hoped to rid themselves of at least one source of competition.[40]

As trade unionists, men were committed to protecting all workers, including women. If they could not exclude them, then they should unionize them, particularly if women shared a trade with them. According to Roberts, "males in the skirt trade found that they couldn't organize until the women were organized."[41] Women have a history of fighting for better conditions in the workplace. Sometimes they did this alone but in some cases the union movement stood behind them.[42]

All three responses by the trade union movement resulted in the differential treatment of women workers. Women were considered to be working only temporarily, that is, until they married. Thus, even when unionized, their lower wages and their segregation into "female jobs" were not high-priority issues for union negotiators. In fact, the segregation of women into low-paying "female jobs" solved major problems for both employers and men workers. For employers, it meant a continued source of cheap female labor. For men, it meant the removal of women from direct competition for their jobs as well as assurance that they would be doing work appropriate to their womanhood.[43] For women, of course, it meant that the jobs open to them in the first part of the twentieth century were severely limited.

Women in the Labor Force since the 1930s

During the Depression years, unemployment soared to levels of almost 20%.[44] Since unemployment insurance did not exist, many families found themselves without any means of support. Working-class wives found it necessary to look for work outside the home despite the prevailing negative attitude toward married women working. Often they could find low-paying, short-term jobs when their husbands could find none.[45]

When World War II began, there were a great many people unemployed in Canada. By 1942, however, the supply of labor had diminished as a result of the recruitment of men for the armed service and the expansion of war production at home. There was renewed demand for labor, and employers and the government had no choice but to turn to women. At first, they did not call on married women; however, as time went on, it became necessary to activate the reserve army of married women.[46] Incentives such as free government nurseries and income tax concessions were provided to attract married women into the labor force. Married women responded to the "call," which was worded in terms of doing their patriotic duty. Other evidence, however, showed that many women entered the labor force because they needed the money.[47] During this time, "there was a continuous movement of women out of low-wage, service and unskilled occupations into the higher-paying skilled positions in manufacturing, where they undertook a variety of jobs formerly performed only by men."[48]

When the war ended, the incentives were withdrawn in an attempt to move married women back into the home. Single women were once again encouraged to return to their former "female jobs." Many married women did leave the labor force, but many stayed and, with single women, found work in the increasing number of "female occupations." However, as a trade union article at the time indicated, "women workers have shown a disinclination to return to domestic service work, and low wage jobs generally."[49]

The decade of the 1950s was a period of rapid economic expansion in Canada. For the first time, there were more people employed in blue-collar occupations (29.4%) and white-collar occupations (32%) than in primary occupations (19.8%).[50] The mechanization of farming had reduced the amount of labor needed for production. At the same time, increases in the use of private cars and the use of machinery and other equipment in factories, offices, and stores gave rise to new occupations. For example, mechanics and repairmen were needed to install, maintain, and service the new equipment and machinery. Both government and private industry began to spend money on construction and on research and development, which stimulated the growth of professional and technical occupations. As more products came on the market, there was a greater need to keep records and to advertise, sell, and finance them. The result was a multiplication of clerical activity.

These shifts in the occupational structure generated a need for a more highly skilled and better-educated labor force. As in earlier times, immigrants were an important source of labor for the expanding economy. Between the years 1951 and 1961, net immigration amounted to 1.1 million of the total population of 14 million.[51] The typical immigrant of the 1950s was likely to be a young male who was better educated than his native-born counterpart. Occupationally, immigrants were concentrated in professional and technical occupations as well as skilled craft and production-type jobs. According to Ostry and Zaidi, since the early 1950s, Canadian immigration policies "have sought to shape immigration to the needs of the economy for more skilled and educated manpower."[52]

The entrance of married women into the labor force. Unlike earlier times, married women now became an important source of labor power. Whereas in the early part of the century, the segregation of women into low-paying, "female jobs" had the effect of limiting the number of jobs open to them, by the middle of the century, it had just the opposite effect. "Female jobs" were on the increase and this created a demand for female labor.[53] Ostry and Zaidi described it this way:

> A factor of overriding importance has operated on the demand side of the market, i.e., the very considerable expansion in recent decades of jobs that are considered especially "suitable" for female employment. It has been argued, and plausibly in our view, that cheapness combined with quality has made female labour particularly attractive to employers in filling the lower and middle-level white-collar jobs, which have been expanding at a well-above-average rate since the end of World War II.[54]

Women entered the labor force in response to this demand and their participation rates rose from 24.1% in 1951 to 39.5% in 1971. More specifically, the participation rates of married women tripled from 11% to 37% during this same period.[55]

Women also entered the labor force in response to the family's dependence on the wage as a means of meeting its need to consume more goods. By mid-century, 77% of Canadians in the labor force were dependent on the wage and by 1961 this figure had risen to 83%.[56] As the economy developed and as labor became more productive, consumer goods became available to more people. At the same time, mass consumption of these goods became imperative for the economic system to continue. New needs were created through advertising, evidenced by the fact that the number of "advertising salesmen and agents" increased by almost 80% between 1951 and 1961.[57]

The impact of trade unions in the modern era. Trade unions fought the battle for a higher standard of living which seemed just beyond the wage earners' reach. Workers argued that they should get a share of the wealth earned by the increased productivity of their labor. An article in a Trades and Congress Journal stated:

> Our contention is that wages shall be based on the ever increasing productivity of the machine in industry, as a result of technological improvements. . . . Organized labor refuses to concede that wages shall only provide for basic needs in order to maintain a minimum standard of living, with no regard to his spiritual or material welfare, his intellectual needs and the necessary reserves for sickness, unemployment and old age. . . . Can it not be that we again are right in our agitation for a higher wage to maintain an ever increasing standard of living for all in Canada.[58]

Trade unions did succeed to some extent in their aims. Real wages did rise, but nowhere near the rise in the rate of productivity. In discussing wages for the manufacturing sector, Ostry and Zaidi stated that "although manufacturing workers' real wages increased by 84% from 1950 to 1972, the rise was far behind that in their productivity, which was over 144%."[59] Cuneo showed that while real wages did go up in the 1950 to 1974 period, the gap between gross weekly income (the dollars that people saw on their pay checks each week) and real weekly income (the actual increase in dollars after the rate of inflation was subtracted) had also increased throughout that period.

> Between 1950 and 1954 the working class lost an average of $5.99 or 11% of its weekly income through inflation. By the 1970-74 period, this loss had increased dramatically to $66.50 or 43% of weekly gross income.[60]

In other words, even though real wages increased, the amount of dollars that they received each week was of considerably less value because of inflation. Moreover, Johnson showed that between 1951 and 1971 there was a growing disparity between higher- and lower-paid wage earners despite the rise in per capita income.[61]

CURRENT CAUSES OF THE INCREASED NUMBER
OF WOMEN IN THE LABOR FORCE

The reality was that a large number of working-class families needed an additional income in order to achieve the new standard of living. At the same time, there also was an increased demand for female labor to fill the growing number of "female jobs." Evidence shows that women who were married to low-wage earners, increasingly entered the labor force. The percentage of families with more than one income earner rose from 43% in 1951 to 65% in 1971.[62]

With two members of the family working for a wage, the new standard of living could be achieved by the majority of families (although some could do it only through the use of credit). The result was an increased dependence on consumer goods and services while young people, particularly young men, stayed in school for longer periods of time thereby "improving the quality of the labor force."[63]

Because men's real wages increased, the standard of living rose, and young people pursued higher educational goals, women appeared to be working for "extras." However, had women not joined the labor force, men's wages, despite the increase, would have been insufficient to maintain and reproduce the family according to the new standard of living. The Department of Labour's publication, Women at Work in Canada: 1964, substantiated the need for married women to work.

> Since the thirties the level of living of the pop-
> ulation, including the "real" incomes that sustain it,
> has risen remarkably. The standard of living—that
> level at which people feel that they are comfortably
> off and not deprived of anything important—has in-
> creased also; the availablity of a wide range of
> consumer goods has assisted in the latter process.
> Yet a considerable proportion of male wage-earners,
> in fact the majority, do not earn the $6,000 or so per
> year that is necessary to move consumption much
> beyond food, clothing, and shelter. For many Can-
> adian families, however, the earnings of the wife
> added to those of the husband just succeed in bringing
> total income up to a fairly comfortable level.[64]

As we move into the 1980s, it is apparent to more and more people that two incomes are necessary just to maintain the standard of living achieved in the previous three decades. For example, the National Council of Welfare's recent report on Women and Poverty evaluated the impact of women's wages by deducting wives' earnings from the incomes of Canadian families and comparing what was left to Statistics Canada's poverty lines. They found that "51% more two spouse families would be poor if wives did not work outside the home."[65] It should be noted that the Council is discussing poverty level and not an average standard of living.

CURRENT CONDITIONS OF WOMEN IN THE LABOR FORCE

While more people find it necessary to enter the labor force, increased technology and automation have decreased the need for labor power. The unemployment rate has risen dramatically, more than doubling between the late 1940s and early 1970s and averaging 7.8% for the 1976 to 1979 period. [66] The labor force is clearly segregated and women and men do not compete for the same jobs. Consequently, "there is no direct relationship between female employment and male unemployment."[67] Part-time jobs are growing in number as a way for employers to save on labor cost. Women fill over 70% of these jobs and most (62%) of these women are married.[68]

It is under these conditions that women, who are still mainly responsible for the work done in the home, continue to seek low-paying "female jobs" on a full-time or part-time basis in order to help keep the family's standard of living from declining. Thus, once again the family and in particular women are compelled to rearrange their resources to fit the needs of the economic system.

CONCLUSION

This paper has attempted to uncover some of the underlying forces which determine the low value, in monetary terms, placed on Canadian women's domestic and wage labor. It has been argued that although an employer hires individual workers, there is an implicit assumption that these workers are members of a family unit and that this unit will combine some amount of wages and domestic labor to meet the cost of raising children or, looking at it from the employers' point of view, of reproducing the new labor supply. Always paying women less than men ensures that women, even when they work for a wage, will remain responsible for the unpaid domestic work done in the home. In this way, employers can depend on a certain amount of flexibility with regard to the cost and use of the labor supply. When women are needed as wage laborers, they can be called on; when they are not needed, they will remain full-time domestic workers. Current evidence indicates that more and more Canadian families require two income earners. At the same time, the number of "female jobs" is increasing and women are entering the labor force on a relatively permanent part- and full-time basis. Yet, "female jobs" still do not provide wages which allow women to maintain themselves and their children at the average Canadian standard of living. Moreover, although men's wages are still on the average higher than women's, an increasing number of men are unable to earn wages that can support a family at this standard. The implications of these facts for immediate action seem evident. Women and men must fight for equal pay for work of equal value. This equal pay must minimally cover the maintenance of the individual and her or his children at the average Canadian standard of living. This, of course, is crucial for the growing number of one-parent families, most of whom are headed by women. For two-parent families, it will begin to provide the basis for the family itself, and women equally with men, to have control over the way in which domestic and wage labor is distributed between them.

NOTES

1. Gelber, 1975; Gunderson, 1976; MacDonald, 1980.
2. Roberts, 1976:8.
3. MacDonald, 1980:23.
4. National Council of Welfare, 1979:18.
5. Ibid.: 31–32.
6. See: Pinchbeck, 1930; Tilly and Scott, 1978; Zaretsky, 1976.
7. Cross, 1974:34.
8. Johnson, 1974a.
9. Ostry and Zaidi, 1979:65.
10. Pentland, 1959.
11. Lipton, 1967.
12. Pentland, 1950.
13. Lipton, 1967.
14. Royal Commission on the Relations of Labour and Capital, 1889:87.
15. Ibid.:87.
16. Johnson, 1972.
17. RCRLC, 1889:87.
18. Scott, 1976:181.
19. Kealey, 1973:326–327.
20. Copp, 1974:31.
21. Cross, 1977:74–75; Roberts, 1976:10.
22. National Council of Women of Canada, (1900) 1975; Scott, 1976.
23. Quoted in MacLeod, 1974:316.
24. Cross, 1974.
25. National Council of Women of Canada, (1900) 1975:95–102. Domestic servants were often expected to live in. As a result, this type of work was impossible for many married women with families.
26. Ostry and Zaidi, 1979:234.
27. Ibid. Because of the difficulty of getting accurate data at that time, this is probably an underestimate.
28. Denton, 1970:19. It is important to note that emigration was also high at this period.
29. Statistics Canada, Census of Canada, 1921.
30. Denton, 1970:48, Table 12.
31. Leslie, 1974:90.
32. Statistics Canada, Census of Canada, 1921.
33. Leslie, 1974:95.
34. Rotenberg, 1974:48–49.
35. Ibid.:49.
36. Kealey, 1973; National Council of Women of Canada, (1900) 1975.
37. Klein and Roberts, 1974:222.
38. White, 1980:14–15.
39. Klein and Roberts, 1974.
40. White, 1980:Chapter 2.
41. Roberts, 1976:41.
42. Kidd, 1974.
43. Johnson, 1974a.

44. Ostry and Zaidi, 1979:145.
45. MacLeod, 1974.
46. Canada, Dominion Bureau of Statistics, 1942.
47. Pierson, 1977.
48. "Women in the Labour Force," Trades and Labour Congress Journal,
        1947:18.
49. Ibid.
50. Ostry, 1967:50, Table 2.
51. Denton, 1970:19.
52. Ostry and Zaidi, 1979:18.
53. Connelly, 1978.
54. Ostry and Zaidi, 1979:42.
55. Connelly, 1978.
56. Ostry, 1967:89, Table 16.
57. Ibid.:21.
58. Buckley, 1947:14.
59. Ostry and Zaidi, 1978:240.
60. Cuneo, 1980:94.
61. Johnson, 1974b.
62. Canada, Department of Labour, 1958, 1965; Gunderson, 1976; Arm-
        strong and Armstrong, 1978.
63. Robb and Spencer, 1976.
64. Canada, Department of Labour, 1965:6.
65. National Council of Welfare, 1979:20.
66. Armstrong, H. 1979.
67. Armstrong, 1980b:14.
68. Collins, 1978.

A STATISTICAL ANALYSIS OF FEMALE PARTICIPATION IN WORK

AND EFFECTS ON THE POPULATION AND SOCIOECONOMIC

DEVELOPMENT: A CASE STUDY OF GREECE

Stylianos K. Athanassiou

Center of Planning and Economic Research
Athens   Greece

The twentieth century has been characterized by truly amazing progress in the sectors of science and technology. From a social and moral viewpoint, the most important feature of this is the independence of women and the lifting of previously held reservations regarding the capability of women to live and to create without depending economically and socially on men, whether husbands or fathers. In fact, the time has passed when women had as their sole occupation in life the cares of the household. Nowadays, women contribute by their labor supply to the production of goods and services as equal members of society. The female supply of labor has caused an increased demand for their skills and services. Women are no longer employed only in jobs traditionally associated with women, but they also work in jobs considered to be nontraditional, for instance, as workers in industry. Consequently, we can say that female economic activity plays an important role in enhancing productivity while affecting the quality and composition of the economically active population. In spite of all this, many factors influence the participation of women in work, primarily restrictively; therefore, there still is a lack of homogeneity in participation rates in the populations of men and women. Finally, the participation of women in work contributes substantially to socioeconomic development, whereas it unfavorably affects population evolution.

Women's increased participation in the labor force has initiated considerable research interest in recent years. Reasons for this interest include: (1) female participation rates are very much lower than the corresponding male figures and they vary markedly among countries, regions within a country, age groups, etc., and (2) female participation in work greatly affects population evolution and plays a vital role in attempts to improve a given country's level of development. The low

regional and age structure variations in female participation rates are
attributable to a range of demographic and socioeconomic determinants
frequently different from those for males.

The aim of this study is to statistically analyze, using census data,
the determinants of female participation in work and to define the effects
of this participation on the population and socioeconomic development for
the case of Greece. The central hypothesis is that female participation in
economic activities depends on a large number of demographic and
socioeconomic determinants and it affects considerably the overall population evolution and socioeconomic development of a country. It is
believed that the results of this study will be useful for planners and
policymakers in making a large number of short- and long-term decisions
regarding the female labor force.

POPULATION AND SOCIOECONOMIC DEVELOPMENT

When we discuss population, we do not simply mean the population
size, i.e., that there are too many people, or its evolution; we also must
mean the changes in the sex-age structure of the population, regional
distribution (rural-urban), differential fertility and mortality rates, and
international or internal migration.[1] Based on census data, we see in all
countries that some population groups increase their numbers more
rapidly than others. Natural growth rates and differentials in migration
vary markedly from one socioeconomic class to another and from one
geographical region to another. These changes in the features of the
population are caused by demographic and socioeconomic factors and it is
evident that their effects on population differ greatly both within and
among countries.[2]

In general, socioeconomic development means a transformation of
life styles, which includes changes both in production and consumption
activities and, more importantly, changes in values and attitudes that
produce a social environment conducive to rising levels of production and
consumption.[3] More precisely, we refer to the process of an improvement
in the standard of living, together with the decreasing inequality of
income distribution and the capacity to sustain continuous improvements
over time. Economic development, a necessary although not a sufficient
condition for socioeconomic development, may be defined as the long-
term process of structural change in the methods of production, resulting
in rising aggregate and per capita income. Although socioeconomic
development could conceivably occur in the short run in the absence of
economic growth, it includes the capacity to sustain itself and it is
difficult to imagine sustained socioeconomic development without economic development, particularly under circumstances of population
changes. Some of the components of socioeconomic development are:
income, income distribution, employment, distribution of population,
quality of labor force, education, health, and consumption. Finally, the
conventional components (measures) of economic development are the
level and rate of growth of aggregate or per capita national income.

THE ECONOMICALLY ACTIVE POPULATION

By definition, the economically active population in a given area or country is that part of the population which contributes by its labor supply to the production of economic goods and services.   Thus, economically active women are those women engaged in the production process. Based on this definition, the economically active population is examined in physical units for statistical analysis.   To quantify economically active persons, we use the "labor force" approach that is based on the activity of each person during a specific reference period, i.e., whether a person is employed or seeking employment.[4]

STATISTICAL ANALYSIS OF FEMALE PARTICIPATION RATES

The Dependent Variable

As previously noted, the female population which engages in economic activities comprises the female labor force. The proportion of this population to the total female population gives an estimate of the female labor force participation rate.   If the rates, FPR, are restricted to females aged 10 to 64, which exclude the elderly economically active and inactive women, they are expressed as follows:

$$FPR = \sum_{10}^{64} \left[\frac{FLF}{TFP} 100 \right]$$

where

FPR: Female Labor Force Participation Rates in the age group 10 to 64.
FLF: Female Labor Force.
TFP: Total Female Population.

The FPR will be the dependent variable in the regression analyses. Obviously, this variable is related to age structure. Indeed, within the age group 10 to 64, there are marked variations in the propensity of women to work, which has resulted in changes in the FPR and the age structure of the female labor force.   Therefore, other correlates of FPR will be identified and incorporated in the analyses.

The Determinants

The FPR are affected by a variety of demographic and socioeconomic determinants and undergo changes as a result of the influences of these determinants. Because there are too many determinants to list, only the most important will be examined to provide some possible explanations for variations in FPR.

Demographic determinants. Marriage and the Role of Women as Housewives - Marriage has been observed as a factor that tends to take women out of the work force, even in the absence of children. Census data show that although the number of women in the work force increases continuously for some countries, the proportion of married women is relatively low. Furthermore, the proportion of women who engage in economic activities is generally lower among married women who have no children than among single women, a fact suggesting that marriage itself and the responsibilities of being a housewife negatively affect the participation of many married women in work.[5]

Fertility - This is a factor which has bearing on the extent to which women take part in work;[6] it also is related negatively to FPR. Obviously, fertility has a greater impact on FPR for women in the childbearing years of 20 to 39, during which the female population can make its greatest contribution to the labor force.[7]

Women with Children - The responsibility of women, married or not, for the care of their children is another determinative factor in the participation of these women in work. Census data show that the correlation is negative between FPR and preschool or school-age children.[8]

Mortality - This demographic determinant also is negatively correlated with female participation rates. However, this variable may be considered as a "standard" one in the sense that its effects on the female labor force can be accurately and readily estimated in the absence of unforeseeable catastrophic events.

Urbanization - The population movements from rural to urban regions (urbanization) are, to a large degree, functions of the labor force and cause variations in the labor force of urban regions. Therefore, these movements also can be considered as another factor affecting the trend of women's participation in the economic activities of urban areas.[9]

International Migration - One of the most important new trends in international emigration that has emerged in the last two decades is the volume of labor migration within Europe itself. This volume of emigration comes from developing countries to the industrialized countries.[10] Obviously, labor emigration decreases the labor force of the countries of origin and, therefore, should be taken into account in the overall labor force evolution.

Repatriation - The noted economic growth of developing countries, and the relative economic decline and the decreasing demand for labor in developed countries in Europe, has resulted in the repatriation of many emigrants to their original countries; therefore, this part of migration also should be taken into account as a determinant of female participation rates.

Socioeconomic determinants. Socioeconomic factors also have an important bearing on individuals who desire or take part in economic activities. The list of factors is longer than that of demographic ones and the available statistical data are very limited. Moreover, because of the complexities and difficulties involved, there have been few attempts at a quantitative analysis of these factors. Thus, only the most important socioeconomic determinants will be discussed.

Husband's Income - As would be expected, wives of relatively well-off men tend to have a lessened financial need to work. Moreover, the wish of married women to stay at home, as stated earlier, is greater than their desire for a relatively low-paying occupation. These facts show that the participation rates of married women vary inversely with the level of their husbands' incomes.[11] However, the husbands' incomes are not always negatively associated with the participation of married women in work and many times this variable may be expected to correlate positively with female participation rates.

Household Income - Most participants in the labor force live with other family members in a household. Thus, an increase of household income provides the possibility of maintaining the family's standard of living by decreasing the number of family members employed. It has been observed that with an increase in family income, women will more readily leave the labor force than other family members.[12] Consequently, it can be concluded that the participation rates of women are inversely related to their family income.

Job Opportunities - Employment opportunities are correlated positively with female participation rates. The more available job opportunities are, the more women enter the labor force. Unfortunately, no source provides a measure of job opportunities within an area, thus, the level of unemployment is used as an index.[13]

Education Level - Education and FPR are closely related. Indeed, a high level of education corresponds with a woman's desire for employment, particularly in urban areas. This relationship is valid for both single and married women.[14]

Student Population - Students of the intermediate and higher levels of education, the 10 to 19 age group, do not engage in economic activities, i.e., they are not included in the labor force. Because this student group undergoes considerable changes during the 10 years of studies, those changes which positively affect the labor force in the 10 to 19 age bracket should be taken into account.

Social Class - The social status of women plays an important role in their decision on whether or not to participate in economic activities. Many middle-class women, for example, need or desire to be employed outside the home; therefore, they want to be educated in order to have the qualifications necessary for a job. Level of education and the level of income together form a crude measure of determining social class.

THE MODELS

Models of different kinds are frequently used in economic demographic analysis and are an indispensable tool in solving quantitative problems.[15] Models are usually expressed by equations which describe the relationships existing between the dependent variable under study and its determinants. By means of these models or equations, we estimate their parameters; that is, we compute the influence of the determinants on the dependent variable and, thus, the appropriate equations are used for making forecasts.[16]

Based upon the determinants of female participation rates, the model in factional relationship is written as follows:

$$\begin{bmatrix} \text{Female} \\ \text{Participation} \\ \text{Rates} \end{bmatrix} = f_1 \begin{bmatrix} \text{Demographic} \\ \text{Determinants} \end{bmatrix} + f_2 \begin{bmatrix} \text{Socioeconomic} \\ \text{Determinants} \end{bmatrix}$$

Following the graphic description of the statistical data and based on the economic and demographic theory, the form of the equation selected is a linear one. The traditional method of least squares is applied for the parameter estimates.

The Female Participation Rates in Greece for the Period 1961 to 1975

On the basis of estimations, the female labor force was comprised of 1,194,000 women in 1961, 905,000 in 1971, and 935,000 in 1975. Thus, there is a decrease of 26.3% in the labor force for the period 1961 to 1971, and a slight increase (4.4%) in the semiperiod 1971 to 1975. The female participation rates are 27.8% in 1961 and 20.2% in 1971 and 1975. The continuous decrease of female participation rates, and the low levels at which they stand, may be attributed to the three factors of the international migration during the period from 1955 to 1970, urbanization, and the relatively low growth rate of the population. The rates of the female labor force also differ by age as can be seen in Table 1. Similar variability also exists in female labor participation rates by either rural or urban region.

In Table 1, we see that the female participation rates decrease considerably in the period from 1961 to 1971 whereas the variations between 1971 and 1975 are relatively small. The greater decrease observed in the age groups 15 to 19 and 20 to 24 can be primarily attributed to the increase of the student population. Indeed, the proportion of female students to total students is approximately double in this period (1961 to 1971) and the number of females who did not complete elementary education is very low. The percentages of economic activity of married women with a spouse and of divorcees/widows are 22.8% and 20.8%, respectively, in 1971. The percentages of women in the labor force by level of education are presented in Table 2. Furthermore, the percentage distribution of educated women in the labor force also has changed greatly during the period from 1961 to 1971, as shown in Table 3.

The distribution of women employed by sector of economic activity in the census years 1961 and 1971 is presented in Table 4. The shifts in the percentages of employment by sector during the decade from 1961 to 1971 parallel the changes for the Gross National Product (GNP) percentage and the rate of growth per sector. The percentage distribution of employment remained fairly constant from 1971 until 1975.

TABLE 1

GREEK FEMALE LABOR FORCE PARTICIPATION RATES
BY AGE, 1961-1975

Age Group	Percentage		
	1961	1971	1975
15-19	46.8	27.7	26.1
20-24	52.2	36.5	37.1
25-29	43.6	33.4	34.2
30-34	48.6	31.9	33.1
35-44	37.1	31.9	32.0
45-54	34.5	28.8	29.0
55-64	24.3	19.8	20.1

TABLE 2

GREEK FEMALE LABOR FORCE PARTICIPATION RATES
BY LEVEL OF EDUCATION IN CENSUS YEAR 1971

Years of School Completed	Women in the Labor Force	Percentage
No year completed	343,000	29.9
Elementary school	382,000	23.1
High school	108,000	29.2
University	40,000	68.9
Vocational education	13,000	52.5

TABLE 3

PERCENTAGE DISTRIBUTION OF ECONOMICALLY ACTIVE
GREEK WOMEN IN THE CENSUS YEARS 1961 and 1971

	Percentage	
Group by Occupation	1961	1971
Professional workers	3.54	6.99
Administrative workers	.16	.23
Clerical workers	3.52	9.08
Merchants	2.29	4.84
Service workers	7.56	9.32
Farmers	65.48	52.97
Laborers	12.97	15.53
Persons not classifiable	4.48	1.04

TABLE 4

PERCENTAGE OF DISTRIBUTION OF EMPLOYED GREEK WOMEN
BY MAJOR SECTOR IN CENSUS YEARS 1961 and 1971

	Percentage	
Sector	1961	1971
Agriculture	39.9	36.4
Mining	5.8	6.2
Manufacturing	31.7	26.8
Electrical	6.8	10.4
Construction	.8	.6
Trade	14.9	21.3
Transportation	3.6	6.8
Other services	34.5	37.4

STATISTICAL ESTIMATION AND RESULTS OF THE EQUATIONS

The 1971 census data for the 52 geographical regions of the country
are considered a random sample for the statistical analyses. The values
of the determinants have undergone transformations because of the unavailability of adequate data and are based on these "gross" data. The

usual tests of significance are employed as a way of highlighting the correlation and regression coefficients of the most important determinants. From these tests, it is evident that many of the determinants are highly intercorrelated; therefore, they are tested separately, and a few determinants have been replaced by new factors. For example, because fertility is positively associated with nuptiality, only the relative frequency of marriages for the critical age groups 20 to 39 is used. In addition to the determinant CFPR, which refers to women with children, we also can use the rates of women 15 to 19 and 40 to 64, who typically are the most economically active age groups and upon whom statistical data are readily available and accurate. Furthermore, the female population movement toward urban areas may also use the degree of urbanization of a region which is the proportion of the urban population to the total population of each region. With regard to international emigration and repatriation, the rates of working-age women who emigrated and then were repatriated to Greece has to be taken into account.

The influence of the husband's income or household income, which are two of the most important economic factors of female participation in work, cannot be determined; therefore, rates of employed persons in the household are used. For the job opportunities factor, the rate of female unemployment is used as a reasonable indicator. The influence of women's educational level on their participation in economic activities has been estimated on the basis of the correlation that exists between rates. Thus, as an educational index, the rates of primary, secondary, or higher education are included in our regression analyses. For the student population, it is obvious that we must use the number of female students aged 15 to 19 who do not engage in economic activities. The correlation and regression coefficients for female participation in work and in relation to its determinants have been estimated on the basis of the sample selected.

To continue, a number of regression equations have been estimated by using a stepwise procedure. The numerical results obtained for the demographic and socioeconomic determinants are presented in Table 5. The determinants are ordered according to the description above. The results show that marriage and the role of women as housewives, fertility, and women with children are significantly and negatively correlated with the participation of women in the work force. The measure of the proportion of all women aged 15 to 19 and 40 to 64 whom, as we have said, are the most economically active age groups, is found to be positively correlated with FPR. For the measures of urbanization, high positive correlation coefficients are obtained, while their regression coefficients are found to be statistically significant and greater than unity. With regard to the emigration and the repatriation of Greek workers, the correlation and regression coefficients also are statistically significant and are in the expected direction—negative and positive, respectively.

Considering the socioeconomic factors, we can say that the results also are satisfactory. That is, the variable of education correlates

TABLE 5
LINEAR FORM AND STATISTICAL CRITERIA OF ESTIMATED EQUATIONS BY SOCIOECONOMIC AND DEMOGRAPHIC DETERMINANTS

DETERMINANT	CORRELATION COEFFICIENT[1]	CONSTANT	REGRESSION COEFFICIENT[2]	D W COEFFICIENT
Marriage (AFPR)	.45	15.30	−.48 (.22)	1.77
Fertility (BFPR)	.48	14.91	−.41 (.15)	1.74
Children (CFPR)	.58	16.21	−.32 (.14)	1.82
Economically active (DFPR)	.44	11.14	.38 (.18)	1.47
Urbanization (MFPR)	.82	20.13	1.15 (.45)	1.65
Urban population rates (UFPR)	.75	13.85	1.05 (.52)	1.82
Emigration (WFPR)	.41	16.61	−.62 (.30)	1.73
Repatriation (NFPR)	.32	10.15	.32 (.10)	1.47
Education (ZFPR)	.54	18.28	.59 (.34)	1.81
Job opportunities (KFPR)	.46	12.14	.84 (.41)	1.78
Student population ages 10–14 (SFPR)	.87	22.47	−.20 (.04)	1.74
Student population ages 15–18 (SFPR)	.59	67.11	−.39 (.09)	1.81

[1] All values are significant at the .05 or greater levels.
[2] Standard errors are included in parenthesis.

significantly with the rates of female participation in economic activities and, particularly, with the educated and skilled occupations. Furthermore, an increase in job opportunities is highly related to an increase in FPR which can be derived from the results of the KFPR equation. Similar explanations can be made for the estimates of the two equations that are included for student populations, aged 10 to 14 and 15 to 19. The continuous decrease in the proportion of female students signifies an increase in the proportion of women in the work force.

The variable substituted for the unavailable husband's and household income is reported as nonsignificant. In this case, however, it was initially suspected that the substitution would be inappropriate and it is believed that, if the necessary data were available, the income of the husband or the household would give satisfactory results.

IMPACT OF FEMALE LABOR FORCE PARTICIPATION ON THE POPULATION

The effects of women's participation in economic activities on the population include the following considerations:

Fertility Changes

The literature suggests that the increased participation of women in the labor force has consistently been correlated with the decline in fertility. This stimulus-response relationship has been proven empirically.[17] In the case of Greece, the census data examined here show that the mean number of children born to economically active women is lower than that for inactive women. The lowest average number of live births for working women occurs in Athens. Furthermore, the analysis of 1971 census data shows that the female participation rates vary not only according to whether or not the woman has had children, but even more so according to the number and the age of the children. These factors obviously affect the woman's decision to stay at home, given that women who live in cities must typically leave home if they wish to work.[18] As a consequence, both female participation in the labor force and other demographic factors, such as emigration, as we will explain later, are associated with the downward trend in fertility from 1960 to 1975. The fertility rates were 18.9 per thousand in 1960 and they decreased to 16.0 per thousand in 1975.

Population Growth

The two immediate components of population growth, fertility and mortality, produce different trends that give the net natural growth of the population. Thus, the decline in fertility causes a decrease in these growth rates and, as an explanation of the changes in population growth rates, is more important than that related to the differentials in mortality.[19] In Greece, since the end of World War II and particularly during the sample period from 1961 to 1975, the population growth experienced a downward trend, primarily as a result of the decline in fertility and the expansion of emigration. The natural increase in population was .98% and .72% on the annual average for the periods from 1961 to 1971 and 1971 to 1975, respectively, compared with the average rate of growth of 1.20% during the previous period, 1951 to 1961.

Emigration and Urbanization

In examining labor force statistics, we observe that economically active persons tend to be the better educated or well trained. These individuals, obviously, have more knowledge of existing employment opportunities and are better equipped to take advantage of them. In Greece, emigration amounted to 715 thousand persons and 440 thousand persons in the periods from 1955 to 1966 and 1966 to 1975, respectively. The majority of emigrants were economically active persons and more than 85% on the average belonged to the age group 15 to 44.[20] Female emigration consisted of 40% of that population or more than half a million women in the 1956 to 1975 period. The proportion of these women in reproductive ages 20 to 39 was more than 50%. In addition to the effects of emigration, we see that female participation in work also affects, indirectly, the population growth by means of its third immediate component, international migration. An analogous thought can be made regarding the population movements within the country, mainly from rural to urban areas. Finally, female participation in economic activities by means of fertility and emigration affects the sex-age structure of the population. In Greece, the proportion of persons aged 65 and over experienced an accelerated increase of 5%, whereas the proportion of the young population (0 to 14) decreased by 4% in the period from 1951 to 1975. The ratio of the population aged 65 and over and 14 and below to the population aged 15 to 64 also rose to 56% for the year 1975 compared with 53% in 1951.

IMPACT OF FEMALE LABOR FORCE PARTICIPATION
ON SOCIOECONOMIC DEVELOPMENT

In modern economies, labor is considered as a basic factor for the production of economic goods. The transformation of the material to be used in the process is made by means of labor and other productive factors. The known factional relationship among national output, capital, and labor (the Cobb-Douglas function) has also been estimated for Greek industry.[21] The statistical results show that the contribution of labor to output is highly significant. Given that women employed in the modern sector of the economy account for more than 15% of the force, it is easy to understand how very important the role of women is in industrial production. Further, the contributions of women in agriculture, services, and trade also are significant.

As is known, technological progress requires more qualifications or skills on the part of those persons entering the labor force. Thus, the participation of better-educated persons also has resulted in an improvement in the quality of the labor force. In the Greek labor force, the proportion of high school graduates is 11% and the ratio of skilled to unskilled workers—another measure of the quality—has increased considerably. The Greek labor force has the capacity to enter up-to-date production units.[22] Moreover, although the participation rates of females in the labor force have experienced a decrease in the decade from 1961 to 1971, the percentage distribution of better-educated women has improved greatly, as presented in Table 3. This fact means that there has been a favorable change in the quality of the female labor force. Analogous changes in the percentage distribution of females in the work force are projected for the period from 1971 to 1978. The contribution of the quality of labor to output can be further used to evaluate the economic benefits of investing in education, which plays a major role in developing the quality of the labor force.[23] Unfortunately, such data are not yet available.

Studies at the individual household level have shown that the contribution made by working women, single or married, to the family's level of living is considerable.[24] Studies on this subject have not been conducted for Greece and the weaknesses of available statistical data in terms of both coverage and accuracy preclude any conclusions. Besides income, unemployment is another index of socioeconomic development and is also associated with participation rates. In Greece, this index is at a very low level. The social and economic conditions in urban areas of the country are much better than those in rural ones. The income is higher, there are more job opportunities, and, in general, these are prerequisites which tend to reduce the traditional role of women as mothers and housewives. All of these factors enable women to participate increasingly in the labor force. Indeed, a consequence of these factors is the increase of participation rates in economic activities and the propensity of rural residents to migrate to urban areas. The development of the urban population as an index of social development from 1951 up to the most recent census in 1971 is as follows: The urban population (towns of 10,000 inhabitants and over) amounted to 4,668,000 in 1971 (53.2% of the total) compared with 2,879,000 (37.7%) and 3,628,000 (43.3%) in 1951 and 1961, respectively. This reflects an increase of 62.2% in the last two decades from 1951 to 1971. Thus, the proportion of people who live in developed and less developed areas, or as they may be characterized in a broad sense, the urban and rural areas, changed significantly.

The economic development in Greece, from the beginning of the decade from 1960 to 1970, experienced considerable acceleration. The rate of growth of the GNP was more than 7.0% and 4.0%, respectively, in the periods from 1960 to 1972 and 1972 to 1978. The per capita income was $3,461 in U.S. dollars at current prices in 1978 compared with $471 in the year 1960. Considering the per capita income as a measure of the degree of economic development and disregarding other characteristics, we can say that Greece belongs among the developed countries. Of course, there is an income inequality which has increased in the decade from 1961 to 1971. The reasons for the increased income inequality are mainly the population movements to urban regions, the diversification of production and the consumer's preferences, and the developments in the labor market, etc.[25] Finally, private consumption, another index of socio-economic development, increased considerably. The rate of growth was more than 7.2% in the period from 1960 to 1977. Moreover, the composition of private consumption changed by an increase of services and a decrease of food. Obviously, all these changes in the per capita GNP and consumption are reflected in the improvement of the level of living of the Greek people and, of course, part of these effects may be attributed to Greek women who participate in economic life.

CONCLUSION

The main findings of the present study, as they are derived from the statistical analysis of female participation in work and its effects on the population and the socioeconomic development, include the following: (1) Female participation rates are lower than those for males. A number of possible explanations of a demographic, social, and economic nature can be suggested for this. (2) Fertility has experienced a decline during the period from 1960 to 1975. In addition to other demographic factors, for instance, the expansion of international migration, urbanization, etc., and the participation of women in economic activities contributed, to a greater or lesser extent, to the downward trends of fertility and to lower rates of population growth. (3) The participation of women in the work force and in paid employment has contributed in part to the national economy and to improvements in the standard of living.

On the basis of these findings and given the working age contingent of the population and the demographic situation in Greece, it can be concluded that an increase in women's labor force participation will contribute to a more rapid socioeconomic development. However, such an increase in participation rates will unfavorably affect fertility trends unless the negative demographic factors of population evolution are considerably restricted and the economic and social conditions of women who are both mothers and labor force participants are further improved.

NOTES

1. Pressat, 1970.
2. United Nations, 1975a.
3. Kocher, 1973.
4. United Nations, 1975b.
5. Koller, 1949.
6. Long, 1958.
7. Friedlander, 1965; McDiarmid, 1977.
8. Wabe, 1932.
9. Koutsoyannis-Kokova, 1964.
10. Livi-Bacci and Haymann, 1971.
11. Mincer, 1962.
12. Lester, 1945.

13. Dernberg and Strand, 1966.
14. Zissimopoulos, 1975.
15. Koutsoyannis-Kokova, 1964.
16. Athanassiadis, 1951.
17. Collver and Langlois, 1962; Ostrorich, 1970.
18. Athanassiou, 1978.
19. Keyfitz, 1971.
20. Athanassiou, 1975.
21. Koutsoyannis-Kokova, 1964.
22. Ministry of Coordination and Center of Planning and Economic Research, 1979.
23. Friedlander, 1965; McDiarmid, 1977.
24. Klein, 1965.
25. Lianos and Prodromidis, 1974.

LABOR FORCE PARTICIPATION AND EARNINGS

OF WOMEN IN ANDALUSIA[1]

Michelle Riboud

University of Orleans and Casa de Velazquez
France and Spain

Over the last two decades, the study of women's behavior both in the labor market and at home has been aided by fruitful human capital and allocation-of-time approaches. With respect to women in the labor market, research by Polachek, Mincer and Polachek, and Landes[2] have shown that observed male-female wage differentials can be largely explained by differences in human capital investment behavior. Similar results are reported for a study using French data.[3] As suggested by human capital theory, these disparities are linked to different patterns of labor force participation. Persons who expect to spend less time—or participate in a discontinuous way—in the labor market will not make the same amount of investments aimed at enhancing their income earning potential as persons who do expect to spend most of their lives in the labor market.

To understand the labor force participation of women, the allocation-of-time approach has proved to be extremely beneficial. First, it has shown that women's behavior is not independent of the main characteristics of their households and that the relevant decision unit is more likely to be the family rather than the woman alone. Second, it has shown that the "world of work" is not restricted to the labor market, particularly for married women. To account for these effects, it is useful to base the classification on the distinction made within the nonmarket activities of women—work at home and leisure.[4]

This paper intends to apply these economic approaches to analyze the situation of women who live in Andalusia, the southern part of Spain, and to answer the following three questions. How does the human capital model work when applied to women? What are the main determinants of their earnings? Which economic factors determine their labor force participation?

THE DATA

Data on education, work history, and earnings, either on an individual or household basis, do not exist for Spain. Because of the absence of this relevant information from official statistical departments, a research program on human resources in Andalusia was initiated in 1978 which was financed by the Casa de Velazquez and the Instituto de Desarrollo Regional of the University of Sevilla. As part of this program, a survey was conducted between August 1978 and August 1979 of 22,000 households located, for the most part, in the province of Sevilla (only 2% from the province of Málaga). Information was collected on the age, educational attainment, work history, earnings, and social background of men and women. Information also was obtained on the date of marriage, number and birth dates of children, and distribution of time for various daily activities (for women only).

Despite the low educational attainment of Andalusian people (only 5% of the male population have attended college and 80% have only attended primary school or have not attended school at all), we decided to include in our sample a fairly equal representation of the levels of education. Approximately one-third of the family heads have 0 to 7 years of school; one-third, 8 to 14 years; and another third, 15 years or more.[5]

WOMEN IN THE LABOR MARKET: THEIR EARNINGS

As is now well known, human capital theory states that the level of earnings of an individual at a given time is a function of the human capital stock previously accumulated.[6] Human capital investment refers to the time, energy, and the amount of sacrificed income devoted to a period of training, either through education or on-the-job experience. This investment is then converted into income-earning potential. Thus, a worker's gross earnings at any point in time are equal to his earnings from his "inborn" stock of capital plus the sum of the returns of his previous investments. His net earnings are obtained by subtracting from gross earnings the amount invested in the current period.

Since our available information deals with school and postschool investments, we can concentrate on the effects of these two types of investments. School investments can be measured by the years of schooling. How postschool investments vary during an individual's working life remains to be defined. In previous analyses of males' earnings,[7] a linearly declining postschool investment profile has proved to be a satisfactory statistical hypothesis, which also is used in the present study.

The test of this human capital model with Spanish data on males' earnings gives results fairly similar to those which have been obtained in previous studies with data from the United States or France.[8] The dependent variable is the logarithm of annual earnings (net of taxes). School investments are measured according to the number of years of schooling and work experience by comparing the difference between the age at the time of the survey and the age at the time of entry into the labor force.

The explanatory power of the model is rather high: schooling and experience explain 51% of the relative earnings inequality and 57% when hours of work are taken into account. The average rate of return to school investments is equal to 8% and work experience appears to be a relevant variable to explain how earnings vary during paid employment.

When the model is applied to the analysis of women's earnings, Mincer and Polachek[9] show that a correct measure of women's work experience and their postschool investments requires taking into account the pattern of their labor force participation behavior. That is, not only are women's participation rates (especially those for married women) lower than men's, but they also spend less hours in the labor market than men and their participation is frequently discontinuous throughout their lives.

With respect to hours of work, we expect to find variability based on the family's demand for the woman's time in nonmarket activities, i.e., according to marital status and number of children. Women without children or without husbands (widowed or separated) will presumably work more hours in the labor force than others. The results obtained with the sample of Andalusian women show the correspondence between observed and expected behavior (Table 1).

TABLE 1

ANNUAL HOURS OF LABOR FORCE PARTICIPATION OF SAMPLED ANDALUSIANS BY GENDER, MARITAL STATUS, AND PRESENCE OF CHILDREN

	SAMPLE SIZE	MEAN HOURS
Married women with children	341	1,493.50
Married women without children	88	1,678.27
Widowed or separated women	43	2,006.14
Unmarried women	243	1,941.71
Men	1,914	2,128.17

The number of hours worked is greater for men than for women. Among women, those without husbands (unmarried, widowed, or separated) work more than married women. Among married women, there still exists a difference based on the presence or absence of children.

With respect to job discontinuity, results of comparisons indicate that only 16.8% of married women report having interrupted their working life once and 1.4% more than once. For approximately 55% of those with one exit and 30% for those with two exits, the motives for the interruptions are marriage or the presence of children. Other reasons include studies, health, or unemployment. As shown in Table 2, the average number of years of nonparticipation since school for this group is equal to 3.03 and since marriage, 1.71. The average number of years of experience in the labor market is 15. These women are on the average 35 years of age and have 2.19 children. Unmarried women are somewhat younger (33 years of age) and have an average number of years experience of almost 13 years.

On the other hand, most women who have worked some time in the past and are not working at present (81%) report having withdrawn either at the time of their marriage or in order to take care of their children.

TABLE 2

HISTORY OF ANDALUSIAN WOMEN WORKING IN 1979

	MARITAL STATUS	
	MARRIED	UNMARRIED
Sample size	429	243
Number of women with children	341	----
Age	35.40	33.24
Years of schooling	11.45	12.74
Age at first job	19.50	20.16
Age at marriage	23.70	----
Years of nonparticipation since school	3.03	2.12
Years of nonparticipation since marriage	1.71	----
Total years of experience[1]	15.00	12.90
Years with current employer	6.51	7.18
Number of children	2.19	----

[1] Present age minus age at first job minus time out of the labor force in the interval.

For this group, with an average age of 37.7, the numbers of years of nonparticipation since school and marriage are 11.80 and 11.22, respectively. Finally, we observe that a large number of married women (36% in our sample) report having never participated in the labor market.[10]

These results suggest that in Spain, at least in this particular region, women who show a strong labor force attachment try as much as possible to adjust their participation in market labor by decreasing the intensity of their work rather than by discontinuing their jobs for periods of time. This is certainly made easier in Spain than in other countries by the greater availability of housemaids and the still moderate prices for such services (33% of the married women who are working report having maids). [11]

As a consequence, the women's earnings function has been first estimated in the same way as for men, i.e., measuring experience by the effective number of years spent in the labor force (the variable of time is defined as the age at the time of the survey minus age at first job minus length of time out, if an interruption occurs).

In the next stage of the analysis two additional variables are introduced: marital status and the number of children.[12] The results reported in Table 3 indicate the following:

1. The average rates of return to school investments are 9.4% or 8.4% after adjustment for the number of hours worked.

2. The variable "age" is inappropriate in explaining how earnings vary over a lifetime. While age is statistically significant in men's earnings function, this result is not found for women. On the contrary, work experience affects significantly the level of women's earnings. This evidence contradicts the view still prevalent in European studies which holds that the earnings profile reflects an age phenomenon such that the relevant information to collect in surveys is age and never experience. For women, earnings are clearly shown to be a function of work experience and not of age.

3. As could be predicted from our previous comments, taking into account the number of hours worked raises the explanatory power of the model; the coefficient of determination rises from .387 to .567.

4. Human capital theory predicts a progressive decline in postschool investments over the working life. The evidence presented with data on women (as well as with data on men) does not contradict this hypothesis. The coefficient of t^2 is negative and statistically significant. Earnings profiles are concave.

5. Although earnings are shown to be a parabolic function of work experience for both men and women, the coefficient of work experience is lower for women than for men which indicates lower postschool invest-

TABLE 3

EARNINGS FUNCTION OF MARRIED AND UNMARRIED WOMEN BY EXPLANATORY VARIABLES[1]

Explanatory Variable	EQUATION									
	(1)	(2)	(3)	(4)	(5)	(6)	(7)	(8)	(9)	(10)
Constant	.189	2.060	2.220	1.790	2.230	-1.790	2.370	-1.680	-1.700	-1.730
Years of schooling (s)	.184 (10.800)	.175 (10.240)	.094 (18.920)	.184 (10.830)	.092 (20.100)	.084 (19.880)	.092 (18.810)	.083 (19.830)	.083 (19.830)	.115 (7.490)
s²	-.004 (-5.540)	-.004 (-5.040)	---	-.004 (-5.530)	---	---	---	---	---	-.001 (-2.140)
Years of participation (t)	.005 (2.220)	.005 (2.290)	.019 (2.760)	.021 (3.110)	---	.020 (3.370)	.024 (3.450)	.021 (3.500)	.020 (3.340)	.020 (3.390)
t²	---	---	-.0004 (-2.490)	-.0004 (-2.510)	---	-.0003 (-2.210)	-.0005 (-3.160)	-.0003 (-2.360)	-.0003 (-2.290)	-.0003 (-2.260)
Age (a)	---	---	---	---	.005 (.326)	---	---	---	---	---
a²	---	---	---	---	-.000008 (-.046)	---	---	---	---	---
Log of hours worked during period studied	---	---	---	---	---	.563 (16.490)	---	.551 (15.490)	.555 (15.340)	.538 (14.540)
Married	---	-.192 (-3.940)	---	---	---	---	-2.440 (4.940)	-.051 (-1.150)	-.067 (-1.300)	-.069 (-1.340)
Number of children	---	---	---	---	---	---	---	---	.009 (.605)	.013 (.880)
R²	.409	.422	.387	.414	.382	.567	.409	.568	.568	.571

NOTE: Values in parentheses are asymptotic t-ratios. Number of observations: 659.

[1] Dependent variable ln Y (logarithm of observed earnings).

ments for women, not only in monetary terms but also as a fraction of wage rates. Even if women had the same amount of work experience as men, they would still earn less because of smaller investments in the labor market; the contribution of men's actual experience to the logarithm of earnings is found equal to .431. The contribution of women's actual experience is found equal to .190.[13] Even if women have the same amount of experience as men, the contribution of experience to the logarithm of earnings would be .258.

6. The dummy variables of "married" and "number of children" are not statistically significant, which confirms what has been found in previous studies.[14] Those variables, which are often used as proxies for differences in labor force attachment, do not have explanatory power once correct measures of the years in the labor force and the number of hours worked are used to explain earnings.

7. The explanatory power of the human capital model appears similar when applied to men's or women's earnings as can be seen by comparing the coefficient of determination found in both cases, which are .571 and .568, respectively.

We may, therefore, conclude from this analysis that the main difference between the work of men and women is that, in addition to the fact that the latter work less hours and, as a consequence, earn less per year, their wage rates are lower because of lower postschool investment rates.

As predicted by human capital theory, lower postschool investments are linked to weaker prospective labor force attachment. Although the proportion of women who have interrupted their participation in the labor market is small, some women in the sample may leave the labor force, either temporarily or permanently, in the near future because of marriage or child care (20% of the sample of married women in the work force have no children). We expect smaller investments from these women. It also may occur that women with husbands tend to make less market-oriented investments than others. Some evidence in support of this hypothesis is observed in the regression of earnings on schooling and experience for a sample of 286 women heads of household (unmarried, widows, or separated). The investment rate rises: the coefficient of the variable t is now equal to .026 which comes closer to the one found for men (.030).

LABOR FORCE PARTICIPATION
OF MARRIED WOMEN IN ANDALUSIA

The labor force participation rate of women (all levels of marital status) is still rather low in Spain compared with other European countries: 26.7% of women ages 14 or older are recorded in the labor force in 1979. In France, for example, the rate is reported at 36.1% in 1968. In the province of Sevilla, the rate is somewhat lower: 17.9% for women ages 14 or older,[15] and for the other seven provinces of Andalusia the rates are between 15.3% (province of Jaen) and 24.3% (province of Málaga). Of course, these rates vary according to age and marital status as can be seen in Table 4.

The proportion of employed married women is very small and falls almost continuously with age after age 35. In many countries, a decrease in labor force participation occurs during the period of childbearing and child care which is followed by a subsequent increase (a return to the labor force after the age of 35 to 40), but this is not the case in the Andalusian region of Spain. This observation suggests a need to examine the factors which explain the labor force participation of married women and the effects of this participation.

Past research and the theory of allocation of time make clear that the appropriate decision-making unit in analyzing labor market supply is the family. This is particularly relevant for the decision of married women on how to distribute their time between market and nonmarket

TABLE 4

LABOR FORCE PARTICIPATION RATES
OF ANDALUSIAN WOMEN
BY AGE AND MARITAL STATUS

AGE GROUP	UNMARRIED WOMEN	MARRIED WOMEN	ALL WOMEN[1]
15–19	.351	.100	.339
20–24	.529	.108	.374
25–29	.599	.088	.208
30–34	.603	.070	.141
35–39	.604	.061	.124
40–44	.570	.056	.119
45–49	.560	.050	.128
50–54	.499	.056	.127
55–59	.457	.048	.125
60–64	.358	.034	.099
Total	.456	.064	.190

SOURCE: Data calculation from "Encuesta de Población Activa," Ministerio de Economia, Instituto Nacional de Estadística, Madrid, 1979.

[1] Includes widows and separated women.

activities, a decision which not only depends on tastes and biological conditions but also on relative prices. Men usually earn more than women in the labor market and, consequently, specialize in working in the labor market. Women, on the contrary, either specialize in home production or work both at home and in the market.

Let us assume that a family combines goods and services, either purchased in the market or produced at home, with consumption time to yield utility. The maximization of utility is subject to two contraints: a budget constraint and a time constraint. The marginal rate of substitution between goods and consumption time must be equal to the marginal product of work at home and to the shadow price of time (in real terms).

Whenever the woman is participating in the labor market, her shadow wage (the value of her work in the household) is equal to the market wage; whenever she does not supply time to the labor market, her market wage is inferior to her shadow wage. The probability that a woman will participate in the labor force is, therefore, a probability that her market wage will be larger than her shadow wage.[16]

As can be seen by the results of the past section, the market wage (or potential wage) is a function of education and experience. We expect to find higher wages for more-educated and more-experienced women. Education and experience should increase the probability of participation in the labor market.

The shadow price of a woman's time is expected to rise when family income (other than the woman's earnings) increases. An increase in income leads to an increase in the demand for the wife's consumption time, and we, therefore, expect a lower probability of participation in the labor market. However, when income from other sources than her work decreases or becomes more uncertain, we would expect the shadow price of the wife's time to decrease which, in turn, increases the probability that she will work in the labor market.

An increase in the number of children, especially of young children, makes it more difficult to substitute the wife's time with someone else's time. It raises the cost of working in the market, raises the wife's shadow wage, and lowers the probability that she will work.

When the number of hours supplied by the husband to the labor market increases, one expects the wife's time to replace her husband's time in nonmarket activities and, therefore, lowers her participation in market labor. To test those possible effects, we fit a logistic function to the sample of 1,614 Andalusian married women. Results of the analysis are presented in Table 5.

Work-Related Variables

Apart from the variables on levels of schooling of both husband and wife (proxies for their potential or actual levels of earnings) and number

TABLE 5

MARRIED WOMEN'S LABOR FORCE PARTICIPATION LOGIT ESTIMATES

Explanatory Variable	EQUATION									
	(1)	(2)	(3)	(4)	(5)	(6)	(7)	(8)	(9)	(10)
Constant	-1.530 (-5.530)	-.964 (-2.950)	-1.360 (-4.880)	-.740 (-2.250)	-1.020 (-3.080)	-8.210 (-7.700)	-1.470 (-4.130)	-1.150 (-3.340)	.009 (.026)	-1.510 (-4.180)
Wife's age	-.014 (-2.270)	-.031 (-4.620)	-.018 (-2.800)	-.036 (-5.510)	-.027 (-3.430)	-.006 (.728)	-.033 (-4.900)	-.037 (-5.530)	-.034 (-5.100)	-.031 (-4.660)
Wife's years of schooling	.172 (11.400)	.176 (11.800)	.194 (10.300)	.201 (10.660)	.173 (11.460)	.189 (8.740)	.219 (11.370)	.212 (11.090)	.205 (10.900)	.216 (11.240)
Husband's schooling	---	---	-.051 (-3.650)	-.055 (-3.890)	---	-.006 (-.357)	-.033 (-2.190)	-.040 (-2.730)	-.057 (-4.000)	-.036 (-2.390)
Husband's annual income	-.006 (-3.490)	-.006 (-3.710)	---	---	-.006 (-3.530)	---	---	---	---	---
Number of children	-.089 (-2.320)	---	-.106 (-2.800)	---	---	-.108 (-2.510)	---	---	---	---
Number of children age <7	---	-.258 (-3.460)	---	-.279 (-3.790)	-.264 (-3.520)	---	-.298 (-4.000)	-.282 (-3.800)	-.275 (-3.700)	-.306 (-4.080)
Number of children age >7	---	---	---	---	-.038 (-.906)	---	---	---	---	---
Work before marriage	---	---	---	---	---	6.330 (6.290)	---	---	---	---
Husband: Unemployed: number of times	---	---	---	---	---	---	.144 (6.150)	---	---	---
Change of employer	---	---	---	---	---	---	---	---	---	.126 (5.660)
Hours worked	---	---	---	---	---	---	---	---	-.00041 (-4.690)	---
Farmer or farm worker	---	---	---	---	---	---	---	.700 (4.030)	---	---

NOTE: Values in parentheses are asymptotic t-ratios. Number of observations: 1,614.

of children, the following variables can be introduced:

Work before marriage. This index is a dummy variable that indicates whether or not the wife had some labor force experience before marrying (36% of the women in the sample did not work before their marriage). Experience before marriage is expected to have a positive effect on the probability of work.

Hours worked. This variable consists of the total number of hours worked by the husband during the year. As explained before, the number of hours worked is expected to have a negative effect.

Unemployment. This variable is calculated according to the number of times the husband has been unemployed during the past five years.

Change of employer. This variable is calculated according to the number of times the husband has changed jobs over the last 10 years.

Farmer or farm worker. This dummy variable simply indicates whether or not the husband belongs to this category. The great majority of farm workers are usually employed only during periods of harvest, and unemployment rates are very high among them.

These last three variables reflect the instability of the husband's employment and the uncertainty which can affect the family and lead to important transitory variations in income. They are expected to have a positive effect on the wife's participation in the labor market.

Results of the logistic analysis confirm the hypotheses. Among the most important results are the strong positive effects of work experience before marriage and the wife's level of education on the probability that she will participate in the labor market. Census data[17] do show that 63.5% of women with a college education are in the labor force as compared to 14.9% of women who have only a primary education (or no schooling at all).

With respect to the number of children, we observe that only young children (ages less than 7) have a negative effect on labor force participation. The coefficient of older children is not statistically significant.

CONCLUSION

Human capital theory predicts that school and postschool investments are important determinants of an individual's level of earnings. The analysis of data on earnings of women in Andalusia does not contradict this hypothesis: a large fraction of the inequality of relative earnings is explained by these two forms of human capital investment. A comparison between male and female earnings shows that an important source of the income differential is the number of hours worked (women work less hours than men and married women work less than unmarried); another reason is that women have lower postschool investment rates, especially married women.

A characteristic of married women's labor force participation often observed in other countries is discontinuity. Although discontinuity exists in Andalusia, it is not very frequent and, therefore, does not seem to be a notable feature of women's behavior. Women who show a strong labor force attachment seem to adjust their supply of labor more by varying the number of hours worked than by job discontinuity. However, we still observe a large proportion of women who have never been in the labor market.

Among the most important factors affecting the probability of a married woman participating in the labor market are education and work experience before marriage. These characteristics promote a strong positive influence while the presence of young children exerts a negative one.

If one realized that, in this part of Spain, only 2.0% of all women have received a college or university education, 16.6% are illiterate, and 70.4% have only been to primary schools, coupled with the fact that fertility is still high (3.76 children per family in 1970), it is not surprising that many women find it unprofitable to shift time from nonmarket to market activities.

NOTES

1. This study is part of an ongoing research project financed by the Casa de Velazquez (a French research institute specializing in topics related to the Hispanic world) and the Instituto de Desarrollo Regional of the University of Sevilla. The author is grateful to Antonio Gonzalez Casas and Jose Fernandez Garcia for their computational assistance.
2. Polachek, 1973; Mincer and Polachek, 1974; Landes, 1974.
3. Riboud, 1977.
4. See Mincer, 1962; Gronau, 1977.
5. This low educational achievement is equally true for the other seven provinces of Andalusia.
6. See Becker and Chiswick, 1966; Mincer, 1974.
7. Mincer, 1974; Riboud, 1977.
8. Ibid.
9. Mincer and Polachek, 1974.
10. This percentage is certainly an underestimate compared to what it must be in the whole population.
11. The proportion rises to 52% for working women with more than primary schooling.
12. Widows and separated women have been eliminated because their sample size was very small (n = 41).
13. The average number of years of experience for men is 24.09, and the

mean for t^2 is 728; for women, those figures are 14.24 and 332.4, respectively.
14. Mincer and Polachek, 1974; Jones and Long, 1979; Hill, 1979.
15. For women between the ages of 15 and 64, the rate is 19%.
16. See Gronau, 1973, 1977; Heckman, 1978; Nakamura, Nakamura, and Cullen, 1979 a.
17. See Instituto Nacional de Estadística, 1976.

ECONOMIC ASPECTS OF FEMALE LABOR FORCE PARTICIPATION

IN THE FEDERAL REPUBLIC OF GERMANY

Wolfgang Franz

University of Mannheim
Mannheim   Federal Republic of Germany

Labor supply decisions are at the heart of economic theory and policy. Many studies, therefore, have been conducted to develop a theory of labor supply and to test the hypotheses empirically.[1] The purpose of this paper is to briefly review the trend and structure of female labor supply in the Federal Republic of Germany and to provide a background for an understanding of female labor supply within the framework of a family life cycle model.

One caveat should be kept in mind. The emphasis of this paper lies on economic aspects of female labor supply. These seem to be important but, needless to say, other aspects may be even more important.

FEMALE LABOR SUPPLY IN WEST GERMANY: TREND AND STRUCTURE

Changes in Labor Market Conditions since World War II

Before describing the trend and structure of female labor supply, a brief overview of some structural aspects of labor market conditions in West Germany after World War II may be helpful. The period 1950 to 1960, usually regarded as the reconstruction phase, is characterized by a high rate of growth limited primarily by the shortage of capital with labor in excess of supply. The increase in the labor force was caused partly by the influx of refugees from Eastern Europe and East Germany which amounted to a new immigration of 3.1 million persons, i.e., nearly 6% of the population. In addition, the female labor supply increased because, as a consequence of the war, many women were the main breadwinners of their families.

In the 1960s, labor market conditions changed rather dramatically. The construction of the Berlin Wall in 1961 stopped abruptly the immigra-

269

tion of refugees. Furthermore, the increasing demand for higher education resulted in a decrease in labor force participation rates. For
instance, whereas in 1960 only 15% of the male population and 9% of the
female population between 15 and 25 years of age attended high schools
and universities, the proportion increased to 25% for men and 18% for
women in 1970. The resulting excess demand for labor during this period
was reduced substantially by the immigration of foreign workers. Employment of these workers rose from 280,000 in 1960 to 1.8 million in
1970, i.e., from about 1% to 7% of the total labor force.

The third decade, 1970 to 1979, was characterized by a recession
which started in 1973. The official male and female unemployment rates
increased to a maximum of 4.3% and 6.0%, respectively. These figures,
however, do not tell the whole story since they include only those
unemployed persons registered at the Labor Office, which tends to
underestimate the true unemployment due not only to hidden unemployment but also to an "export of unemployment." The number of foreign
workers is determined in part by employment policy and, therefore,
declined to 1.9 million workers in 1978.[2]

In the long run, labor force participation of men and women showed
a similar pattern in that participation rates increased approximately 6%
since 1907 to 68% for men and 36% for women in 1939, but declined after
World War II to 56% and 31%, respectively, in 1978. With regard to more
recent trends of female labor supply, it was noted that this long-term
pattern was repeated after World War II. Female labor supply increased
from 31.3% in 1950 to 33.4% in 1961 and decreased in the following years.

The Significance of Changes in the Population's Age Structure

Such figures give only limited information because of changes in the
age structure of the population. At the very least, participation rates
should be based on a labor force and population between 15 and 65 years
of age. Usually, primary school ends at the age of 14 years, which
represents the first possible age of working life. About 65% of those who
leave primary school request apprenticeship training, 30% attend full-
time vocational training schools, and only 3% seek jobs.[3] The duration of
apprenticeship training is between 3 and 4 years. Youths attending high
school leave these schools usually at the age of 19. About 70% of them
change to universities and finish their studies at the approximate age of
25.[4] The date of leaving the labor force is the retirement age of 65 years
for men and 63 years for women, but they may claim a flexible retirement
starting at the age of 63 and 60, respectively.

A more careful analysis, therefore, should be based on participation
rates for different age groups. Table 1 is a presentation of the age
distribution at 5-year intervals. These 5 time frames from 1957 to 1977
have been chosen to account for an approximate business cycle of 5 years,
thereby avoiding an inadequate comparison between labor supply in a
recession and in a boom. As can be seen, total female labor supply in the

TABLE 1

RATE OF WOMEN'S[1] PARTICIPATION IN THE LABOR FORCE BY AGE AND MARITAL STATUS, 1957-1977

AGE	1957		1962		1967		1972[2]		1977[2]	
	TOTAL%	MARRIED%	TOTAL%	MARRIED%	TOTAL%	MARRIED%	TOTAL%	MARRIED%	TOTAL%	MARRIED%
15-20	76	55	72	60	62	55	60	57	45	59
20-25	76	50	71	52	69	50	66	57	70	64
25-30	52	40	51	40	49	40	52	45	59	52
30-35	45	36	45	37	42	35	47	47	52	47
35-40	44	36	46	39	43	37	47	43	51	47
40-45	42	34	47	39	47	40	49	44	52	47
45-50	39	32	44	36	47	39	50	44	50	44
50-55	36	29	40	33	42	36	46	40	47	40
55-60	32	26	34	27	36	30	36	30	39	33
60-65	23	20	22	19	24	20	18	14	14	11

SOURCE: Schwarz, 1978:473.

[1] Employed women divided by population according to each subgroup.

[2] These figures include German women only, thereby excluding immigrant women.

first age group declines across the years by about 30%, whereas the supply increases in the age group 30 to 40 years by approximately 7% and in the age group 40 to 50 by about 11%. Due to recent regulations concerning a flexible retirement age, the participation rates of women older than 60 years decrease substantially. The reason for the decline among younger women is primarily an increasing demand for higher education whereas the increases are attributed to an increment in the labor supply of married women. Also found in Table 1 is some support for a theory which describes the timing of female labor supply using three periods.[5] In 1971, according to Handl and his associates,[6] there is a peak (58%) in labor supply for married women at the age of 20. This peak is followed by a decreasing labor supply to approximately 35% at age 32. Starting at the age of 32, there is a slight increase from 35% to nearly 40% by age 40. The most important reason for this pattern is child care. Married women leave the labor force in order to rear children, and many return when their children become older. This effect, however, is not as strong as pointed out in some studies and is clearly not the sole reason for a re-entry into the labor market. Other reasons may be the desire for a higher standard of living (for instance, to buy a house) or a preference for working outside the home.

The Significance of Children on Labor Force Participation

The influence of child care on labor supply is shown in Table 2, which presents some evidence using recent census data. For example, if a married woman has no children (less than 18 years old), her participation rate is 46.5%. If she has one child less than 15 years old, her participation rate decreases to a value of 43.5%; if the child is less than 6 years old, it decreases to 35.6%. Although the first decrease of three percentage points may not be significant, the latter of 10.9% is substantial. In general, the rate of decline in participation increases with the number of children to be educated.

Table 3 shows that there is a substantial difference between participation rates of women with a primary school education and those with a university degree. In comparing the values in Tables 2 and 3, the implication is that there is a negative association between level of education and number of children.

Changes in the labor supply may be influenced by factors other than age, such as a change in the number of households or the number of persons per household. A decrease of the latter is equivalent to an increase in the former under the assumption of a stationary population. The resulting increase in labor force participation rates is not necessarily based on economic factors, but perhaps more on changing attitudes toward family life. For an economic analysis, it might be appropriate to separate these effects.

TABLE 2

PERCENTAGE DISTRIBUTION OF EMPLOYED WOMEN[1]
BY MARITAL STATUS AND NUMBER OF CHILDREN,[2] 1978

MARITAL STATUS	NUMBER OF CHILDREN				
	0[3]	1	2	3 OR MORE	TOTAL
Single	57.4	77.4 (65.4)	0	0	57.7
Married	46.5	43.5 (35.6)	35.4	31.3	42.8
Divorced	71.0	76.0 (54.0)	55.2	28.4	69.6
Widowed	29.2	45.2 (34.3)	33.3	0	31.1
Total	50.1	42.2 (32.7)	36.0	31.1	46.6

SOURCE: Statistisches Jahrbuch, 1979, p. 99.

[1] Employed women aged 15-65 divided by population aged 15-65 years for each subgroup.

[2] Less than 15 years old (less than 6 years old in parentheses).

[3] Less than 18 years old.

TABLE 3

FEMALE LABOR FORCE[1] PARTICIPATION RATES
BY LEVEL OF EDUCATIONAL ATTAINMENT, 1971[2]

	EDUCATIONAL LEVEL				
	PRIMARY SCHOOL	VOCATIONAL TRAINING SCHOOL[3]	HIGH SCHOOL	COLLEGE	UNIVERSITY
Participation rate	30.9	38.2 (48.1)	45.2	41.4	69.4

SOURCE: Handl, 1978, based on census data from 1971.

[1] Women more than 14 years of age.

[2] The categories are approximate due to the difference in school systems between the U. S. and Germany.

[3] The first number is for training in craft and trade, and the number for business administration appears below it in parentheses.

The Female Labor Force and the Influence of Foreign Workers

The female labor force is still heterogeneous. One reason that heterogeneity is still prevailing is the different participation rates with respect to German and foreign workers. This structural effect has been in operation since 1962. Since that time, the number of female employees has grown substantially. For foreign workers, participation rates are extremely high, averaging 98% for men and about 85% for women since 1972. This is not very surprising because foreign workers, for the most part, do not intend to become permanent residents, but plan to earn as much as possible in order to secure financial means for emigration. Hence, leisure is regarded as an "inferior good" and attitudes toward leisure may be quite different than those held by native inhabitants.

The intersectoral distribution of female employment is reported in Table 4. As can be seen, there is a substantial decrease in employment in the agricultural sector and a corresponding increase in the service sector. This picture is in accord with a similar trend for male employment and represents the decrease in the share of the agricultural sector in the Gross National Product as well as the remarkable progress in labor productivity in the agricultural sector.

DIMENSIONS AND DEFINITIONS
OF THE GERMAN LABOR FORCE

Labor supply has several dimensions, among which participation and hours of work are most important. Using workers in the manufacturing industry as an example, there is a downward trend in weekly hours for male and female workers, and hours of work are lower for women due to part-time work. Taken at face value, the trend noted for 1960 to 1978 suggests that the proportion of male hours of work to female hours remains constant (0.93) over the entire period. This result, however, may not be representative of all industries.

Finally, some remarks about the definition and measurement of the German labor force are in order. This definition differs substantially from others and sheds some light on various concepts of measuring hidden unemployment. All of the preceding tables (except Table 4) are based on a 1% and 0.1% census ("Microcensus"). According to the Microcensus, employed persons "comprise (a) all those, including unpaid family workers, who worked as much as one hour during the survey week and (b) all those who had a job or business at which they had previously worked, but from which they were temporarily absent during the survey week because of illness or injury, industrial dispute, vacation or other leave of absence, or temporary disruption of work for reasons such as bad weather or temporary breakdown. Persons on temporary layoff and career military personnel are also considered to be employed."[7] Unemployed persons in the Microcensus are defined as persons "14 years of age and over who are not at work in the survey week and who state that they are unemployed or that they are looking for work."[8] Using this restrictive definition results

TABLE 4

INTERSECTORAL DISTRIBUTION OF FEMALE EMPLOYMENT, 1962-1978

Time[1]	Total Employed Women	PERCENTAGE BY SECTOR			
		Agricultural	Manufacturing	Trade and Transportation	Service & Other
1962	9,825,000	18.8	32.8	19.7	28.8
1964	9,785,000	16.4	33.5	20.1	30.0
1966	9,700,000	15.4	33.7	20.4	30.5
1968	9,412,000	14.5	32.5	20.5	32.5
1970	9,582,000	12.4	34.1	20.6	32.9
1972	9,613,000	11.2	33.2	21.3	34.3
1973	9,734,000	10.7	33.3	21.3	34.7
1974	9,627,000	10.4	32.7	20.8	36.0
1975	9,366,000	10.3	31.3	20.7	37.6
1976	9,269,000	9.7	30.9	20.4	39.0
1977	9,638,000	8.2	30.6	20.4	40.8
1978	9,695,000	8.0	30.0	20.6	41.4

SOURCE: Statistisches Jahrbuch: 1979, p. 97; 1978, p. 96; 1977, p. 96.

[1] 1962-76: averages of each year; 1977-78: April.

in higher unemployment percentages than those reported by the Labor Office. The figures compiled by the Labor Office are most commonly used as a measure of unemployment in Germany and, therefore, unemployment rates are not comparable internationally. For instance, an 8.5% U.S.A. unemployment rate and a 4.7% F.R.G. unemployment rate result in a 6.5% and 4.4% rate, respectively, after adjustment according to Labor Office criteria.[9]

With respect to female labor supply, the Microcensus of 1976 reports 31,000 married and 28,000 not-married "unemployed" women more than those registered at the Labor Office.[10] This outcome can be questioned seriously, however, since the concept is not very operational. To put it differently, virtually all women "would like to work in principle," if wages are high enough and work places are convenient enough, regardless of actual conditions with respect to wages and work places. A question like this should be based on something pragmatic, such as a given wage rate, in order to obtain a more realistic estimate. Besides this, no claim can be made that a woman who would like to work (more) now will not reallocate a given lifetime labor supply only intertemporarily. A more sound basis perhaps is to estimate the number of those persons not in the labor force because of poor demand conditions ("discouraged worker"). Note, however, that the estimated number represents a net effect only, due to "additional workers" who enter the labor market in such economic situations as those needed to maintain family income.

In an econometric study, Egle[11] obtains an estimate of 185,000 women who might have been in the labor force in 1978 if a corrected unemployment rate actually would have been 0.7% instead of 4.3%. Among these women were 94,000 married women aged 45 to 65 years, but no significant "discouraged" married women could be found beyond that age. This finding is not consistent with that of an earlier study by the author.[12] Within a framework of a macroeconometric model of the German labor market, the highest and most significant long-run and short-run elasticity of labor supply of married women with respect to the (one-period lagged) unemployment rate was -0.028 and -0.015, respectively, within the age group 35 to 45 years of age. There might, however, have been a change in the behavior of discouraged workers. In fact, Pauly recently estimated a rather low elasticity (-0.0086) of female labor supply with respect to excess supply of labor.[13]

THE INTRAFAMILY LIFE CYCLE MODEL OF LABOR SUPPLY

How, then, might such a phenomenon be explained? A possible solution may be found in the intrafamily life cycle model of labor supply. According to this model,[14] the family determines current labor supply and consumption in a life cycle setting. The members of the family have preferences over present and future work as well as consumption which are formally expressed by a utility function. The problem is to maximize lifetime utility subject to a budget constraint that includes endowments of time available (valued by expected wage rates), wealth, and consumption goods (available at expected prices). With respect to female labor supply, the woman has to decide how many periods (e.g., years) she wants to work in her lifetime, how many hours in each period, and when within her life cycle this work should take place.

Leaving aside other than economic factors for a moment, the woman will be more likely to work as long as the market's reward for this behavior is not less than her preference for nonmarket activities. An increase in the wage rate at some phase in her lifetime, for example, may cause an entry into the labor market then or a higher amount of hours supplied to the market, but may also lead to a reduction of periods of work offered in other intervals of her life cycle. Hence, she substitutes work intertemporarily. On the other hand, if the increase in the wage rate is not large enough to induce her to enter the labor market, it will not alter her lifetime labor supply decisions at all.

This model indicates that people work harder in some years than in others because real wages are higher in these years than in previous years. "People are not tricked into extra work in a boom; they find the work desirable because the return, in the relevant intertemporal sense, is unusually high. Slumps are just periods when a lower level of effort is economically efficient."[15]

The decision-making unit is the family. In most families, income is pooled and the relevant income variable which influences consumption and leisure is family income. One consequence of this view is that there may take place an intrafamily substitution of labor supply. For example, assume an increase in the male wage rate, ceteris paribus. Even if the husband does not work more, family income and, therefore, demand for leisure increase, i.e., female labor supply decreases. This negative effect on female labor supply becomes stronger, if one takes into account that the husband is now more productive in market activities. To put it differently, he is more costly to the family in producing, say, home services, and, as a consequence, he is encouraged to participate in greater market activities. He reduces housework which is now done by the woman, thereby reducing her supply of labor to the market.

CONCLUSION

The decision problem itself is based on several choices which have been described by market and nonmarket activities. The latter covers a variety of activities such as housework, schooling, and leisure. Treating these activities as one composite good (called "leisure") is an oversimplification and can be justified only by the lack of sufficient data such as housework wage rates. As a consequence, the influence of higher incomes on the supply of labor may be overestimated: The increase in the demand for leisure may reduce the time devoted to housework instead of reducing the supply of labor.

An elaboration of the model and discussion of its theoretical implications are beyond the scope of this chapter. Such information, however, is included and discussed at some length in an expanded version of this paper which is available on request from the author. It should suffice to say that the theoretical considerations suggest a female labor supply model in which labor supply depends on wealth and male and female permanent and transitory real net incomes. The presence of wealth has a negative influence on the woman's labor supply as well as the permanent income of her husband. On the other hand, a higher female permanent and transitory income increases her labor supply. Moreover, it is well known that the total effect of a higher female wage rate on her labor supply is ambiguous since it depends on two contrary effects. The results of the empirical part of this study indicate that the so-called "substitution effect" on labor supply is positive, i.e., a higher wage rate makes the consumption of leisure more "costly" and, hence, the woman will work more.

NOTES

1. Most of the more recent work on female labor supply draws on the pioneering work of Mincer (1962). He has developed a model of the timing of labor supply decisions over the life cycle which has been extended in later work by several authors. At the present time, Heckman and MaCurdy (1980) have estimated a highly sophisticated version of a life cycle model by generating the entire set of parameters of the life cycle labor supply function. For a more recent survey, see Smith, 1980:3-23.
2. For a theoretical and econometric study of international factor mobility and the German labor market, see Franz, 1977.
3. For more details on youth employment and unemployment, see Franz, 1979.
4. The transition rates between high school and universities decreased in the 1970s from about 90% to 70%.
5. See Myrdal and Klein, 1956, for example.
6. Handl, Mayer, Müller, and Willms, 1979.
7. U.S. Department of Labor, Bureau of Labor Statistics, 1979b:105.
8. Ibid.:101.
9. Koller and König, 1977:184.
10. Brinkmann, 1979:8.
11. Egle, 1979.
12. Franz, 1974:131.
13. Pauly, 1978:160-165.
14. An incomplete list of studies of this type of labor supply models includes: Hall, 1979; Heckman and MaCurdy, 1980; Heckman and Willis, 1977, 1979; Mincer and Ofek, 1979; Lucas and Rapping, 1970; MaCurdy, 1980; and Mincer, 1962.
15. Hall, 1979:3.

PART IV

WOMEN AND THE WORLD OF WORK:
ROLE INTEGRATION AND REVISION

INTRODUCTION

Mary P. Wagner, Anne Hoiberg,

Camille Kim Cook, and Lawrence A. Palinkas

Naval Health Research Center
San Diego, California   U.S.A.

One of the most fundamental difficulties women face as they enter the world of paid work is that of role integration and revision. The authors included in this section address this issue directly by examining women's role and responsibilities in the family and how these are affected as the character of women's work undergoes change. In all of the contexts described, from rural, agrarian Turkey to industrialized America, one critical fact is evidenced: the changes in family organization and attitudes result from economic and ideological shifts within the society. While these changes are considered beneficial to society as a whole, the status of women is undermined by an incompleteness of transition. That is, whatever "wobble" obtains in the economic, political, and ideological realities within a society is borne most heavily by women in that culture. These shifts, in turn, can create an imbalance in women's sense of "wholeness," which is defined as an integrated collection of roles and expectations that provides a paradigm for both self-identity and social interaction.

The chapters in the other sections, which focus primarily on the economic, political, and ideological factors related to women's increased participation in the labor market, assess the prevailing conditions as unnecessarily unfavorable to employed women and present strategies which could rectify this situation of inequality. Throughout, there is a veiled hope that with time and concerted effort the position of women in the work force can be elevated to one of equal opportunity and equal remuneration.

Although several authors present solutions that might enable some women to overcome some of the obstacles, the few instances of optimism expressed elsewhere are difficult to detect in this section. As many of the authors point out, more and better research would clear away some of the myths and groundless assumptions about women and work, and more comprehensive legislation and enforcement would support and effect changes for the betterment of women. However, the bonds of family and children are, because of their subjective and emotional nature, resistant to enforced change from without. This realization necessitates a different locus of change and effort which will be discussed following an overview of the chapters and presentations. Obvious throughout is the crying need for justice, although it is recognized that a ferocious battle would ensue if equity were seriously demanded.

WOMEN'S WORK IN DEVELOPING COUNTRIES: FAMILY STRUCTURE

Perhaps the most poignant example of this need for equality is presented by Deniz Kandiyoti in her paper, "Women's Work: A Critical Appraisal within the Context of Rural Transformation in Turkey." The author critically examines the issue of women's productive and reproductive roles in relation to the structure of developing economies. Using the situation of women in the rural areas of her native country as a paradigm, Kandiyoti clarifies the connection between domestic and capitalist production and examines the dynamics of labor in Third World countries.[1]

A main pivot in Kandiyoti's discussion of the transformation of women's work in this rural setting is the nature of the context in which it is taking place. It is to be expected that the position of women would be radically different if changes occur within the context of the traditional patriarchy rather than within a system that allows women a greater, though relative, autonomy.

This comparison requires a description of the dynamic and gender roles of the traditional household and the labor process. Unlike sons, who continue the family lineage and inherit the family property, daughters at an early age are married off into another lineage and enter their new families as the lowest status members. The function of these young women is to produce sons and contribute labor to agricultural production. It is the function of the older women to perform household tasks and to supervise and coordinate the activities of their sons' brides. This role shift, which occurs as a woman's sons marry, reflects the norm of a woman's possible status achievement in classic, patriarchal, agrarian cultures: the opportunity to withdraw from agricultural drudgery to engage in household duties is attributed to the presence of her sons' brides who are lower in the family hierarchy and, thus, are subject to her supervision.

Several factors are brought into sharp relief in considering this tradition. First, the only possibility of escape from toil in the fields is the presence of other women; therefore, a woman's elevation in status must always be at the expense of other women. Second, in order to have these women to dominate, she must produce sons to whom she will be subservient. And third, paradoxically, the young female family members who are valued least are the ones who do the hardest work. That these young women are willing to accept such circumstances can be explained not only by acknowledging their conditioned submissiveness, but also by recognizing that they perceive following in their mother-in-laws' footsteps as their only avenue for advancement within the family and the family as their only possibility for security or survival.

Using this description of the traditional family as her reference point, Kandiyoti observes that the actual changes in rural Turkey are based on the interrelation of the household as both a domestic and production unit. When there is a shift in the labor process within the household, the family social organization will necessarily change to accommodate it. A major impetus of this change has been the penetration of capital resulting in a consolidation of land ownership and, even among the wealthy landowners, a diversification of income sources. The typical pattern of the extended family deriving its income solely from the individual family property has been gradually dissolving; the traditional standard of residential extendedness is giving way to the nuclear family.

The inability of parents to exert the traditional rigid control over the young is a direct result of this transition in the basic economic structure. Being less financially dependent on the family patriarch, many young men now choose their own marriage partners, participate in family decision making, and establish their own households after reimbursing the family for the price of their weddings.

Kandiyoti emphasizes, however, that this movement toward a greater independence for the young, nuclear family does not necessarily provide greater freedom for the young wife. Although many more young married women head their own households when they normally would still be subservient to their mother-in-law, their exploitation emerges in other forms. Oftentimes the young husband will supplement the family income by temporarily migrating to find work, leaving the wife with total responsibility for the family's agricultural production. This migration, however, is seasonal, at most several weeks per year. During the time that the husband is _not_ working away from home, the wife alone still performs the tasks which the couple originally shared, in addition to remunerated tasks, such as rug weaving, and the domestic chores. In short, when a husband's labor force participation takes him away from traditional duties, the wife is expected to assume his duties in addition to her own. When he returns, he does not reassume his traditional duties, but requires that his wife continue to perform them while he remains idle. In effect, a new, prevalent norm has been established for the rural, nuclear family which allows the husband to retain his traditional control over family resources (goods, money, high-quality food) and policies without performing his traditional duties. Even though the women are conditioned to be submissive, the increased dependence on women's labor without extending them greater autonomy has been made possible only by means of harsher and more violent subjugation.

The crux of Kandiyoti's presentation, that is, the context of the transformation of the rural family in Turkey, reflects clearly both the penetration of capitalism and the traditional patriarchy. The former emancipates the young woman from subservience as the lowest status member of the extended family. The latter, far from dead, continues to deprive women of opportunities for education and mobility, excludes them from family decision making, and places the burden of labor production squarely on their shoulders without commensurate remuneration. This system of making up the difference of the leftovers from the transition is another aspect of what Clair Brown calls "bringing down the rear": whoever is last in order of priority is encumbered with the greatest burden.

WOMEN'S WORK IN THE HOME:
FAMILY STATUS PRODUCTION

In another paper, prepared by Hanna Papanek, an effort is made to clarify the importance and significance of women's work by examining the trade-off between women's paid work outside the home and the various activities within the household. As pointed out by Kandiyoti, the division of labor and rewards within the household is not a static situation, but rather is subject to renegotiation as conditions change. This flexibility is of special importance to the study of women's work because women in most societies are more likely than men to shape their work around the demands of the family setting.

Papanek's basic thesis posits the importance of status in the eyes of others as a given value in most cultures; within the society a degree of consensus obtains regarding the accepted criteria for the conferring of status. Her theory of family status production as an alternative to paid work outside the home proceeds from the concept of status relations and focuses not on the content of women's work, but on its intended and unintended consequences for the family unit in the context of the community.

Family status represents the collective interest of the household to which the woman belongs in relation to the social order. It is judged in terms of both controlling and maintaining assets; thus, a contribution to family status usually has an effect on income or control of assets. A woman's family status production work, therefore, in some way maintains or improves the status evaluation of her household by the community, and any loss of status can have serious effects on future negotiations for resources between the family and the community.

Types of Family Status Production

Family status production is an alternative way for women to make economic contributions to the family by expending their time, energy, and skills to enhance the ability of other household members to increase the family income as well as by maintaining and enhancing the family's secular and religious status to enlarge the family's control over assets. Four categories of family status production work are described to illustrate this concept. [2]

Support work is generated by the conditions of another family member's paid work. It is often requested directly by the person, but is rarely explicitly required by the employer. Examples include: feeding co-workers, washing and ironing work uniforms, performing secretarial or other professional work, making speeches, engaging in political activities, appearing at functions, and counseling parishioners.

These examples of support work share certain common features. The work is not directly paid, but is rewarded indirectly in the form of appreciation and approval from those family members who benefit from it and who have integrated its performance into their expectations. In some families, there is an expressed assumption that because the men's earnings pay for women's food and lodging, women are amply rewarded and men are entitled to expect such services. Indirect rewards also may include access to "benefits" (e.g., health insurance) that are features of certain middle-class jobs. The work is usually gender-specific and contextually integrated into the wifely role, being part of a "hidden contract" between members of the household or couple.

Status-appropriate training for children is especially important for middle- and upper-class families and includes: supervising homework; educating children in status-appropriate language, behavior, and presentation of the self (i.e., choice of clothing, hairstyles, ornaments, etc.); and training in particular skills, such as sports. This category also includes encouraging children to participate in formal education, motivated in part by the parent's anticipation of their children's role in providing them with old-age security.

Political activities related to status are explicitly associated with the community: exchanging gifts, obtaining and conveying information, and preparing feasts are often very important in maintaining a family's social position. The most important activity in this category, which overlaps with status appropriate training, is the making of marriage arrangements, a crucial parental function in many societies. Papanek emphasizes the importance of this datum by reminding us that in societies where women's employment opportunities are limited, the arrangement of a good marriage may be the only contribution the parents are capable of making to a daughter's future economic security.

Rituals, the fourth category of family status production work, pertains to those rites that maintain the family's relationship with God or gods. The performance of these rites also indirectly affects status within the community. Women often assume greater responsibility for these ritualistic practices than men, especially as men of the culture increasingly adopt public and global roles which require a more internationally standardized mode of behavior. Papanek points out here that the pressure on women to maintain family and national cultural traditions, though serving many functions, ensures that the social position of women remains unchanged. Again we see the paradox described in Kandiyoti's presentation: As the opportunities for breaking with past traditions and establishing new norms open for the men of a culture, pressures on women to embody the traditional mode intensify.

Family Status Production: Underdeveloped versus Developed Countries

An important condition for the prevalence of family status production over the paid work of women is the norm of shared income. For women who work outside the home, the contribution of status production may not be significant; whereas for those who have no alternatives to living and working within the household, it is very important. Papanek identifies several Asian countries in which the options for women are so few that, as we saw in the Turkish case, the family is the only possibility for women's survival and security. For the three countries of India, Pakistan, and Bangladesh, the extreme of this condition is manifested in the phenomenon of purdah, the virtual segregation of women from unrelated men as a means of meeting the required norm of modesty. These women work: Their work in the home is enmeshed with status and emotion, and the fact that they perform the task is often more significant than the work itself in securing the desired status. But because of a dependence on family support, their family status production work may be highly exploitative, enhancing the family's position without improving the women's status within the family. Thus, once again, women are left at the mercy of marriage and household arrangements wherein they have no bargaining position or education or employment.

An examination of industrialized western society initially looks quite different. Status evaluations are increasingly dependent on life style, ownership of consumer goods, and types of leisure activities. Because the "technology of leisure," consumer services, education, and health and social services are all high-cost items, the four categories of status production work may be declining in status productivity in comparison with earned income. This shift in consumption patterns, which is based on money income rather than the "unpaid" services of women in the home, accounts to a great extent for the massive movement of women into paid work.

At first glance, this attachment to the labor market, as contrasted with the locked-in position of many Asian women, looks rather rosy. Just below the surface, however, are many concerns which Papanek addresses through a critique of the adequacy of the conventional social sciences. Basically, her criticism centers on two related issues: First, the social sciences today reflect the life experiences of white, middle-class, educated men in Western Europe and North America in the late nineteenth and early twentieth centuries; and second, the misperceptions of women's roles are linked to the social vision that underlies conventional research and planning, a vision that sees men primarily as workers and women primarily as homemakers. According to Papanek, it was important for men of that era to assume that women did not "work." Their narrow vision, generalized from private experience, was imposed and accepted as the norm.

The Undervaluation of Women's Work

Today all evidence is to the contrary, but this vision is so deeply rooted in the assumptions of social researchers that even in Eastern Europe and the USSR, where women in great numbers have participated in paid labor for many years, researchers addressing the problem of women's work overload continue to ask women why they work. Although this view is based on the primacy of women's family responsibilities, the study of women's work in the home has not been a prime target for social scientists. In fact, women in paid work and housewives have both failed to be absorbed into the theoretical models and research methods of sociologists concerned with status and stratification.

Papanek's theory of family status production provides us with a key to understanding this situation. As she states above, women in most societies are more likely than men to shape their activities around the demands of the family setting; thus, when the family requires more monetary income, women enter paid labor. Social norms, however, continue to stress that women's work is—or should be—in response to family needs. The dilemma Judith Stiehm points to in her paper on women elected officials is a clear demonstration: While pictures of male candidates with their young children are believed to generate votes, women officials avoid being shown with their small children (or delay running for office until they are grown) because the assumption of the electorate is that holding office will cause her to shirk her primary familial duties. That family needs do, or should, determine the nature and character of a woman's work (and that her personal undertakings, there-fore, are ancillary) are basic to the expectations of not only social scientists, but employers, legislators, educators, lenders, and, especially, other family members.

Papanek's application of her theory of family status production emphasizes two very clear and critical observations: First, women work, and the undervaluation of household and family tasks overlooks their importance in maintaining and enhancing the status of the family; and second, family status production work is often extremely exploitative, especially in societies where it represents the woman's only work option. The most crucial finding is that, in the accepted vision of the culture, women have no recourse to their responsibilities for family status production work whether they work in the home or are remunerated in the labor market. Women employed with men continue to be seen as "homemakers" in an extended sense which perpetuates the traditional evaluation of their work as transient, supportive, and adaptive at the expense of its real worth and integrity.

This traditional perpective was not without representation at the symposium. A paper delivered by Horst Helle presents a traditional view of sex roles and family types. He supports his separate, different, but somehow-still-equal stance with a number of arguments, both biological and cultural. The contrast between Helle's traditional perspective and the more prevalent, progressive views sparked lively and interesting discussion at the session on family considerations and women's work.

IMPACT OF FAMILY STRUCTURE ON WOMEN
IN THE WORLD OF WORK

The obvious emphasis that the papers place on the structure of the family is relevant and central to the concern here of investigating the role of women as family members and their relation to the world of work. In Kandiyoti's presentation, the transition from the norm of residential extendedness to the nuclear family is attributed to an economic transformation within the society. Papanek postulates that a shift in the values within a given culture evokes a change in the division and focus of labor within the family and, consequently, changes the shape of the culture itself. Although the content of women's work may be altered due to these societal changes, the presumption remains intact that the family's needs dictate the nature of women's work.

Papanek's criticism of the conventional social sciences as perpetrators of a narrow and illusory vision of women as workers was echoed and expanded upon by Elisabeth Beck-Gernsheim in the paper she presented, "Where Experts Fail: Nontraditional Family Patterns and How Professional and Popular Knowledge See Them." Beck-Gernsheim's view is that social scientists have failed to perceive the family structure as a fluid one that can adapt in response both to economic change from without and to changes in expectations and goals by family members themselves. The "experts," she claims, have held up the traditional nuclear family and the typical roles and responsibilities as the unalterable norm, against which all innovations in response to changed expectations and desires are seen as deviant, problematic, and dysfunctional. For example, growing numbers of women and men, who view the traditional structure and its division of labor between the sexes as too narrow and restrictive, look to social scientists with hope for clear-cut directives and are discouraged by biased and partial advice that continues to give responsibility to women in matters of both child rearing and marital stability. This advice, she asserts, is implicitly oriented toward reestablishing the traditional nuclear family.

Once again, the recurring theme surfaces with a new face. With the economic and ideological changes evidenced in Western industrialized society, women's labor force participation has become crucial to the maintenance of the economy and vital to the family, regardless of socioeconomic class. Although the natural response of women is to seek a revision of their familial role that would take into account their new responsibilities, this role transition has met with rigid resistance. Women are being encouraged both to change and not to change, that is, to become full-time (though not full-fledged) labor force participants without expecting that any support will be forthcoming to ease the load of traditional responsibilities.

ESCAPING THE DOUBLE BURDEN

Beck-Gernsheim's central point is that family patterns which are creative alternatives to the "double burden" must be found, and that they will most likely be discovered in spite of, rather that with the help of, the "experts." The two possibilities she examines in depth are the choices to postpone childbearing or to remain childless; both of these options have traditionally been denigrated as either selfish or dangerous or both.

The traditional view, which posits that children both hold a marriage together and enhance the couple's happiness, is being challenged by many contemporary couples for the following reasons. (1) Children no longer represent clear economic benefits; on the contrary, they constitute a considerable financial burden. (2) The problem of "who will care for the children" creates a serious and difficult tension for couples aspiring to intimacy and equality in marriage. (3) Children can limit drastically the couple's options in the realms of travel, social activities, career development, and economic success. The importance of taking these factors into account has been borne out by recent studies which show that parents indicate they are subject to greater pressure and stress, feel significantly less healthy, and find their marriages restrictive whereas couples without children are less likely to report these adverse circumstances and are more likely to speak of happiness and general satisfaction.

Despite these findings, most researchers continue to classify as deviant behavior family decisions that would limit the number of children and allow couples more options, such as the decisions to bear children at an older age or to remain childless. By choosing to investigate these two nontraditional family patterns, Beck-Gernsheim indirectly highlights the issues of the other papers in this section: child care and the responsibilities of motherhood.

Child Care and Motherhood: New Studies and Ideas

As Harriet Presser states in her chapter, "Child Care Use and Constraints in the United States," women do in fact have major responsibility for the rearing of children, regardless of one's convictions about the appropriateness or inappropriateness of this situation. Therefore, increased female labor force participation is contingent on a corresponding expansion of substitute care for children.

Presser's study examines the extent to which the scarcity of quality, low-cost child care constrains the labor force participation of both employed and nonemployed mothers of preschool-age children. Her findings support the contention that much hidden unemployment exists within this group; many women who are not presently participating in the labor force report that they would probably do so if they could find and afford satisfactory child care. Further, women in this group who are presently employed part time indicate that they would most likely work more hours if suitable and affordable child care could be obtained.

Presser's research also points up the association between specific occupations—both the wife's and the husband's—and the use of child care. Her finding that fraternal care is proportionately higher when the mother does shift work provides a basis for two tentative conclusions. (1) Couples seeking alternatives to the traditional mode can share the responsibilities of child care and family income more equally by adjusting their work hours to meet these needs. (2) While this is a constructive step, it is obvious that the adaptation, both in terms of type of employment and hours of the day worked, is being made by the mother who adjusts her work hours to the father's more regular work schedule.

The most important aspects of Presser's study are the questions she raises and the light she sheds on the need for further research. By posing the fundamental query—who is caring for the children and what arrangements are currently being made—she calls attention to the issue which, if not addressed, constitutes a primary limiting factor to the equal participation of women in the world of paid work.

An interesting complement to Presser's study and her unanswered questions is Henning Transgaard's chapter, "The Attitude of Danish Mothers to Child Care, 1975 to 1979." He presents a detailed analysis of child care attitudes in Denmark, a country where child care has official government priority and receives substantial support from government funds.

Transgaard's study shows that an individual's attitude toward child care in general, and toward the specific forms available, cannot be isolated from other factors. He contends that the mother's decision about what she wants to do with her life, a decision enmeshed in a conflux of attitudes about appropriate sex roles, is strongly associated with her attitudes toward specific forms of child care. Therefore, the variable of the mother's preferred occupational status is inextricably entwined with the issue of child care. Through his use of a wide range of variables, Transgaard reveals the extent to which attitudes toward child care reflect other societal values. This finding is germane to the themes of the other papers in that it both reiterates the primary need to clarify and examine the values upon which choices are made, and discourages us from attempting to treat the illness without treating the cause, as in the effort to provide forms of child care not consonant with the life choices of those involved.

In addition to this study of attitudes, Transgaard investigates the impact of Danish child care on women's employment. In this respect, his presentation is similar to Presser's. However, because he examines a wider range of variables and attitudes, he succeeds to a great extent in filling the research gap described by Presser and others.

Sandra Tangri's chapter, "Research on Women's Work and the Family at the Urban Insitute," also addresses both the child care issue and those associated with women and the family. Her work, and that of other researchers at the Urban Institute, is directed toward identifying problems and providing solutions or clarifications where obscurities remain in order to supply solid factual bases for public and private decision making. Her paper surveys such topics as displaced homemakers, teen-age childbearing, the consequence of remaining childless, and other issues central to women's labor force participation and the ramifications of that participation.

Specific issues that Tangri raises are the inequities in the income tax and Social Security systems for married women who are employed outside the home. Several solutions, which have been presented before Congress, also are discussed. Another interesting point that Tangri makes is that occupational training in high school may be harmful for women because the preparation received is directed only toward placement in "women's jobs." She also stresses that men employed in jobs that are predominately women's occupations receive lower wages than in jobs employing few women.

The final chapter in this section is by Barbara Bergmann who surveys the problematic issue of the relation between women's labor force participation and the economic support of children. Somewhat similar to Brown's thesis, Bergmann's objective provides a philosophical and historical overview of the trends in economic flows to children from their mothers, fathers, and the state and clarifies the problems women must resolve as the structure of the family and society undergoes change.

Bergmann's research approach has a dialectical quality: Women are suffering both economically and personally because of society's assertion that when the traditional family breaks down, women are expected to fulfill double roles—mother and father as well as provider and caretaker. But, in the face of the American version of what we have seen as a chronic dilemma for women, Bergmann halts those on the path of retreat to the old, overwhelmingly oppressive social institutions. She frankly discusses the traditional disadvantages of marriage and applauds the presently possible and more conscious options.

Most important, however, is her realistic study of the contemporary economic bind that many women experience and her assessment of both causes and possible solutions. Legislative and judicial decisions could have a powerful impact on the enforcement of child support payments by centralizing the process and attaching the distribution of payments to the Social Security system or the Internal Revenue Service. The economic support of many single-mother families, moreover, directly affects all of us through paid taxes that are funneled into the welfare program.

Although the money distributed through the welfare system is a constant source of controversy, there seems to be little awareness of the fact that the major cause of the high cost for this program is attributable to the delinquency and irresponsibility of the fathers.

Another issue discussed by Bergmann is the serious problem of work overload for married, employed women which could be significantly alleviated if the major organs of communication and education were utilized to influence changes in the distribution of household and child care. And, of course, low-cost, high-quality, child care must be made readily available. Bergmann is another herald for creative and innovative solutions to the real and essential problems we have traced as foremost to women today.

CONCLUSION

At the outset, it was observed that women who are family members must resolve a major role conflict as they enter the world of paid work: the problem of role intergration and revision. In addressing this basic issue, it would seem that, on the face of things, women are without resource. As many authors emphasize, the assistance that could be given by "expert" advice and legislative support has, for the most part, not been adequately extended. Morover, recent political trends bode ill for those who would seriously undertake to stretch into the household the equality that has been accomplished (although far from optimal) in the workplace. This goal is by no means one that we can presume will be achieved by ordinary efforts or ordinary means.

Initially, it appears that certain realities have to be acknowledged. As Deborah Freedman, the symposium discussant for the session on "Poverty Viewed as a Woman's Issue" stated in her remarks, ". . . it is important to remember that for better or worse, most women spend most of their lives in traditional families, divorce notwithstanding, and their answers to survey questions indicate that they value that kind of life." It also should be noted that most women consider marriage and the traditional family as desirable objectives, which is reflected in the high marriage rates recorded during recent decades (reaching a peak of 95% for the 1950s generation) and currently showing a strong resurgence.[3] Nevertheless, less than 16% of all families in the United States today, according to Reagan, conform to the traditional model in which the father is the breadwinner and the mother remains at home to provide household and child care. The remaining 84% consist of single persons, couples, one-parent families, other household units that include children or other dependents, related adult units, nonfamilial adult groups, and institutions.

The efforts of the women's movement have been strong and consistent in endeavors to achieve absolute equality for women in our society. But these efforts have not, until now, firmly addressed the fundaments of the issue at hand. Women today are confronted with the task of integrating the changes in the criteria for social role performance and the conflicting expectations of their position within the family. Such integration is crucial to the maintenance of a sense of "wholeness" which enables an individual to operate effectively in a social context. If women's lives are to be reshaped to include both equality and wholeness, then we are faced with difficult vistas of action.

The first of these vistas focuses on the need for education. Women should be informed of various employment opportunities, provided with the means for determining their occupational objectives, and trained in the skills necessary to accomplish these goals. Career counseling, workshops, and job training programs are particularly relevant. Most important, women must be convinced, as many of the symposium participants conclude, that they have equal need, along with men, to be educated and trained for jobs. Early on, girls should be taught the importance of making the difficult commitment to educational preparation, because the "Prince Charming" view of marriage is both currently unrealistic and overwhelmingly oppressive.

By encouraging this commitment, however, only half of the educative issue will have been addressed. If left at that, another generation of women will have been exhorted to assume the double burden. The dichotomy that presently obtains in the culture must be brought home: the traditional mode is valued in itself, but women who embody it are comparatively undervalued. Women are faced with an untenable imperative to both maintain the traditional norm that is abstractly prized and settle for being undervalued because of their participation in it. Also within education is a need to question the criteria by which women judge themselves and challenge the assumption that their real personal worth is measured by the degree to which they meet the traditional norm.

The second vista for action turns this challenge on women themselves. Transgaard's study showed that child care choices proceed from basic personal values. In analogizing from this, it is suggested that choices in family structure and career reflect the values of those participating. Because values determine decisions, the outcome of changing the form without transforming the content will lead to eventual failure. Therefore, the radical and basic changes must begin in the individual, facilitated by education and role models. Support systems of any kind, be they legislated programs, educational training, stricter child support enforcement, or whatever, can simply provide a framework. These efforts supply only a limited, temporary solution—unless individuals desire the changes and believe in their fundamental value.

With the help of organizations devoted to equality, women and men can pull together, assiduously demanding the educational, legislational, and judicial support which must aid their own efforts. More fundamentally, they need the courage to exercise those options which directly or indirectly contribute to their wholeness and well-being. As human beings reaching for wholeness, they cannot afford to exclude family and intimacy from their aspirations by categorizing them as inherently oppressive, but they cannot allow the momentum of past expression to swallow them up.

The lack of exuberant optimism in these chapters is in no small part attributable to the intrepid courage that is required to surmount the problems they delineate. The needs for both earning capacity and rewarding work have to be recognized by women's families and elevated to a position of priority equal to that of men. Domestic tasks and child rearing can be defined in the family self-concept as household duties that are performed by household members, a shift in the division of labor away from rigid role orientation.

The challenge for women is to be relentless in reiterating these essential needs as governing the choices they individually make. Also of importance is the belief that women can have their deepest longings for love and union fulfilled without abdicating their need for equality. Further, women cannot permit their needs to be compromised nor allow themselves to be swayed by the empty guarantees of security and the opportunity to evade the worries of the world that Prince Charming promises—with his fingers crossed behind his back. Women have to, and they can, keep the vow to themselves to stand their ground and educate their loved ones in becoming whole.

NOTES

1. See Kandiyoti, 1974, 1977.
2. For an elaboration of the concept of family status production, see Papanek, 1979.
3. U.S. Department of Commerce, 1979.

CHILD CARE USE AND CONSTRAINTS IN THE UNITED STATES[1]

Harriet B. Presser

University of Maryland
College Park, Maryland   U.S.A.

The study of child care is a critical aspect of both understanding and changing women's status in the labor force and at home. Regardless of one's convictions about the appropriateness or inappropriateness of women having major responsibility for the rearing of children, the fact that they do represents a major constraint on female achievement outside the home and on leisure time within the home. This responsibility for child care is particularly relevant for industrialized societies which have been experiencing an overall increase in female labor force participation, most notably among mothers with young children. This phenomenon is contingent on an accompanying expansion of substitute care for these children. A related trend is the increased use of child care among nonemployed mothers who leave their children in someone else's charge in order to attend school, do volunteer work, participate in recreational activites, or perform household tasks, including the care of other children. Who are the individuals watching the children, what are parents doing with their free time, and are child care needs being met?

As basic as these questions are, our answers are limited due to the lack of data. Moreover, results of studies that have been conducted are not readily comparable because samples are very different. Let me briefly review what we know about child care use and constraints in the United States.

OVERVIEW OF LITERATURE

Much of the research on child care focuses on the types of arrangements that are made.[2] A general finding is that children of employed mothers are cared for mostly by relatives, neighbors, and baby-sitters, and that such care is much cheaper than licensed, institutionalized arrangements such as nursery schools and day care centers. Without

295

informal low cost arrangements, many women would not consider it economically feasible to work—particularly those with preschool-age children.[3] There is evidence that the presence of nonemployed female relatives in the home is associated with relatively high employment rates among women with young children.[4] Such families, however, constituted only 4.8% of all households with children under 6 in 1970.[5] Ditmore and Prosser[6] have argued that, for low-income mothers, the expansion of child care facilities would increase the number of women in the labor force as well as the hours worked among the currently employed. Economists have also considered the cost of nonfamilial child care arrangements in relation to demand.[7] While there has been some consideration of the cost to the U.S. government of increasing the child care deduction, there has been no assessment of the effect that the recent tax credit for child care has on demand.

The extent to which child care is a perceived constraint on employment has received only minimal attention and the information that we do have comes from two studies reported in the United States—Dickinson,[8] using data from the 1973 National Panel Study of Income Dynamics, found that among nonemployed mothers with children less than 12 years of age, 68% felt that, if they wanted to take jobs, some child care arrangements could be made; 16% were uncertain. In the Westinghouse study of families with annual incomes of $8,000 or less, 18% of the nonemployed mothers said they were not employed because they could not make (or afford) satisfactory child care arrangements.[9]

In this chapter, we will be looking at both child care use and child care unavailability as constraints on employment for employed and nonemployed mothers with preschool-age children. First, the characteristics of the sample are briefly described and, second, the preliminary findings are discussed.

SAMPLE CHARACTERISTICS

The June 1977 Current Population Survey (CPS) was based on a sample of approximately 53,500 interviewed households. Child care questions were asked of all women aged 18 to 44 with children under 5 years of age residing in the household. (The 5-year-olds were excluded because most attended kindergarten and, thus, were not truly preschoolers—an important restriction when analyzing child care arrangements.) Women's adopted and stepchildren were included, along with their natural children. The subsample consisted of 8,331 women: of these, 2,996 were employed and 5,335 were nonemployed (that is, either unemployed or not in the labor force).

On the basis of this national sample for June 1977, an estimated 35.4% of all women with children less than 5 years old were employed, 5.5% were unemployed, and the remaining 59.1% were not in the labor force. Of the employed women, 65.1% were employed full time and 34.9% were employed part time. Employed and nonemployed mothers in this sample differed very little with respect to selected social and demographic characteristics: mother's age, marital status, race, and education. Family income was only slightly lower among the nonemployed than the employed. Also, for married women with a husband present, there was little difference by employment status in the husband's socio-economic characteristics: his education, labor force status, and occupation. The critical distinction between employed and nonemployed mothers related to their stage of family formation, not their socio-economic status. Nonemployed mothers, as compared with employed mothers, tended to have larger families, more children less than 5 years old, and very young children under 5 years of age.

CHILD CARE USE

Let us turn now to the child care data. To what extent do mothers of children under 5 years of age in the United States utilize child care? The questions on such arrangements were asked differently for employed and nonemployed mothers. For those employed, child care was broadly defined to include any care while the mother is working, day or night, unlike the usual definition of child care, which is essentially day care. The specific question asked was as follows: "Who provides most of the care for your child/youngest child/second youngest child while you are working outside the home?"

As shown in Table 1, percentage distributions of responses relating to the youngest child are separated into whether the mother is employed full time or part time.[10] For the sample as a whole, relatives care for the child only slightly more than nonrelatives. (Child care centers and nurseries are included in the nonrelative category.) Relatives, however, are more likely to care for the child when mothers work part time rather than full time. This finding is primarily attributable to the increased role of fathers as caretakers of the child when mothers are employed part time (over one-fifth is father care). Indeed, it is interesting to find that, overall, 14.0% of all fathers are primary caretakers. As we shall see, this percentage increases substantially to 24.7% when we consider married couples only and wives working part time. Father care is clearly not the predominant mode of child care, but it is an especially important one, given the complexities of role sharing between husbands and wives that this implies.

How can fathers, who are mostly employed full time themselves, be the primary person watching the child when their wives are employed? It may be that many of these couples work split shifts; for example, one works days, the other works evenings or nights. Data on the hours of employment of husbands and wives are not available from the June 1977 CPS, but the occupation of the wife is available—unfortunately, this same information is not available for the husband. Taking a one-sided view of the split-shift hypothesis, it was possible to consider whether women whose husbands care for the child are more likely than other employed women to be in shift-work occupations, which appears to be the case.

TABLE 1

CARE OF YOUNGEST CHILD LESS THAN FIVE YEARS OF AGE FOR WOMEN EMPLOYED FULL OR PART TIME, U. S., JUNE, 1977

PERSON CARING FOR YOUNGEST CHILD	TOTAL $N = 3,899$ [1] PERCENT	FULL TIME $N = 2,589$ PERCENT	PART TIME $N = 1,310$ PERCENT
Child's father	14.0	9.8	22.3
Child's sibling	2.4	2.4	2.4
Other relative	29.5	32.1	24.4
Nonrelative	42.7	47.7	32.6
Mother watches child at work	3.5	2.2	6.0
Mother works at home	7.5	5.4	11.9
Child watches self	0.4	0.4	0.4
Total	100.0	100.0	100.0

[1] Weighted *n*.

Table 2 reports on married couples only, husbands present and employed. We see that father care is most characteristic of professional nurses (when part time, 39.7% of the fathers watch the child), sales-workers (part time, 42.6%), waitresses and other food service employees (full time, 31.7%; part time, 34.8%), and practical nurses and other health service employees (full time, 32.1%; part time 40.7%).

Unfortunately, we do not have data on the time of day women work in these occupations, but most occupations appear to be types that disproportionately entail shift work. Given the remarkably high pro-portion of fathers who watch the children and the impact that shift work may have on family life, this is an area that clearly seems worthy of further research.

Thus far we have been looking at employed mothers. Nonemployed mothers were asked about child care use, but in a different way. The specific question was: "In the past 4 weeks has your child/youngest child/second youngest child been cared for during the day in any regular arrangement, such as a day care center, nursery school, play group, baby-sitter, or some other regular arrangement?" The emphasis here was on day care to eliminate the reporting of baby-sitting for social reasons in the evenings.

TABLE 2

CARE OF YOUNGEST CHILDREN LESS THAN FIVE YEARS OF AGE
BY FATHER FOR EMPLOYED WOMEN,
HUSBAND PRESENT AND EMPLOYED, U. S., JUNE, 1977

	PERCENT CARED FOR BY FATHER			
OCCUPATION OF MOTHER	Mother Employed Full Time		Mother Employed Part Time	
	Percent	Number	Percent	Number
Teacher	2.6	13.6	12.6	70
Nurse (RN) and other health professionals, excluding physicians, dentists, and other practitioners	14.9	86	39.7	99
Other professional	13.3	98	13.8	50
Managers and administrators	9.5	73	21.6	39
Sales worker	14.0	50	42.6	144
Bookkeeper	6.1	79	5.3	49
Office machine operator	9.8	61	*	12
Steno/secretary	5.3	258	10.1	66
Other clerk	10.0	307	36.2	144
Craftsperson	7.3	30	*	19
Operative, durable	21.8	168	*	18
Operative, nondurable	18.4	173	17.9	39
Operative, other	11.0	45	4.8	25
Laborer	12.7	26	*	15
Private household worker	7.0	24	10.4	44
Cleaning service	14.6	31	10.4	46
Waitress and other food service	31.7	68	34.8	89
Practical nurses and other health service	32.1	67	40.7	83
Personal or professional service	5.5	94	7.2	83
Farmer and unpaid family laborer	0	50	8.3	34
Total[1]	12.0	1,926	24.7	1,124

*Percentage not computed; base less than 20.

[1] Weighted *n*.

The data indicate that 9.6% of nonemployed mothers with children under 5 years of age have a regular child care arrangement. The prevalence of such care varies according to the age of the youngest child from 4.2% for a child less than 1 year of age to 18.8% for a child between 4 and 5 years of age. Nonemployed women were also asked whether they regularly participated in any specific activities while their child was under such care. This survey is the only source of data of this type and provides an interesting perspective on the use of child care among the nonemployed. Overall, 60% of the women questioned say they regularly participate in nonfamilial activities during the scheduled child care time. As shown in Table 3, the primary regular activity of nonemployed child care users is recreation. This form of activity is noted for all education levels, except for women who have 13 to 15 years of schooling; these women are most likely to be attending school (although recreation is a prevalent secondary activity).

TABLE 3

REGULAR USE OF CHILD CARE BY NONEMPLOYED MOTHERS OF CHILDREN UNDER FIVE YEARS OF AGE BY EDUCATIONAL ATTAINMENT AND SELECTED NONFAMILIAL ACTIVITIES, U. S., JUNE, 1977

SELECTED ACTIVITIES[2]	EDUCATIONAL ATTAINMENT[1]				
	Total[3] (N=511) %	<12 (N=66) %	12 (N=187) %	13–15 (N=116) %	≥16 (N=142) %
Going to school	14	14	10	26	9
Other instruction or training	8	11	9	7	5
Looking for work	8	14	11	5	3
Volunteer work	10	3	5	13	18
Recreational activities	19	9	18	18	27
Other regular activities	16	19	10	16	21
No regular activities	40	33	46	36	39

[1] Highest grade completed.
[2] Women may participate regularly in more than one activity; thus, the percentage for each column exceeds 100.
[3] Weighted n.

CHILD CARE CONSTRAINTS

As previously noted, over one-third of all women with children less than 3 years old were employed in 1977.[11] This high employment rate might lead one to expect that all mothers with young children who want, or need, to work are able to find satisfactory low-cost child care without much difficulty. The availability of child care, however, does not appear to be the reason for the employment or nonemployment of women.

Mothers who were nonemployed and not looking for work were asked the following question: "If satisfactory child care were available at reasonable cost, would you be looking for work at this time?" Approximately 18.2% said yes, and an additional 6.1% did not know. Thus, approximately one out of five mothers with preschool-age children who are not in the labor force said they would be looking for work (or would be employed) if suitable child care were available (almost one out of four if the "don't knows" are included). Of course, attitudes do not necessarily predict behavior, and it is possible that women overestimate their readiness to seek employment. On the other hand, there may be an underestimate of readiness for employment, since greater availability of child care might increase demand for its use and stimulate interest in employment. In the absence of data for either position, we shall regard the actual response to the question as our best estimate of child care as a constraint on employment. This response is supported by estimates previously cited for different samples of nonemployed mothers.

Employed mothers were asked, "If you could find additional satisfactory child care at reasonable cost, would you work more hours?" The percentage of employed mothers who answered affirmatively was 15.9% and 3.3% did not know. As might be expected, those employed part time were much more likely than full-time employees to feel restrained from working more hours by the unavailability of suitable child care. About one out of four part-time employed mothers indicated they would work more hours, compared with about one out of eight full-time employed mothers. Although relatively low, the prevalence of child care constraint among full-time workers is surprising; these women usually work at least 35 hours a week.

Family formation factors (age of youngest child/children and number of children in the household less than 5 years old) are not relevant correlates of child care constraint on employment among either part-time or full-time workers. As noted earlier, such life cycle variables are found to be associated with whether or not a mother is employed. Apparently, they are not related to the extent to which child care unavailability keeps the nonemployed from seeking employment or the employed from working more hours.

Other demographic and social variables (which are not related to whether or not mothers are employed) show some association with child care contraint on employment. Women who are most in need of

employment are most likely to report that the unavailability of satisfactory child care at reasonable cost affects their labor force participation: the young mother (18-24), the unmarried mother, the black
mother, the woman who did not graduate from high school, and the woman
whose family income is less than $5,000.

It may be that the most relevant factors affecting child care
constraint are structural, relating to employment opportunities and the
types of child care available as well as cost. While we cannot assess the
relevance of employment opportunities, we can examine the type of child
care arrangements currently made by employed mothers.

Forms of child care vary in the extent to which they allow
flexibility in hours, add commuting time to the day, are costly in terms of
money or concern for the child's welfare, or create indebtedness with
relatives. Do the types of child care currently being used by employed
mothers differentiate the prevalence of constraint from working more
hours? As shown in Table 4, there appears to be some relationship.
Institutionalized child care arrangements are associated with the lowest
report of constraint on employment, followed by care by a nonrelative in
a private home. The latter (often referred to as family day care) is also
the most prevalent single form of child care arrangement made. Among
full-time workers, those using a nonrelative in the child's home show the
highest prevalence of constraint; among part-time workers, those using a
relative in a private home report the highest. These findings imply that
payment for care is not generally the critical constraining factor, since
care by a relative tends to be inexpensive and care by a nonrelative in the
child's home and institutionalized child care tend to be the most expensive
forms of care.[12]

While the cost of child care is related to the type of care which, in
turn, is related to feeling constrained from employment, there is no
consistent relationship between paying for care and feeling constrained
(Table 4). The types of care that have the highest and lowest percentages
of women reporting constraint almost all require payment and are the
most expensive. For those types of care where some women pay and some
do not, there is no clear pattern of paying and feeling constrained. Full-
time workers who have a relative in the child's home are more likely to
report constraint when the relative is not paid; when a relative cares for
the child in another home, they are more likely to report constraint if the
relative is paid. Among part-time workers, there is no difference in
constraint according to whether a relative in the child's home is paid or
not; among women who have a relative care for the child in another home,
those who pay are less likely to report constraint—the opposite of the
relationship found for full-time workers. The complex relationship may
be affected by other considerations such as the amount paid or the nature
of the relationship with relatives (e.g., quality, availability, location).

TABLE 4

EMPLOYED MOTHERS WITH CHILDREN LESS THAN FIVE YEARS OLD
WHO ARE CONSTRAINED IN THEIR LABOR FORCE PARTICIPATION
BY THE UNAVAILABILITY OF CHILD CARE,
BY TYPE OF CHILD CARE AND PAYMENT,
SEPARATELY BY EMPLOYMENT STATUS (%), U. S., JUNE, 1977

| TYPE OF CARE AND PAYMENT | WOULD WORK MORE HOURS IF CHILD CARE WERE AVAILABLE | | | | | |
| | TOTAL | | FULL TIME | | PART TIME | |
	%	N	%	N	%	N
Nuclear family	19.3	497	15.8	240	23.0	257
Mother	16.9	356	11.8	169	21.4	187
Relative, child's home:	18.8	291	15.4	199	26.1	92
Pay for care	16.8	131	12.6	87	25.0	44
Do not pay	20.6	160	18.9	112	27.1	48
Relative, other home:	16.7	511	11.7	366	29.3	145
Pay for care	16.8	316	14.8	243	23.3	73
Do not pay	16.4	195	4.9	123	36.1	72
Nonrelative, child's home:	21.9	195	19.2	119	26.3	76
Pay for care	22.7	185	20.5	112	26.0	73
Do not pay	*	10	*	7	*	3
Nonrelative, other home:	12.4	667	9.4	498	21.2	169
Pay for care	12.2	648	9.6	490	20.3	158
Do not pay	*	19	*	8	*	11
Institutionalized:	10.6	349	8.2	267	18.3	82
Pay for care	10.5	324	7.9	252	19.4	72
Do not pay	12.0	25	*	15	*	10

SOURCE: Presser and Baldwin (1980).

*Percentage not computed; base less than 20.

CONCLUSION

In sum, this paper has presented some basic descriptive data on child care use and constraints among mothers of young children, both employed and nonemployed, in the United States. The data are limited, and there are numerous child care issues we have not raised here.[13] Nevertheless, the data strongly suggest that we pursue in depth such areas as specific occupations (both male and female) as they affect child care use, the role of fathers in caring for children when both spouses are employed, and child care use among the nonemployed—including what mothers do with their child-free time. The substantial level of hidden unemployment due to child care constraint is especially noteworthy and has important theoretical and policy implications. Why does a substantial minority of

American women feel constrained in their labor force participation because of the unavailability of child care, when so many women do arrange such care? It is time to give serious attention to these and other child care issues if we seek to enhance women's achievements both inside and outside the home.

NOTES

1. A similar version of this paper was presented at the Research Conference on Women: A Developmental Perspective, November 1980, which was sponsored by the National Institutes of Health.
2. Duncan and Hill, 1975:Vol. 3; Kurz, Robins, and Spiegelman, 1975; Lajewski, 1959; Lave and Angrist, 1974; Low and Spindler, 1968; Ruderman, 1968; Shortlidge and Brito, 1977; U.S. Department of Health, Education, and Welfare, 1976; Waite, Shortlidge, and Suter, 1974; Westinghouse Learning Corporation, 1971. For a review of findings from the major studies, see Woolsey and Nightingale, 1977. There is a considerable body of literature on child development and administrative aspects of nonfamilial child care, which goes beyond the focus of this paper.
3. Child care is typically viewed as a means of facilitating female, not male, employment, and the costs relative to earned income are related to her salary, not the child's father's. This perspective may well change in the near future.
4. Sweet, 1970; Waldman and Gover, 1971.
5. Waldman and Gover, 1971.
6. Ditmore and Prosser, 1973.
7. Duncan and Hill, 1975; Heckman, 1974; Kurz, Robins, and Spiegelman, 1975; Steiner, 1971.
8. Dickinson, 1975:Vol. 3.
9. Westinghouse Learning Corporation 1971.
10. For women who had more than one child less than 5 years old, there was little difference in type of use when considering the second youngest child rather than the youngest. Also, there is little difference when looking at women who had only one child versus two or more, or age of youngest child.
11. This section summarizes findings previously reported in Presser and Baldwin, 1980:1202-13.
12. U.S. Department of Health, Education, and Welfare, 1976.
13. For an elaboration of issues, see Presser, 1978:167-80.

THE ATTITUDE OF DANISH MOTHERS TO CHILD CARE,

1975 TO 1979

Henning Transgaard

The Danish National Institute of Social Research
Copenhagen   Denmark

With ever-increasing numbers of women entering the labor force, the question of who will care for the children becomes an important one, not only for the working mother, but for society as a whole. Different programs in various countries have been developed in response to the child care question. In Denmark, the answer has been to heavily subsidize outside forms of child care. Under the Danish system, parents pay only 20% to 25% of the actual operating costs. Because of this low expense, the number of children on the day care waiting lists is between 53,000 and 55,000, out of a population of 500,000 children less than 6 years of age. About 40% of all Danish children below school age are being cared for by people other than their parents in some form of public child care. Furthermore, the majority of these children come from families in which both parents live at home. For example, the results of a 1974 survey on the family background of kindergarten children show that only 16% are from one-parent families, whereas 82% are from families with married or cohabiting parents.[1]

The apparently widespread use of Denmark's child care program leads us to consider its impact on the labor force participation of married women as well as their attitudes toward the different forms of child care available. Has the increased availability of child care influenced women's willingness or desire to enter the labor force? Are these forms of child care perceived as incentives to seek employment outside the home? Are some forms of child care preferable to others?

In an attempt to provide at least partial answers to these questions, this chapter will specifically address the attitude of Danish mothers toward this pervasive child care program. There are two important reasons for examining this attitude. First, the program itself has raised several interesting theoretical and methodological issues with respect to attitude theory, issues which will be outlined below. Second, the resolution of these issues is believed to be essential in the process of answering the questions outlined above.

To these ends, multivariate analyses of two cross-sectional attitude surveys will be presented, the first conducted in 1975 and the second conducted in 1979. Danish mothers, married or cohabiting, with one or more children below the age of 10 were asked to report their attitudes toward the forms of child care available, their beliefs about the differences between them, the conception of their role (whether housewife or fully employed), and their preferred occupational status. Focusing upon the distinction between care at home by a parent versus any outside form of care, the various determinants of this attitude will be explored crosssectionally as well as over time.

THEORETICAL PERSPECTIVE

The standard concept of attitude is that of an affective charge associated with the attitude object on a general positive-negative continuum. The most direct operationalization of the concept would seem to be the semantic differential:[2] the location of the object on a series of scales, defined by pairs of general evaluative objectives. This concept of attitude, however, contrasts with that implied in other studies in which the key problem considered is that of establishing the dimensionality and identifying the dimensions of the evaluative space involved.

A second problem related to attitude theory is that it contains very little about the notion of an attitude object. In fact, it is common to assert that practically anything may be an attitude object. In effect, this means that no defining attribute of attitude objects, as compared with nonattitude objects, is being proposed or employed. This problem also applies to the concept of an attitudinal dimension, whose main defining attribute would merely seem to be its evaluativeness.

The consideration of these current limitations of attitude theory is motivated by the fact that the availability of several forms of child care represents the existence of several attitude objects, which raises the problem of the number of attitudes involved. Do we deal with a single attitude or an attitude structure? The presence of several forms of care, of several attitude objects, would seem to suggest that several attitudes must be involved--an attitude toward nurseries, kindergartens, family day care, etc. On the other hand, all of these forms of care do belong to a single domain and have a single function. Thus, even if several attitude objects are involved, and consequently also several attitudes, these attitudes might be so closely associated that for all practical purposes only a single attitude would be involved.

With regard to dimensions, Lieberman[3] provides a fascinating account of the various evaluative concepts, concerns, or themes to which day care has been linked in American opinion, past and present. Lieberman considers the following in relation to child care: (1) the dichotomy between the affluent and the poor; (2) the ambivalence toward women— mothers first, but also as a reserve labor force; (3) a pervasive ambivalence toward the poor; (4) day care as an instrument of social change, as a service for the underprivileged, or as an institution providing relief in national emergencies; (5) cognitive stimulation through day care attendance; (6) changes in the accepted definition of what constitutes good mothering; (7) reconciling the rights of mothers with what is good for children; (8) the potential that day care may emotionally damage the child due to maternal separation and deprivation; (9) paying professionals or the welfare mothers to assume child care responsibilities; and (10) the costs versus the gains of such programs, such as the cost to society of childhood deprivation as contrasted with the cost of subsidizing child care centers.

Analyzing the views of the politicians involved in the "first explicit federal authorization of day care," Greenblatt[4] summarizes his policy analysis in terms of the following 12 factors: (1) violating traditional assumptions of the family and of the maternal role; (2) labor market changes (increased job competition, male unemployment); (3) welfare statism; (4) pejorative images of day care; (5) confused professional identity on the part of day care workers; (6) male sexism; (7) fiscal cost; (8) preventing neglect of children of employed mothers; (9) helping the child achieve his maximum potential; (10) national benefit; (11) freedom of role choice for mothers; and (12) providing relief from the social stigma and financial burden of welfare.

Had this list, or a similar Danish/Scandinavian itemization, been available earlier, it might have been possible to include this aspect of attitudes in a more systematic and comprehensive analysis of employment of women outside the home and of child care alternatives.

Another problem has to do with feasibility or level in the sense of a potential discrepancy between what is ideal and what is realistically possible. A related issue is that of the costs of the various forms of child care. What comes to mind is not primarily weighing of the costs and the benefits, but rather how this is reflected in the demarcation of the forms of care between those that are realistically possible and those that are not thought to be so.

Furthermore, the attitude objects (the forms of care realistically available) may be secondary to another attitudinal element, namely, the mother's chosen or preferred occupational status. To some respondents, the decision regarding this status may be more important than selecting the form of child care to be adopted. And whose decision is this? Is it the mother's, the husband's, both, or an element in a more extensive negotiation? It would seem highly probable, then, that attitudes toward forms of care are not isolated, but are themselves embedded in a more comprehensive attitude structure or value system which involves the division of labor between men and women—both in the family and in society at large. The married mother's preferred occupational status would seem to constitute some important, perhaps even the main, elements of this comprehensive attitude structure or value system having to do with the proper position of women in society and the family.

An earlier Danish study of fertility also was concerned with attitude structure.[5] In the study, a national sample of women was presented with various so-called "family types," which consisted of all combinations of the paid employment of the husband and the wife (i.e., both either working full or part time or either the husband or the wife staying at home) as well as different kinds of child care: full-time or part-time kindergarten or care at home by the mother or father. Six different "family types" were produced which were then ranked by the respondents. Analyzing the rankings from this study, two aspects—the wife's preferred occupational status and the preferred form of care for the children—were not sufficiently associated to establish a single dimension. This result was in accord with other analyses.

A final problem concerns the link(s), or rather the conceptualization of the link(s), between attitudes toward forms of care, on the one hand, and the respondent's self-conception, role conception, norms, motives, relative deprivation, etc., on the other. Of these, only the attitude-self link seems to have received any attention in the literature.[6] As an aspect of this issue, one might also point to the interplay between the mother's own attitude and her perception of others' attitudes concerning the climate of opinion.[7]

METHOD

The approach adopted for the analysis of these issues involved the following, all of which pertain to the standard concept of attitude: (1) the forms of care were each considered as a separate attitude object; (2) three forced-choice comparison tasks were employed instead of separate semantic differential ratings; (3) the problem of level (i.e., ideal versus realistically possible) was built into the comparison tasks; (4) only an indirect assessment was attempted of the implications for the respondent's self-conception (or role conception) of her attitude (unfortunately, for reasons of cost, these responses were only obtained for the 1975 study); and (5) an assessment was made of the beliefs about some of the attitude objects as well as the perception of the climate of opinion. These will be described below.

Analysis and Results of the Attitude Variables

Table 1 presents the 1975 and 1979 responses to the three attitudinal questions: (1) "For smaller children (0 to 3 years), which form of child care do you prefer, nursery or family day care?" (2) "For older children (4 to 6 years), which form of child care do you prefer, kindergarten or family day care?"   (3) "Assuming that one parent is at home during the day, which form of child care do you prefer for older children (4 to 6 years): care at home by a parent, half-day kindergarten, family day care, or full-day kindergarten?"

TABLE 1

PERCENTAGE DISTRIBUTION FOR PREFERRED FORM
OF CHILD CARE BY AGE OF CHILD, 1975 TO 1979

Form of Child Care[1]	Age of Child					
	Percentage					
	0-3 Years		4-6 Years		4-6 Years	
	1975	1979	1975	1979	1975	1979
Care at home by parent	....	....	....	....	38.7	25.7
Family day care	71.5	72.7	26.6	20.4	2.3	2.4
Nursery	22.1	20.4	....	....	....	....
Kindergarten	....	....	66.0	77.1	....	....
Half-day	....	....	....	....	52.1	68.2
Full-day	....	....	....	....	2.8	2.0
Don't know	6.4	6.9	7.4	2.4	4.0	1.6
Total	100.0	100.0	100.0	99.9	99.9	99.9

[1]In the 0- to 3-year-old group, only the options of nursery or family day care were offered; in the 4- to 6-year-old group, the options were kindergarten or family day care; columns 5 and 6 also consisted of 4- to 6-year-olds with the added condition that one parent was at home during the day and the child care options included care at home by a parent, family day care, and half- or full-day kindergarten.

A striking feature of this table is the shift in preference from one age group to the next; the preference expressed for children 3 years and younger is reversed for the 4- to 6-year-old group, indicating that the semiprivate form of day care (family day care), preferred for smaller children, is replaced by an institutionalized form of care. These forms of care, at least in their all-day varieties, are in turn replaced as the majority choices, when care at home and half-day kindergarten are added as options. In comparing data from 1975 with data from 1979, the preference for kindergarten as compared to family day care and the preference for half-day kindergarten as compared to care at home by a parent have both increased.

The values in this table suggest the appropriateness of the Coombsian conception in which the various forms of child care may be viewed as located on a continuum that also contains an ideal point and the attitude or preference expressed is based on the distance of the options from this ideal point, as conceived by the respondents. The majority choices for older children, care at home by a parent and half-day kindergarten, would seem to come closest to this assumed ideal point, if not actually constitute it.

The three preference variables of "nursery vs. family day care," "kindergarten vs. family day care," and "care by a parent vs. any form of outside care" were dichotomized and entered into a Pearson product-moment correlation analysis. Results of the analyses for both 1975 and 1979 reveal that the three attitudes are clearly not independent of each other. However, considering the fact that we are dealing with a single domain, a single function, the attitudes in question do not seem so closely associated that it is possible to speak of a single attitude, but rather, three different attitudes seem to be involved. The pattern obtained is not without interest in that there are two different clusters: "family day care" and "care at home by a parent," on the one hand, and "nursery," "kindergarten," and "any form of outside care," on the other hand. Another more elaborate analysis of the structure of the attitudes also was performed. Instead of just looking at the pair-wise associations, all three attitudes in their original nondichotomized version were combined into a single matrix. The structure revealed indicated that no dominant order of preference for the various forms of child care could be identified. This, of course, is not to be interpreted as evidence against the Coombsian conception, which may describe the individual preference order or attitude structure adequately. The above analysis merely suggests that in the aggregate such a structure cannot be identified.

Table 2 presents additional information on the nature of the aforementioned change in preference. The respondents were asked to indicate which of the two forms of day care was more accurately described by each item. There also was no difference between the two with respect to a particular item. The two columns, showing the proportion of respondents answering "kindergarten" to the particular item in 1975 and 1979, are remarkably similar, except for the final item. This one exception may possibly be traced to the so-called "indoctrination debate" in Denmark which began in 1975. This controversy had to do with the possible discrepancies between the social and political values of the parents and those of the personnel of the day care institutions. It was thought that the values held by the latter might, in some instances, be promoted too actively among the children. The debate was rather short-lived and was no longer current in 1979. The results, therefore, probably are a reflection of a specific issue at a particular point in time rather than a true difference in beliefs.

Aside from this one item, there obviously has not been a change in the beliefs about the differences between the two forms of care, although the attitude toward them has changed, as shown in Table 1. Further analyses of this apparent discrepancy show that these items do not seem conclusive for the attitude—as would be argued by such models of the attitude-belief relationship as Fishbein's.[8] Faced with a similar state of affairs, Stipak drew the same conclusion, arguing that "cognitive or

TABLE 2

BELIEFS ABOUT DIFFERENCES
BETWEEN KINDERGARTEN
AND FAMILY DAY CARE BY PROPORTION ANSWERING
"KINDERGARTEN" TO THE VARIOUS ITEMS, 1975 AND 1979

Item	Percentage	
	1975	1979
More attentive to child's health	32	28
More responsive to individual child	18	21
Safer place	24	22
Close and stable contact with same adults	22	19
More responsive to individual mother	19	19
More space, more possibilities for child	79	82
Teaches the child to interact better with other children	79	82
Provides to a greater extent the experiences and stimulation the child requires	61	66
More qualified care	41	41
Can better handle minor emergencies	49	46
Less interference with parents' upbringing	43	34

rational processes of attitude formation" (such as Fishbein's model) "require a stronger perceptual basis." This statment implies that the attitude is "grounded in perceptions or knowledge of the actual characteristics of the attitude referents" (i.e., the attitude objects).[9] This implication is not supported by Stipak's data either; he suggests "adoption from reference groups or generalization from other attitudes" as more plausible theoretical approaches.[10]

Other items provide further evidence of a change in attitude. For example, approximately 28% of the 1979 respondents have changed their attitude toward kindergarten. Another item shows that a change in what is assumed to be the majority view or the climate of opinion also has occurred. This interpretation of the question pertaining to a perceived trend toward an increased or decreased recognition of kindergarten is obviously vulnerable to the counter argument that the responses to the item do not necessarily reflect any change in the climate of opinion, but are merely a reflection of the respondent's own attitude.[11]

Some independent, although direct, evidence is available. First, one may point to the increasing demand for day care institutions in Denmark in spite of a large increase in supply. Second, a general law concerning the Danish Social Security system contains a paragraph stating that it is the responsibility of the local governments to make available the necessary number of day care institutions. The proper legal interpretation of this paragraph is currently being decided by the Supreme Court after the local government of Copenhagen was sued by the parents of a child for not providing needed child care facilities. Third, some mothers and others in the population clearly felt pressured by the climate of opinion with respect to the proper form of care for children, even to the point of organizing an effort at changing this climate of opinion. "Ideological traditionalists" is Duncan's apt term in the area of sex typing represented by this movement.[12] As another example of this trend, a Swedish study may be cited in which mothers in the capital of Sweden were asked about their use of, and attitude toward, the forms of care in connection with the choice between paid employment and work at home. For some mothers, an insecurity was observed with regard to whether or not the woman as mother is really allowed to like or enjoy being at home. Consequently, even if the responses involve an attempt at reducing the ambiguity of one's attitude by calling upon its social support, whatever one's attitudinal position, they are, nevertheless, viewed as mainly reflecting the climate of opinion.[13] Fourth, the analyses of Noelle-Neumann—employing both measures of the respondents' own attitudes and of various aspects of the climate of opinion as perceived by the respondents—indicate that the two responses of attitude and perceived climate of opinion are not necessarily identical.[14]

The results of a Danish Gallup poll[15] may be cited as evidence of the attitude change as well as the change in the climate of opinion. The poll compared responses in 1973 and 1979 to the following question, addressed to a national sample of men and women: If we consider a married woman with small children under the age of 10, what do you think would be her natural activity: working as a housewife at home or employed in outside employment, either full or part time?

The proportion who answered full-time employment was the same (3%) in 1973 and 1979 for both men and women. However, the proportion for part-time employment increased from 35% to 54% among the women interviewed (most likely all women in the sample, married and unmarried, with and without children). The line connecting these two figures is parallel to the one connecting the 39% and the 26% (from Table 1, columns 5 and 6). Thus, the rates of change on the two items, occupation and forms of child care, correspond to one another.

As more women express a preference for part-time employment outside the home and for institutionalized forms of child care, there has been an increasing demand for child care outside the home. One political response to this increased demand, in a period otherwise characterized by cuts in public spending, has been the discussion of the possibility of a larger emphasis on the less-expensive alternative, family day care. However, this proposal may have the long-term effect of stimulating demand. Recalling the associations between the attitudes toward the various alternatives of child care or the distance between them, care at home by a parent and family day care are relatively close to each other, suggesting that the latter may serve as a bridge between care at home and the institutional forms of care.

Association of Other Variables with Changes in Attitudes

The term "other variables" means not only the usual survey background variables but also attitudinal variables that would be difficult to order causally with respect to the attitude. Since a comprehensive analysis of the 1975 data had already been performed, including multivariate analyses with various combinations of independent variables with a dichotomized version of responses as the dependent variable (care at home by parent versus any form of outside care), and since this item also reflects the largest amount of change, the responses listed in Table 1, columns 5 and 6, were again selected to serve as the dependent variable to provide the best single indicator of the attitude toward form of care. The analysis of the 1975 responses mainly relied upon the Automatic Interaction Detection (AID) program developed by the Institute for Social Research.[16] The number of interaction effects which appeared in the tree diagrams supported the mode of analysis employed. Most of the 1975 analyses were run within the following three occupational status groups: full-time employed, housewives, and part-time employed. Interaction effects were present even at the zero-order level within these three groups as exemplified by the fact that education, satisfaction with one's occupational status, and occupation were the three different variables most strongly associated with the attitude separately within these three status groups. This accounts for 13%, 6.5%, and 23.4%, respectively, of the variation in attitude. The grouping by occupational status (housewife, full-time employed, part-time employed) of the 1975 responses, accounts for 5% (eta = .05) of the variation in attitude. With the 1979 sample only about one-half of the 1975 one, it was not deemed advisable, however, to analyze the 1979 data in the same way as those from 1975. Instead, one- or two-way analyses of variance were employed to determine how much of the variation in attitude (care at home by a parent preferred to any form of outside care) is accounted for by five variables. In the 4-year period under study, the background variables of occupation, age, education, and urbanization, account for less of the variation in attitude, whereas the link to preferred occupational status has increased. One implication of the first part of this conclusion is that the change in attitude has occurred in all of the subgroups of the four background variables referred to, but differences between the subgroups of these variables have become smaller. The particular fact of the decreasing association between occupation and attitude speaks against any simple account of the change in attitude as due to the increasing rate of employment of married women which seems common in the Western industrialized countries in the 1970s.

Both aspects of this observation, i.e., the initial low as well as the increased association between attitude toward form of care and preferred occupational status, also are in accord with analyses of the change in American women's sex-role attitudes.[17] The responses over time to a number of sex-role attitude items constituted the data of these analyses. In general, the various sex-role attitude items were only weakly associated with one another. Only two clusters of items could be identified; they were labeled "job" and "family" by the investigators. In 1970, these two clusters were themselves unrelated, but when followed over time, a moderate assocation between the two clusters emerged.

A more detailed examination of the link between the two clusters of "job" and "family" was conducted. More than 72% and 76% of the mothers indicated that they would prefer to be employed either part of full time as determined respectively from the responses in the 1975 and 1979 surveys to the item asking for the respondent's preferred occupational status. The two marginal distributions are highly similar. The stronger link between this variable and attitude is clearly not caused by any change in the marginal distribution of this preference. Rather, as is implied by the change noted above, the difference in attitude between the preferred occupational status subgroups has widened; in this sense an attitudinal polarization is emerging.

ATTITUDE AND SELF/ROLE CONCEPTION

The self- or role conception of the respondents was assessed by presenting them with a set of reasons or motives along with a request that they indicate for each of the items whether or not this was one of the most important reasons for their occupational status.

The items are not to be interpreted as being relatively comprehensive and stable systems, involving a kind of local energy or directive source and influencing attitudes and behavior in a fairly direct way. Rather, we are dealing with a vocabulary of motives, reasons, and outcomes relating to a particular role. And as such, they are partly based upon the vocabulary embodied in the list of reasons and motives generally employed in social science analyses of the motive factors concerning women's employment, role choices, etc. The study by Weiss and Samelson[18] of the contribution of women's roles to a sense of usefulness and importance was one of the major references in the formulation and selection of the role conception items employed in the present study.

Though the items refer to two different roles, full-time employed and housewife, an attempt has been made to secure as large an overlap between the two sets of items as possible. It did not seem possible to develop a related list to be presented to the dual-status respondents, the part-time employed. This difficulty with the vocabulary or motives for the part-time employed suggests that this status is not as autonomous or primary as the other two, that it is indeed a dual-status one, being derived from the other two rather than an equal level one. No distinction will be attempted between the self- and role conception aspects of the responses. Though the repondents are asked for a "yes" or "no" answer to the question of whether the particular item is important for them as a housewife or as a full-time employee, it would not seem possible to decide whether the interpretation actually adopted by the respondents referred to their occupational status only or to the more comprehensive concept of their self, roughly comparable to "Is this important to you or not?"

Tables 3 and 4 represent the zero-order association of the role or self items with attitude. The variation accounted for by the full AID-analyses is, respectively, 14% for the housewives and 7% for the full-time employed, both relatively small figures.

Note that the task of child care, as represented by the items referring explicitly to this, does not manifest itself as especially influential among these determinants of attitude. These "children-items," therefore, clearly do not imply the attitudinal position that care at home is the best form of child care.

Consider the two role items having the strongest overall association with the attitude: "I am best able to be something (of use or service) for the family as a housewife," and "I am more respected and esteemed, by husband and children, as employed." Also, note the similarity, the almost mirror-like quality, of the items. The direction of the association is also important; in both cases, it is the group of "yes-sayers" who express the highest preference for care at home. This finding suggests that the same attitudinal position may be embedded in two different social identities, which is constrained by the relatively weak links (in 1975) between attitude and the actual or the preferred occupational status of the respondents.[19]

INFLUENCES ON PREFERRED OCCUPATIONAL STATUS

In this section a change in perspective will be effected, allowing for a more comprehensive view and an opportunity for assessing the presence of children and child care as two of several potential influences on the preferred occupational status. This change (unfortunately only available for 1975) will be initiated by substituting the preferred occupational status for the attitude toward form of care as the dependent variable and by considering all married or cohabiting women in the (1975) sample. This consideration will allow the presence of children below the age of 10 to be used as one of the independent variables. The dependent variable, the preferred occupational status, is dichotomized as housewife versus employed (full time or part time). Role satisfaction, mentioned previously, was actually operationalized as the correspondence (satisfaction) or the discrepancy (dissatisfaction) between the actual and the preferred occupational status. This means that for housewives the dependent variables of role satisfaction and the dichotomized version of the preferred occupational status (housewife or employed) are identical. Consequently, either may be interpreted in terms of the other for this group. For the other two occupational status groups, the two variables are merely related. Taking all married or cohabiting women in the (1975) sample, aged 16 to 63, the presence of children below the age of 10 accounts for less than 1% of the variation in the wish to be a housewife.

TABLE 3

PERCENTAGE OF VARIATION IN ATTITUDE AND IN WISH TO BE A HOUSEWIFE ACCOUNTED FOR BY INDIVIDUAL ROLE CONCEPTION ITEMS FOR HOUSEWIVES, 1975

Role Conception Item	Attitude	Wish to be a Housewife			
		Percentage			
		Age			
		16–25	26–35	36–45	46–63
Impossible to get a job	0	4	5 *	9 *	3
Cannot get the children cared for	1	3 *	0	0	0
Child care is too expensive	0	2 *	1	1 *	1 *
It will not pay for us if I work	0	2	0	2	0
It is not necessary for me to work	1	22	0	13	0
I have never considered working	0	3 *	0	1	0
One is less tied down and freer as a housewife	2	3	17	0	6
I could not manage to take care of husband, children, and the home if employed	0	1 *	0	2	0
Work at home is more satisfying to me	5	26	19	7	10
I am indispensable at home	0	19 *	6	5	2
I am best able to be something for the family as a housewife	7	14	7	7	3
My husband and children deserve for me to be a housewife	4	8	5	13	2
My husband and children expect me to be a housewife	4	4 *	1	17	1
I am more respected and esteemed, by husband and children, as a housewife	5	2	1	12	7

NOTE: AID zero-order BSS * 100/TSS coefficients; starred entries are the comparable eta coefficients.

TABLE 4

PERCENTAGE OF VARIATION IN ATTITUDE AND IN
WISH TO BE A HOUSEWIFE ACCOUNTED FOR BY
INDIVIDUAL ROLE CONCEPTION ITEMS
FOR FULL-TIME EMPLOYED, 1975

Role Conception Item	Attitude	Wish to be a Housewife			
		Percentage			
		Age			
		16–25	26–35	36–45	46–63
Contact with other people	1	0	1	16	5
To keep a job and an income in a period of unemployment	1	0	2	0	0
Outside employment is more satisfying to me than work at home	1	10	10	11	7
In order not to stagnate	0	6	6	9	2
My income is necessary for us	0	0	2	1	1
I have a larger economic surplus and more to give family as employed	0	3	2	6	2
One is less tied down and freer when employed	0	3	1	3	3
My work is really interesting and important to me	0	4	3	22	0
To be able to afford more than just the necessities	3	0	0	3	2
To become economically independent of my husband	2	2	0	1	2
To have a richer, more varied life	0	9	3	13	3
I feel more useful and more of an accomplisher	1	5	4	5	4
I am more respected and esteemed by husband and children as employed	4	1 *	0	4	1 *

NOTE: AID zero-order BSS * 100/TSS coefficients; starred entries are the
comparable eta coefficients.

The presence of children was also included as an independent variable along with the role conception items in the AID analyses (one set including the housewives, the other the full-time employed) of the wish to be a housewife. These two groups were further classified by age and the AID analyses were performed within the resulting eight groups, as shown in Tables 3 and 4. The presence of children, though, was still too weakly related for it to be drawn into the analysis. Since the presence of children is not in itself a strong influence, the relatively weak effects of the two child care items shown in Table 3 are not surprising. The change in perspective effected, therefore, would seem to corroborate the previous analyses.

SUPPLY, DEMAND, AND ATTITUDE CHANGE

It is sometimes asserted that supply of child care invites or even creates demand. Indeed, a comprehensive Swedish survey[20] of the parents of 190,000 Swedish children documents that the demand for child care (operationalized as the response to an item asking for the number of day care places needed in 1980) often increases in well-supplied communities.

A specific analysis[21] with a much smaller n focused upon these two problems: (1) Does demand increase when supply increases (holding the proportion of children with working or studying parents constant), and (2) what is the effect on demand of an increase in the proportion of employed mothers (holding supply constant)? Demand—operationalized differently from the large survey mentioned above—entered into the analysis both as total demand and demand for new places.[22]

Analyzing these links, both in the aggregate and cross-sectionally, by means of multiple and partial correlations, the results depend upon the type of demand considered, either total or new. As to the latter, however, there was no effect on supply, holding the employment rate constant. The existence of a latent demand, manifesting itself when supply allows it, is assumed to be one of the causes of this outcome.

A fine-grained analysis of the interplay between supply, demand, and attitude change, however, is not possible with the data available from the two surveys reported here, nor are any specific conceptual/theoretical analyses available of the process or processes which might be operating. Crosby's analysis of relative deprivation seems to be a fruitful starting point for a suggestion of one such process. Crosby argues "that for an individual to feel resentment, five preconditions must be met. The person who lacks X must: (1) see that someone else (other) possesses X, (2) want X, (3) feel entitled to X, (4) think it is feasible to obtain X, and (5) lack a sense of personal responsibility for not having X."[23]

Now let X be "enrolled in publicly supported child care," even as a purely practical arrangement without any attitudinal support, along with the assumptions that (1) the greater the number of other parents using publicly supported child care, the greater the resentment; and (2) the greater the resentment, the greater the demand (and the greater the pressure on the individual's attitude).

It is also possible, referring back to both Weiss and Samelson as well as Rudebrant and Thorn, that X might be the social recognition and esteem previously awarded self-evidently by the climate of opinion to mothers who cared for their children at home. This social reward is now more likely to be perceived as going to the unmarried mother; coupled with this is a new, more ambivalent view of the housewife.

However, these suggestions are partial in the sense that they only account for attitude change in the aggregate, and even assume some aggregate attitude change up to a critical point; the differential susceptibility to change at the individual level as well as the initial (aggregate) attitude change assumed are not accounted for.

CONCLUSION

The purpose of this paper encompassed two specific objectives: (1) to outline and address some theoretical and methodological issues with respect to attitude theory raised by the attitudes of Danish mothers toward various forms of child care, and (2) to employ the examination of these issues in an attempt to answer some questions concerning the impact of Denmark's pervasive child care program on women's employment.   The primary data for the analyses were derived from two nationwide surveys conducted in 1975 and 1979. The analyses, first of all, revealed that the forms of child care represented several attitude objects and were differentiated in the attitude structure of Danish mothers. An examination of three specific attitude objects, family day care, care at home by a parent, and other forms of child care, revealed them to be different, in spite of their dependence on each other.

Second, its was observed that attitudes toward the various forms of child care have changed over the recorded 4-year time span while beliefs have remained relatively stable. A change in the climate of opinion also was observed. It was noted that this change corresponded with an increasingly favorable attitude toward part-time employment outside the home.

Third, a specific analysis of the determinants of these attitudes indicated that certain background variables such as age, occupation, education, and urbanization account for less variation over the years while the link between attitudes on forms of child care and preferred occupational status has increased. This link was verified by reversing the variables and examining the influence of children and child care on preferred occupational status. While the difference in attitude between the preferred occupational subgroups has widened over the years, an analysis of the link between attitudes toward forms of care and self-conception/role conception demonstrated that the same attitudinal position may be embedded in two different social identities.

Finally, it was noted that while the data for an analysis of the impact of the availability of child care on demand and attitude were limited, the relationship might be characterized through the use of relative deprivation theory.

It would appear that the data confirmed the hypothesis that attitudes toward forms of care were not isolated but embedded in a more comprehensive structure. An important element of this structure was preferred occupational status. A close association was found to exist between this variable and attitudes toward forms of child care; even current occupation did not account for as much of the variation in attitude as did preferred occupational status. It can be concluded, therefore, that there was a connection between preferred occupational status and attitudes toward child care, reflecting an increased "ideological" linkage of the two attitudes: one (child care) was beginning to imply the other (the preferred occupational status), i.e., the choice of child care was now not only concerned with the interests of the child, but also with those of the mother—at the level of general, "self-evident" principles.

NOTES

1. Kjerkegaard, 1976.
2. Osgood, Suci, and Tannenbaum, 1957.
3. Lieberman, 1978.
4. Greenblatt, 1977.
5. Bertelsen and Ussing, 1974.
6. Ostrom and Upshaw, 1968; Hooper, 1976.
7. Noelle-Neumann, 1977.
8. Reported in Transgaard, in press; Fishbein, 1967.
9. Stipak, 1977:50-51.
10. Ibid.:53.
11. Noelle-Neumann, 1977.
12. Duncan, 1979.
13. Rudebrant and Thörn, 1979:70.
14. Noelle-Neumann, 1977.
15. Berlingske Tidende, September 16, 1979.
16. Sonquist, Baker, and Morgan, 1971.
17. Mason and Bumpass, 1975; Mason, Czajka, and Arber, 1976.
18. Weiss and Samelson, 1958.
19. See also Hooper, 1976.
20. SCB., 1980.
21. SOU., 1978.
22. SCB., 1980.
23. Crosby, 1976:90.

RESEARCH ON WOMEN'S WORK AND THE FAMILY

AT THE URBAN INSTITUTE

Sandra S. Tangri

The Urban Institute
Washington, D.C.   U.S.A.

In 1975, with assistance from The Ford Foundation, the Urban Institute established the Program of Research on Women and Family Policy. The aim of the Program is to provide objective analyses of the ongoing changes in women's economic and social status and the implications of these changes for individuals, families, and economic and social institutions. It seeks to provide insight into the changing roles of women and men associated with the increase in the percentage of women in the U.S. labor force, and to offer a factual and conceptual basis for both public and private decision making. Because of this broad objective, the staff represents a wide range of social science perspectives including economics, sociology, demography, and social psychology.

Since 1975, the work of the Program has dealt with the following issues associated with women and the world of work: women's labor force participation, occupational segregation, displaced homemakers, income tax and social security, women in federal programs, institutional sexism at work, sexual harassment at work, child care, childlessness, teen-age parents, and the role of high school experiences in the attainment process. This review of the Program's research will focus on each of these topics.

WOMEN'S LABOR FORCE PARTICIPATION

During the past two decades, an unprecedented increase in the employment of women outside the home has profoundly affected the economic and social status of women and is fostering a fundamental change in general social and economic conditions. The division of labor, in which men work outside the home for pay while women specialize in unpaid housework, is breaking down. As more women work outside the home, the pressure will intensify for equal treatment in the job market and shared responsibility for work in the home. Institutions concerning the role of women in all aspects of society are being questioned, thereby generating conflict and change.

The continued growth of the female labor force will result in an increase of two-earner families. This change affects consumption patterns, living standards, and needs for purchasing home services in the market. In general, this increase in two-earner families is leading to an exertion of pressure to change public and private institutions that are still largely predicated on the notion of a society dominated by one-earner families.

Contrary to popular belief, the expansion of the female labor force is not likely to mean the end of the American family; it will, however, lead to important changes. For example, women who work are more likely to delay marriage. A careful review of the evidence indicates that maternal employment does not necessarily harm children; the effects depend on a number of factors, including the mother's reasons for working, the quality of alternative care, and the psychological needs of the child.[1]

The growth in women's labor force participation during recent years can be traced to the expansion of job opportunities and the liberalization of attitudes toward women working outside the home. The increase in labor force participation was facilitated by a decline in the percentage of women who are married and have young children. Projections of women's labor force participation in 1990 were made based on estimates of the trend in participation rates for groups of women disaggregated by age, marital status, and presence and age of children.[2] The estimates assume that the trends of the period 1964 to 1977 will continue through 1990 for these detailed demographic groups.

The projections indicate that 68% of all women 16 to 54 years of age will be in the labor force in 1990 compared to 58% in 1977. Married women in this age group with children 6 to 17 years will increase their participation rate from 57% to 70% and those with children under 6 years of age will increase their participation from 39% to 55%. In terms of numbers, Smith estimates that about 46 million women 16 to 54 years of age will be in the labor force in 1990 (compared with about 35 million in 1978)—an increase of almost one million each year. Two-thirds of these women will be married and more than half will have children under 18 years of age.

OCCUPATIONAL SEGREGATION

Even though the size of the U.S. female labor force has increased substantially, the majority of women are employed in stereotyped female jobs. These jobs are often low paying and offer limited opportunities for advancement. Four potential sources of these sex differences in occupation have been identified: differences between men and women in expected length of careers and in job-related skills; differences in tastes and preferences; lack of information or misinformation on the part of employers about the availability and qualifications of women or on the part of women about jobs; and discrimination by employers, employees, and clients. Sorting out the relative importance of these factors could aid in the focus of policies to provide a wider range of jobs for women.

With funding from the Women's Bureau of the U.S. Department of Labor, Vanski has completed a preliminary investigation of the factors leading to the occupational concentration of women.   A range of occupations has been categorized according to their degree of male/ female integration, while other variables have been constructed to permit the analysis of factors leading to the choice of nontraditional careers by younger women.

DISPLACED HOMEMAKERS

Our research interest in the economic problems of elderly women has been expanded by a study of the employment problems of women who are forced to re-enter the labor market in mid-life. Policy analysts are becoming increasingly aware of the detrimental effects of marital dis-solution on the economic well-being of women. In 1975, federal and state level policymakers began to lobby for the passage of legislation which would help a particular group of these women, "displaced homemakers." The displaced homemaker is generally defined as a woman who has devoted herself to family responsibilities and, because of divorce, sepa-ration, or loss or disability of a spouse, has been forced to re-enter the labor market.

Currently, Vanski is designing a demonstration program that will aid in the reduction of barriers to employment for this population.   For instance, if women with certain characteristics have an easy time obtaining a job, they may only be in need of minimal vocational counseling and assistance in finding a job. If outdated or limited job skills are a primary obstacle, then the appropriate policy prescription would be the provision of specific training. If age has an independent adverse effect on labor force entry, then employer discrimination against elderly women cannot be ruled out, in which case programs dealing with employer prejudices would have to be developed. An assessment also will be made of the need for programs to improve self-esteem and confidence as well as for programs that would provide short-term financial assistance to enable displaced homemakers to have sufficient time for selecting an optimal career path.

Before the specific size and type of programs can be determined, two initial studies need to be undertaken.   First, because current estimates vary considerably, it will be necessary to determine the size of the displaced homemaker population. Historic trends in this population will be reviewed in order to assess the relative size of programs needed for the future.   In addition to total estimates, the population will be disaggregated by age, race, education, urban/rural residence, work experi-

ence, marital history, and family income in order to determine the extent
to which various factors affect the level of need for this population.
Second, a structural model will be developed of the displaced homemaker's transition to paid employment which will test the extent to which
formal education, training, age, length of time out of the labor force, skill
level, and attitudes affect labor market success. Results from this model
will be used to indicate the types of program initiatives which promise the
most impact and, therefore, most effectively aid the displaced homemaker.

EFFECTS OF THE SOCIAL SECURITY AND
FEDERAL INCOME TAX SYSTEMS ON WOMEN

The basic frameworks for the Internal Revenue Service and Social
Security Administration were developed in the 1930s and 1940s before the
great increase in the labor force participation of married women. Both
systems were designed for a society in which married women are
essentially homemakers and, therefore, dependents of their husbands in
terms of access to money income. Both systems, in effect, provide
subsidies to the full-time homemaker. Considerable attention is now
being paid to the equity issues raised by these subsidies as well as to the
disincentives to paid employment.

The current tax treatment of families and single persons has been
criticized as creating a disincentive to paid employment, particularly for
women who choose to marry as well as a disincentive to marry for couples
in which both the man and the woman choose to work. In addition, it may
have undesirable effects on the way real disposable income is allocated
among families and individuals because of the way the tax burden is
distributed.

As outlined by Gordon,[3] the problem stems from the following three
features of the income tax structure:

1. Income tax is applied only to money income, not to income from
unpaid work in the home. Families in which the wife is a full-time
homemaker, therefore, will have an extra real income that is not
reflected in their tax burden. A bias against market work, therefore, is
built into the system.

2. Married couples are allowed to file a joint return in which the
combined income of both spouses can be split in half, with taxes paid on
each half regardless of the distribution of earnings between the spouses. If
one spouse has little or no income, income splitting results in a significantly lower tax burden because the progressivity of the tax system
insures that the sum of the tax paid on two halves of an income will be
less than the tax applied to the total income. The gains from income
splitting disappear as the money incomes of both partners become more
nearly equal. Income splitting, therefore, reinforces the bias in the tax
system, which provides strong financial incentive for the wife to be a
homemaker.

3. Since 1971, the tax schedule for singles is lower than the tax schedule for married individuals who file separate returns. Therefore, when two earners marry, they generally pay a higher combined tax, unless one quits work.

In efforts to change the tax system, O'Neill,[4] in congressional testimony, outlined three options: (1) mandatory individual filing where each person with a money income would pay a tax based only on his or her income without regard to the spouse's income; (2) optional individual filing enabling married persons to file on the single's schedule; and (3) tax deduction or credits granted to the lower-earning spouse to offset somewhat the loss of services in the home. Further analysis of labor supply effects is planned.

The Social Security system has some analogous problems. The wife of a covered worker is entitled to a retirement benefit equal to 50% of the husband's benefit and inherits 100% of his benefit as a widow. No additional tax contribution is paid, however, for these "dependent's" benefits. The wife who works in covered employment, on the other hand, does pay taxes. She will receive the standard worker's benefit if she earns one that exceeds what she could be paid as a dependent. (She cannot keep both.) Her worker's benefit, however, is not likely to be worth the extra cost. In fact, estimates made by O'Neill[5] indicate that the implicit rate of return to Social Security taxes is considerably lower for two-earner couples than for one-earner couples and that the longer the wife works in covered employment, the lower is the return. Thus, the system subsidizes one-earner couples at the expense of two-earner couples and single individuals without dependents. In addition, a work disincentive is imposed with respect to the married woman's covered employment.

The Social Security system has been criticized for its treatment of wives as dependents. It should also be pointed out that although the homemaker wife in one sense gets a free ride, she is placed in a vulnerable and demeaning situation. Should she be divorced before 10 years of marriage, she receives no protection as a dependent. Several proposals for restructuring the Social Security system have received considerable attention. In The Subtle Revolution, Gordon[6] discusses earnings sharing and estimates of the benefits and costs of converting to such a system. Under earnings sharing, a husband and wife would split their combined earnings credits and spouse benefits would be eliminated. Thus, a homemaker wife would obtain a record of her own, which would be portable in the event of a divorce and could be increased in the event she entered covered employment.

WOMEN IN FEDERAL EMPLOYMENT PROGRAMS

Although most federal employment programs appear to be sex-neutral, they often have procedural rules built into them which, in effect, adversely affect women's participation in them. There are programs that expressly give priority to male family heads over other people (e.g., the

Work Incentive Program); others set a limit of only one job or training allotment for each family (e.g., the Public Service Employment programs); and some programs place a family-income limitation on eligibility to participate.[7] The Job Corps did not initially include young women at all. When they were introduced into the program, the sex-segregated centers provided fewer openings for women (there being fewer women's centers) and they have never been equal in scope of benefits to those of men.

INSTITUTIONAL SEXISM AT WORK

Public policy approaches to sex discrimination in employment have relied heavily on legal prohibition against differential treatment of women and men in hiring, promotion, and salaries. Although these laws and regulations provide recourse to those experiencing discrimination, the process of filing and carrying through legal suits—either as individuals or as a class action—is tedious and, even when successful, does not completely compensate for the time, money, and psychological losses sustained. In addition, these legal remedies can only scratch the surface of entrenched private policies, practices, and attitudes that shape the conditions of employment adversely affecting women. It is the pattern of these harmful policies, practices, and attitudes that is referred to as institutional sexism.

A conceptual paper on this topic, prepared by Laws and Tangri,[8] describes and analyzes the processes of institutional sexism at three levels. At the institutional level, the authors examine the cultural and organizational context for women's employment, i.e., the normal working arrangements for which no one is held responsible but which apply to everyone and are detrimental to women. At the interpersonal level, the paper deals with behavior toward others based on attributions, perceptions, position, and numbers. At the individual level of analysis, the authors explore men's and women's motivation, stereotypes, and affective dispositions. Some of the phenomena examined include: the time structuring of work, sex-typing of occupations and markets, the impact of federal employment programs and the Social Security system, the culture of formal organizations, interpersonal interaction patterns and group dynamics, leadership, the psychology of tokenism, and the effect of stereotypes on evaluations of persons and their performance.

SEXUAL HARASSMENT AT WORK

The problem of sexual harassment has been raised to the level of public consciousness only during recent years. Although several books and articles have been written on the subject, there has been no way to estimate the extent of the problem. To remedy this state of affairs, a study designed by Tangri, Burt, and Johnson is now being carried out by the U.S. Merit Systems Protection Board. The study covers 1% of the civilian federal male and female work force, in all agencies, at all levels, and over the continental United States and Alaska. This study will be the first on sexual harassment to cover an entire national work force.

Of major methodological concern was the designing of a questionnaire which would be equally acceptable to both sexes. Returns indicate that we succeeded in solving this problem and in generating an exceptionally high return rate. Preliminary results reveal that 42% of the women and 15% of the men report some form of sexual harassment within the last 24 months; 30% of the incidents experienced by women were severe. Harassers were overwhelmingly reported as male (80%) and of coworker status. Further analysis will deal with the consequences of sexual harassment for both the victim and the harasser (such as age, sex, race, position, work relation to each other, and tenure on the job). It also will be possible to identify the various views held about sexual harassment, the working conditions that appear to foster its occurrence, and the beliefs and knowledge about the source of the problem and what can and should be done about it.

THE POTENTIAL DEMAND FOR DAY CARE BY WORKING MOTHERS THROUGH THE 1980s

Day care is a critical issue for working mothers and for those who would like to work. Hofferth[9] shows that the rate of growth of the number of preschool children with working mothers over the next decade is expected to accelerate, both because of the increase in the total number of children and because more of their mothers will be employed. At the same time, the supply of individual providers of home care is expected to shrink as higher wages and expanded job opportunities attract these providers into other occupations. Most evidence points to increased parental dependence on group care, not only for 3- to 5-year-olds, but also for infants and toddlers. Thus, by 1990 there will be 24.1 million preschool children in the United States, of whom 10.4 million (about 43%) will have working mothers. On the basis of past trends, Hofferth estimated that by 1990, over one-half million children under the age of 6 with working mothers will be attending a day care center, and another 3.2 million will be cared for by other nonrelatives. If present preferences and trends do not change, the excess demand for the services of informal child care providers will result in overcrowding and rising prices at these facilities as well as feelings of frustration and desperation among parents.

Because a woman's need for child care is influenced by her labor force participation, fertility, and marital status, we are now planning to develop a model that determines the probabilities that young women will move into and out of employment, bear children, and marry and divorce as well as the time intervals between such life events. Using the National Longitudinal Surveys of the Labor Market Experience of Young Women, these probabilities subsequently will be entered into simulations of a complete sequence of employment and marital-status changes and the birth of children for the decade 1981 through 1990. The final result of this procedure will be annual projections for the decade of the number of children by age as well as the marital and employment status of their mothers. This model will permit us to test alternative assumptions concerning future changes in marriage, divorce, childbearing, and employment.

THE CONSEQUENCES OF REMAINING CHILDLESS
OR HAVING ONLY ONE CHILD

As fertility in the general childbearing population beyond the teen years continues to decline, the consequences of the increasing numbers of small families assume greater importance, both to the family members themselves and to society. The first phase of Hofferth's 2-year study undertaken for the National Institute for Child Health and Human Development found that disparities in the workloads of families increase with the number of children: families with several children spend much more time in paid work and in housework than those with few or no children, thereby reducing the time for leisure activities. Wives with children are more likely to obtain additional training or schooling than those without children; however, they have less work experience and receive slightly lower wages in the early years of marriage. Although there are small differences in the levels of living of childless families and those with up to three children, families with more than three children have substantially lower levels of living. The presence of children, particularly young children, appears to depress the geographic mobility of families as well as the risk of divorce and separation.

TEEN-AGE CHILDBEARING

The failure of teen-age fertility to decline as significantly as that of the rest of the childbearing population has created concern about the social and economic viability of teen-age parents and the degree of public dependency this implies. In their paper on teen-age motherhood, Moore and her associates[10] indicate that mothers who give birth to their first child while still in school tend to complete fewer years of formal education than those who delay childbearing. Since they also tend to have much larger families and lower family incomes than other women, they also are more likely to be in poverty and to require public assistance. Results of a related study show that generous government welfare assistance does not lead to more frequent early childbearing, while employment and educational opportunities for women seem to be only weakly related to less frequent teen-age childbearing.

THE ATTAINMENT PROCESS OF
NONCOLLEGE GIRLS AND BOYS:
THE ROLE OF THEIR HIGH SCHOOL EXPERIENCE

Using data from the National Longitudinal Surveys of the Labor Market Experiences of young men and women, Hofferth[11] designed a study to answer the following three questions:

1. How do noncollegiate youths develop their educational and occupational goals and aspirations and select their high school curriculum? Are curricula selected on the basis of ascribed characteristics such as sex, race, and social class or on the basis of personal preferences?

2. What part do high school experiences play in determining the later labor force outcomes, occupations, and earnings of young noncollegiate men and women relative to the part played by parents and by postgraduate experience?

3. How does the process of occupational choice and wage attainment differ by sex and by race?   Are there substantial differences between college-bound and non-college-bound youth?

Differences between the choices of young noncollegiate men and women in high school are primarily due to differences in aspirations and goals predating entry into high school rather than to specific developments during their high school years.   Parents critically influence the early ambitions, goals, and course selections of their children.   Even the sex-typicality of the occupations children select reflects the sex-typicality of the occupations of their parents.   In this sample of noncollegiate youth, teachers were found to have little direct influence on aspirations and curriculum selection compared with the effects of family background, individual ability, and sex.   Peers exert some influence through their ambitions and goals.   Although individual preferences play some role, they themselves are to a large degree determined by sex and social class.

High school expriences do affect the postgraduate choices of young men and women, especially the work experience they obtain, the occupations they enter, and the hourly wages they earn.   However, the contribution of high school mainly consists of instilling and/or reinforcing values, goals, and aspirations, rather than in providing direct employment skills.   The attitudes and values that youths develop are important in employment.   However, there is a close relationship between the attitudes and values that schools reward and sex-typed aspirations and outcomes.   Individuals who have internalized the work norms of society are likely to also accept and follow the sex-role norms as well.

Young men and women appear to develop along completely separate life tracks. The consequences of the differential tracking of the sexes are serious. Women who work full time, full year earn 60% of what men earn. The fact that men and women are concentrated in different occupations contributes in large part to the difference in their wages. Men who work in occupations employing a large proportion of women also receive lower wages than those employed in occupations employing fewer women. No differences were found between white men and women in basic characteristics (years of schooling, IQ, parental background) that might lead to wage differences.   Forty-six percent of the difference in wages is attributable to occupational segregation. The remainder of the difference is attributable to discrimination and to other factors not included in the model. In contrast, 70% of the wage difference between white and black men is attributable to differences in basic characteristics, none is associated with occupational segregation, and 30% is left unexplained.

Finally, the effects of high school experiences were not found to differ for noncollegiate and college-bound students. Occupational training in high school has only short-term positive effects on the labor force outcomes of noncollege men and women. In the long run, the effects of such training may actually be harmful to the well-being of young women since they lead to placement in "women's jobs." By the tenth year after high school, the same factors associated with going to college and, thus, with higher wages and earnings (e.g., ability, a college preparatory curriculum, studying) also predict the later labor market successes of those not attending college.

OTHER RESEARCH

Other research projects in the Women and Family Policy Program concern: sources of income to the elderly, determinants of private pension income for women, determinants and policy implications of changes in black family structure, barriers to suitable housing for minority and nontraditional families, long-run economic consequences of living in a single-parent family during adolescence, estimates of the cost of children in the United States, and the impact of child spacing on family level of living.

Results of the reviewed projects as well as those briefly noted should form the basis for developing techniques and aids to enhance women's participation in and contributions to the world of work.

NOTES

1. Moore and Hofferth, 1979.
2. Smith, 1979a.
3. Gordon, 1979.
4. O'Neill, 1980.
5. O'Neill, 1976.
6. Gordon, 1979.
7. Underwood, 1979.
8. Laws and Tangri, 1979.
9. Hofferth, 1979.
10. Moore, Hofferth, Caldwell, and Waite, 1979.
11. Hofferth, 1980.

WOMEN'S WORK PARTICIPATION AND THE

ECONOMIC SUPPORT OF CHILDREN

Barbara R. Bergmann

University of Maryland
College Park, Maryland   U.S.A.

The Urban Institute
Washington, D.C.   U.S.A.

In all of the developed countries, and in many of the developing ones, more women every year are participating in the work force. This trend is gradually transforming in many ways the economies of these countries and also is broadly transforming the social and economic relations among women, men, and children.[1] In the course of these transformations, the assumption of the economic burden of caring for children has shifted toward the mother and the state--and away from the father. This paper considers the causes and policy implications of this "burden shifting" in the context of developments and institutions in the United States.

PATTERNS OF ECONOMIC FLOWS TO CHILDREN

In thinking about the traditional form of the family, in which only the husband works for pay, the common locution is that the husband supports or is the sole support of the wife and children. This terminology gives the impression that in such a family the entire economic burden of child support is on the father. What is obscured, of course, is that in most such families the mother makes a considerable economic contribution both to the husband and to the children in the form of domestic services provided.[2] The economic form of the traditional family is diagrammed in Panel A of Figure 1. The mother is the source of all domestic services (ds) to the husband and to the children, while the father is the source of all goods and services purchased for money (mg). We have included the role of the state, which receives taxes (t) from the father, and provides educational services (es) to the children. If we were to use conventional methods of attaching a monetary value to the services rendered, the

331

FIGURE 1

PATTERNS OF ECONOMIC FLOWS
AMONG PARENTS, CHILD, AND THE STATE

A. TRADITIONAL FAMILY

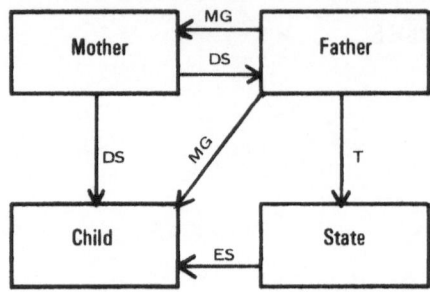

B. "ANDROGYNOUS" FAMILY

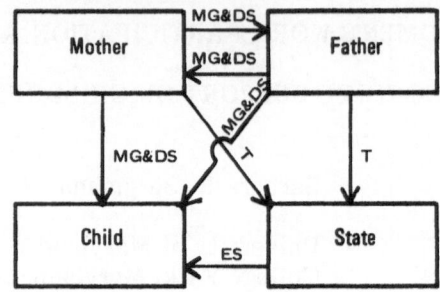

C. "RUSSIAN" FAMILY

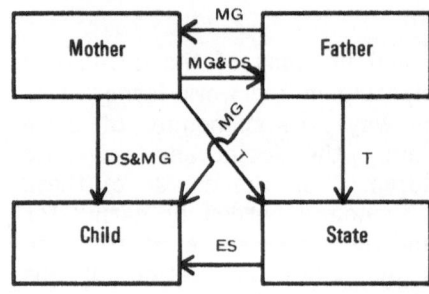

D. NO FATHER'S CONTRIBUTION
   MOTHER ON WELFARE

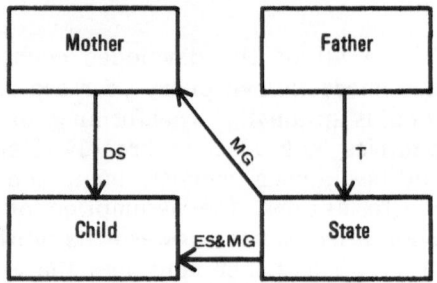

E. NO FATHER'S CONTRIBUTION
   MOTHER WORKING

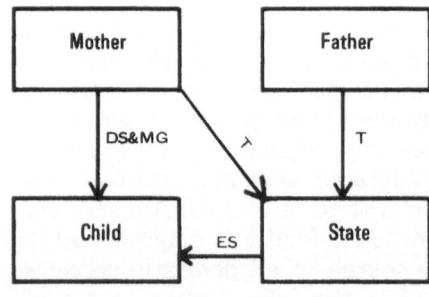

F. FATHER PAYS CHILD SUPPORT
   MOTHER WORKING

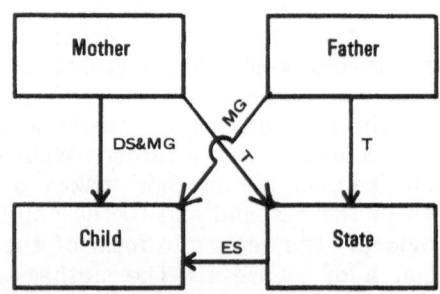

CODE: MG = Market Goods; DS = Domestic Services; ES = Educational Services; T = Taxes.

father's economic contributions in the traditional family would exceed those projected for the mother's services.[3]

When the wife enters the labor force, the pattern of flows of goods and services provided to family members changes. One possible pattern that may establish itself is the one exhibited in Panel B, which shows the pattern of flows that has been dubbed by the author as the "androgynous family." In this example, both husband and wife engage in paid work, providing market goods and services and domestic services to each other and to other family members, and each shares these two major tasks about equally. Surveys have indicated that only a small portion of American families with two earners currently follow the androgynous pattern, but it remains the ideal of most American feminists.[4]

A second pattern that may emerge when the wife works for pay is one in which the sharing of the burden of market work by the wife does not result in any substantial sharing of the burden of providing domestic services by the husband. This pattern, diagrammed in Panel C, has been dubbed the "Russian model," because of its well-publicized prevalence in that country. The Russian model family is by no means restricted to Russia and is a common mode of operation for two-earner families in the United States. Those women moving from the traditional family model to the Russian family model, or even to something halfway between the Russian model and the androgynous model, assume a larger share in the economic provision for children and for family life in general.

Divorce (or illegitimacy) may produce the pattern of economic flows shown in Panel D, where the state has taken the place of the father in the traditional family. It supplies welfare payments, while the woman stays at home and provides domestic services to the child. The standard of living of the woman and her children in this situation is on the average much reduced from what it would be in the intact traditional family.

The increasing proportion of mothers of young children who are working and the rise in women's wages have probably caused a reduction in the public's willingness to contribute public assistance to nonworking mothers. Up to now, welfare policy has been based on the idea that welfare mothers should have the same right as middle-class women with husbands to choose child rearing and homemaking as their sole occupation. Such a claim is bound to sound outlandish to an increasing proportion of citizens, particularly as more married mothers take jobs.

The result will be a greater number of families exhibiting the pattern shown in Panel E, where the mother singlehandedly takes on the economic roles played in the traditional family by both mother and father. Of all the patterns shown, this one puts the greatest share of the economic burden of child rearing on the mother.

Finally, in Panel F, we show a separated father contributing child support, which will ususally be supplemented by the mother's wage and/or welfare payments. As will be shown below, the pattern in F is much rarer in the United States than that shown in D or E.

THE IMPACT OF THE ATROPHY
OF SOCIETAL INSTITUTIONS

Social systems, at least in Western countries, have tradionally dealt with the problems of encouraging men to provide economic support for children by attempting to limit the number of "fatherless" children. The major mechanisms for doing so have been: the institution of marriage; a taboo on sex for unmarried women; and disgrace and deprivation not only for women bearing children out of wedlock, but also for the children themselves.   The problem of male economic support for children "left over" after a marriage ended was solved in part by avoidance of the problem—by making divorce difficult for all but the rich—and also through the extended family.

While the institution of marriage obviously confers certain benefits on men, the major beneficiaries, nevertheless, are women and children. If many or most women are going to have children anyway, the economic strain of child rearing has required, until recently, that each woman attach to herself and her children a man who would make contributions to their sustenance. The disadvantages that women have traditonally suffered in marriage—more or less severe domination by the husband, a requirement (in the absence of servants) that she do all the housework, a double standard with regard to sexual fidelity—can be viewed as the price she has to pay for this support.

Until recently, traditional social institutions and mores have been fairly successful in keeping a high proportion of children with, and receiving economic support from, adult males. It is obvious that some of the most important of the traditional habits and arrangements upon which past generations relied to perform this function, such as a taboo on premarital sex for women, a stigma on illegitimate births, the all-but indissoluble marriage of unequals, and the extended family, no longer operate for large segments of the population in the West. In the United States, it may well be true that they have ceased to function for a majority of the population.

We may speculate that one vital element in the decline of traditional institutions has been the development and accessibility of contraceptive devices and of safe and legal abortions.   Contraception and abortion must surely have contributed significantly to the atrophy of the taboo against premaritial sex for women, since their availability reduces the probability that a woman will have to unwillingly bear an extramarital child as the result of any single act of sexual congress. We can further speculate that indulgence in premarital sex by more women lowers the incidence of marriage by reducing the incentive for men to marry. A greater availability of unmarried female sexual partners also may increase the incidence of divorce by providing a husband with increased opportunities for extramarital involvements that could have a destabilizing effect on marriage. An increase in the divorce rate due to these involvements for husbands makes a divorce a more common occurrence, which in turn could serve as an example for other couples who might have stayed together had divorce been less prevalent.

A second major cause for the breakdown of the traditional system for insuring economic flows from adult males to children is the long-term, upward trend in women's wages. Since the start of the Industrial Revolution, the West has been in a period of continued technological change, which has had the effect of raising the productivity of work and raising year by year the amount of goods and services that can be purchased with the average daily wage earned by both men and women. Women workers' wages (corrected for price changes) have been rising along with men's wages. In the United States, these earnings, however, have remained at about 60% of men's wages, while the increases in productivity, on which rises in real wages are based, have continued at an average rate of 2.3% per year since World War II.[5]

The rise in women's real wages may well have an important effect on both child support and employment practices in our society. The most obvious effect of this increase has been for women to weigh the relative advantages of staying at home against those of entering the paid work force. This process, in turn, has altered traditional thinking concerning the suitability of paid work for married mothers, and caused more and more married women with young children to seek and occupy paying jobs. The rise in the labor force participation of married mothers and, in particular, the increase in the number of women who maintain a fairly continuous attachment to the labor market seem to have stimulated new thoughts about appropriate occupational goals for women and new opinions (and laws) about the rights of employers to treat women differently than men with respect to job availability and wages, which may in time succeed in narrowing the gap between men's and women's wages.

The increasing orientation of married women toward paid work, while reducing the birth rate of within-wedlock children, probably has had the effect of increasing the divorce rate and, thus, the number of children living apart from their fathers. A major reason for this higher divorce rate is that the lessening of economic dependence on the husband has given wives more courage to terminate bad marriages. Also, women's work force participation probably has reduced the guilt that husbands feel about leaving their families and failing to contribute to their subsequent support. It also may have increased overt conflict within the marriage, in proportion to the extent that the unchallenged dominance of the husband, based partly on his monopoly of access to money wages, reduced overt conflict in the past.

Because of rising real wages, a large proportion of women in the United States are now capable of earning enough money to support themselves and their children at above the poverty level (as officially defined), unaided by a man. In 1976, the median money wage of a woman who had a full-time, year-round job was $8,099, which is far below the 1976 median male money earnings of $13,455.[6] However, the rate of productivity increase has been such that employed women now have real wages that are about what men's real wages were on the average 22 years ago, wages which allowed men to be the sole financial support of their families. As women's financial capacity rises, it no doubt will be more and more tempting for separated fathers (and for governments) to continue to place on single and divorced mothers the burden of supporting their children without help from them or the state.

CHILD SUPPORT IN THE UNITED STATES

As of 1978, more than one in six children in the United States (10.7 million children) was living with a woman who had no husband present, a dramatic increase from the ratio of one in nine observed in 1970 and one in twelve in 1960. Approximately 75% of this increase since 1960 was due to divorce or separation while about 25% was attributed to an increase in births to never-married women.[7] More than 40% of these "fatherless" families had a money income below the official poverty line in 1977.[8]

A survey conducted in 1975 by the U.S. Bureau of the Census found that of currently divorced, separated, or never-married women with children, only one-fourth received any child support payments at all in the survey year.[9] Of those who did receive some payments, only 9% received as much as $1,000 per child during the year (see Table 1). Those who received no child support payments were almost three times as likely to be below the poverty line as those who received some payment.

Of the 4.9 million women who were not living with the father of their children and of whom 2.0 million had remarried, 3.1 million had an income from earnings in the survey year. Of the 2.0 million who reported having received government transfer payments, 0.9 million indicated that these payments were their only source of income.

An important reason that more of the "fatherless" children in the United States receive little or no economic support from their fathers lies in the poor enforcement of child support court orders. When a father is ordered by a court to make child support payments, the court in most jurisdictions sets in motion no governmental mechanism which sends him bills, notes his payments and delinquencies, or moves against him if he is delinquent. The woman, who is to be the recipient of the payment, must herself initiate legal action in the case of delinquency and, until recently, has in most cases had to seek the help of a lawyer to do so. In the United States, not only is the court procedure archaic and slow, but the problem of the legal pursuit of a deliquent ex-husband may be further complicated by the man's move to another state. Recent changes in this procedure have removed the need to hire a private attorney, but the process of collection remains tedious.[10] Even if, after a long legal process, the

TABLE 1

MOTHERS SEPARATED OR DIVORCED FROM THEIR CHILD'S FATHER BY RECEIPT OF CHILD SUPPORT PAYMENTS, 1975 (UNITED STATES)

Extent of Child Support Payments	Number of Women	Percentage Distribution
Received no support payments	3,676,000	75
Received less than $1,000 per child per year	800,000	16
Received more than $1,000 per child per year[1]	446,000	9
Total	4,922,000	100

SOURCE: Current Population Reports, No. 84, p. 23, Table 8.

[1]Mothers of more than three children were included in this category if they received at least $3,000.

delinquent father has been ordered by a judge to "make good," he may shortly thereafter become delinquent again and must be legally pursued anew. The process is so cumbersome that obedience to these court orders can almost be said to be voluntary, and the proportion of volunteers among separated fathers is very low.

It is commonly conjectured that most of the fathers defaulting on child support payments have low incomes, but surveys do not bear this out. Data from the Michigan Panel Study of Income Dynamics show that in 1973 the proportion of fathers who paid nothing varied little with the father's income.[11]

NEW INSTITUTIONS FOR SHARING THE ECONOMIC BURDEN OF CHILDREN

During recent years, mothers have been carrying an increasing share of the burden of providing economic resources to their children, and a high and growing proportion of these mothers is assuming an unfairly large share. Working mothers who are married obviously have the problem of "overload," deriving in part from the persistence of the tradition that men should do little or no domestic work even if their wives have jobs outside the home. This problem needs to be attacked through the organs that influence public opinion: radio, television, newspapers, books, magazines, and the schools. Even with a more equitable sharing of domestic work, the two-earner family has less time and energy available for domestic work and needs to be able to purchase some of the services previously performed by the wife on an unpaid basis. Conveniently available, high-quality child care is the most obvious of such services.

The situation of the unmarried, working mother is made difficult by the relatively low wage to which discrimination is likely to condemn her. She also suffers from her household's shortage of adult person-hours available for domestic tasks.[12] She cannot solve her problem, as her male counterpart can, by acquiring a "wife" who would do all of her domestic work in return for room, board, and a modest clothing allowance. And, as we have documented above, American women suffer from their inability to effectively compel the father of their children to contribute to the economic burden of child support.

Many unmarried American mothers, facing this combination of adverse circumstances, go on welfare and live a poverty-stricken version of traditional family life. An even higher percentage of these women combine the economic functions of both mother and father in the family. With inadequate resources in time and money, they suffer from unceasing toil, responsibility, and inadequate rest.

It is plausible to suggest that the government should consider the implementation of three provisions for unmarried mothers: (1) subsidized, high-quality child care to enable the single mother to buy the economic services of a "wife" as inexpensively as a man can; (2) cash payments to raise the child's standard of living to a minimal level of decency if either of the parents is judged incapable of providing adequate financial support; and (3) perhaps most important of all, a mechanism for collecting, monitoring, and disbursing child support payments from fathers living apart from their children.

The mechanism for collecting from absent fathers should work on the principle of the withholding tax, and could be administered by the Internal Revenue Service and/or the Social Security Administration. Entry into such a system should not await delinquency, but should start at the very beginning of the separation.[13]

CONCLUSION

The drying up of the flow of resources from adult males to children is reflected in the welfare burden which has been placed on the taxpayers, but the burden is borne also by millions of women and children who never see a welfare dollar. The burgeoning welfare roles and the dissatisfaction felt with the amount of tax dollars going to welfare are really just the tip of the iceberg of the social problem created by the atrophy of long-standing human institutions which kept children under the economic support of their fathers.

The social institutions which in the past served to channel resources to children from their fathers are unlikely to be revived in their old forms, nor should we wish them to be. These institutions kept women in a grossly inferior status and ruined many lives by preserving unhappy marriages which should have been dissolved. The problem is that we have not yet created and initiated new patterns and institutions which serve some of the positive purposes of the old institutions and which are felt to be just and adequate. The development and implementation of such patterns and institutions is of high priority.

NOTES

1. For an extensive review of the effects in the United States of the increasing labor force participation of women, see Smith, 1979b.
2. Ignoring the economic contribution of the mother in the traditional family probably results from two causes: the fact that her contributions are not exchanged for money and the fact that in a small but highly visible minority of middle- and upper-class households mothers may be relieved from providing such services through the employment of paid servants.
3. At least part of the reason for this result is that discrimination has reduced the rate of pay of the kinds of services that women are relegated to delivering; a second reason is that the value of parental care is scanted by assigning a monetary value.
4. For a review of the literature on the sharing of housework, see Smith, 1979b:111-116.
5. U.S. Department of Commerce, 1977.
6. U.S. Bureau of the Census, 1978b.
7. U.S. Bureau of the Census, 1979e.
8. Placement of "fatherless" families above or below the poverty line as defined by the U.S. government gives an incomplete and biased picture of their material situation. It leaves out of account the fact that a single parent who is working full time is going to experience an acute shortage of adult person-hours available for domestic tasks—such as shopping, cooking, and going to the doctor as well as being with and supervising children (See Vickery, 1977:27-48). Even more important, the rationale on which poverty lines are based includes no allowance whatever for cash payments for child care services.
9. U.S. Bureau of the Census, 1979e. The survey shows that remarriage did not increase or decrease the probability of receiving child support payments; one-quarter of remarried women received some payments.
10. Cassetty, 1978; Gates, 1977.
11. Jones, Gordon, and Sawhill, 1976.
12. Vickery, 1977:27-48.
13. Cassetty, 1978.

CONCLUDING COMMENT

The objective of this symposium was to examine several areas associated with women's increased participation in the world of work: contributions and progress; physical, mental, and economic well-being; socioeconomic factors; and role integration and revision. In comparing the research reported on these topics, a commonality across countries is revealed which indicates that women throughout the world face the same fundamental problems regardless of differences in status, education, utilization, and wage scale. Many solutions also are presented in this volume, and the final issue to be addressed is to summarize these into a few action-oriented recommendations. This final comment, therefore, endeavors to highlight the work that still has to be accomplished in our efforts to realize the "exciting possibility of creating a much improved society."

Of all the issues associated with women's participation in the world of work, the most important is economic: Women enter the labor force because they need the income. For most women, however, their jobs offer relatively few opportunities for advancement or for making substantial increases in pay. Women's work in all occupations tends to be undervalued and, as a result, the wage is considerably less than is paid to their male counterparts or to men in jobs of equal value.

These inequities exist despite the fact that equal pay laws have been enacted in almost all European and North American countries. The earliest recorded equal pay laws date back to 1871 in Germany and 1882 in France; more recently, all of the Member States of the European Community have adopted the Treaty of Rome in which Article 119 sets the principle of equal pay for men and women. The U.S. Congress passed into law the Equal Pay Act in 1963 while the Equal Pay Act of 1970 in the United Kingdom stipulated that all employers had 5 years during which collective agreements, employers' unilateral pay scales, and arbitrated adjustments were to reach full equality. Although it may be too soon to observe a substantial narrowing in the wage disparity between men and women, the reported comparisons in 1978 among 17 European and North American countries showed that the United Kingdom had the widest wage differential (women's wages were on the average 55% of those for men), followed by the United States with 60% which has since widened to 57%.

While these laws apparently have had little impact on eliminating the inequities in pay and treatment in the workplace, the passage of re-

341

cent legislation will pave the way for promoting equality in the future. In December 1979, the 34th Session of the United Nations General Assembly voted unanimously to approve the International Convention on the Elimination of All Forms of Discrimination against Women. This monumental Convention will become law and part of the official U.N. Documents when ratified by 20 governments; as of July 1981, seven countries had ratified the Convention. Approving countries agree "that all contracts and other private instruments that restrict the legal capacity of women shall be deemed null and void."

On February 10, 1981, the sitting of the European Parliament was opened by President Simone Veil for debate on the position of women in the European Community. After the debates and discussion, the Members of the European Parliament voted and passed the Resolution of 59 specific directives that encompass proposals to improve existing Community measures and recommendations for new Community action. The "new" directives center on: the social and economic involvement of women; educational and vocational training; health care (including research on improved birth control methods and a request for Member States to enact legislation whereby women can obtain a safe and legal abortion in their own countries); the special categories of women such as migrant women and women working in family firms; and the way in which the position of women in developing countries is influenced by western cultural principles. One of the directives is especially pertinent to the purpose of this symposium; it states that consideration should be given to guaranteeing for each individual "the freedom to attain his or her own aspirations and to combine family tasks and a role in society according to his or her own choice."

In North America, legislation has been passed in both Canada (the Human Rights Act in 1978) and the United States. The U.S. Congress passed the Equal Rights Amendment (ERA) in 1972 which will not become law until ratified by 38 states, a goal that is being hard fought by both proponents and opponents alike. If the ERA is not ratified, the view of the United States as a progressive country primarily concerned with human rights will no longer be held by peoples throughout the world. Without ratification, the disheartening thought that comes to mind is the image of the United States that then will be conveyed abroad.

In order for these human rights and equal pay laws to be effective, individuals have to have the dedication, energy, courage, and time to ensure their enforcement. The burden of this legal labor falls on the shoulders of the individuals involved, namely women. During the past decade, increasing numbers of women have initiated the legal steps to obtain equal pay—even though the struggle requires considerable energy and time to learn the rules and procedures to become successful combatants in the fight-for-equality arena. The time element is an important factor because most employed women work an average of 71 hours per week in paid and unpaid work, as noted by Reagan in her chapter. As a result of their legal pursuits, moreover, many women probably have experienced a variety of retaliatory unpleasantries in the workplace which will detract from the battle for equality and, more important, from their work and career goals. Further, the judicial process requires that individuals seeking justice have patience to endure this frequently time-consuming procedure. Given these circumstances, it is understandable why such little progress has been made by women in achieving equality on the job. It is only through the court system, however, that women will reach full equality. Therefore, to be taken seriously, which in turn will lead to consideration for jobs of increased authority, responsibility, and pay, women who are victims of discrimination will have to employ the legal process as the first step in the evolution to equal status with men.

Another facet of the struggle for equality concerns women's role in the family. Most women who work outside the home also perform the household and child care tasks as well as the work of "family status production" which was conceptualized by Papanek and described in the previous section. For most employed women in all countries, this double duty has been accepted as their "lot in life."

Many other women, however, perform triple duty whereby they assume responsibility for their own tasks in the home, the absent father's chores, and remunerated work outside the home. For example, in Turkey (as reported by Kandiyoti), Greece, Italy, and Portugal, many women have become operators of a family-owned farm or business, paid workers, and household and child care providers while their husbands emigrate to work in highly industrialized countries. Single mothers who are employed also face similar work demands and time constraints in that they, too, have had to take on the triple role obligations of his, hers, and remunerated work. Attaining assistance from members of the extended family has become more difficult because of the greater mobility of most families, the restrictive policies of the social services, the increased financial independence of members of the extended family, and the construction of smaller homes that only have space for members of the nuclear family. Current trends suggest that in the future most women assuming double or triple duty will be unable to depend on members of the extended family for either economic or social support.

To effect changes in the aforementioned circumstances and to increase equality in the family, several specific recommendations were formulated by the symposium participants: (1) promoting the establishment and expansion of low-cost, quality child care programs; (2) implementing the division of household and child care tasks among all household members (including boys who worldwide spend the least amount of time, if any, engaged in such chores); (3) using all forms of the media and the educational system to encourage men to become more active participants in family activities and work; (4) legislating the automatic withholding of child support payments from absent fathers' pay checks; (5) providing training programs for boys and girls to prepare them for assuming the roles of responsible adults and parents; and (6) forming support networks for women to work for the fulfillment of these goals.

At the community level, efforts should be reinforced that promote a stronger attachment to the family and the community which in turn will contribute to a strengthening of such individual attributes as cooperation, compassion, and responsibility. Further, as more and more large multi-family housing projects are constructed in our cities, endeavors to foster the values of helping, caring, and sharing will become an essential part of each occupant's responsibility in order to ensure the safety and survival of all inhabitants. The results of these efforts will permeate throughout the community and bring about positive changes in creating environments that are conducive to the realization of each person's own growth-promoting goals and aspirations.

The third area of consideration addressed in this volume is that of women's physical and mental well-being. Being members of the labor force has had a positive effect on women workers' health in that women who are employed report better health than homemakers. According to most women, moreover, a job is perceived as a source of social support, self-esteem, and self-confidence. Most women also find that a job offers them the opportunity to belong to a health care program that may include medical care benefits for other family members. Having more women in the workplace also leads to health-promoting changes in the work environment because the policies and laws that are designed to protect women, especially those who are pregnant, ensure safer working conditions for men as well. In these diverse ways, women have helped to humanize the work setting by providing the impetus for improved working conditions, and therefore better health for all workers, and by requesting employee health care plans that provide medical care for their family members. As more women assume positions of authority in the workplace, another facet of this humanizing effect perhaps will be reflected in increased rates of productivity and improved workers' health.

Other issues that should be noted in these concluding comments include: the need for girls and women to make a training or educational commitment, an increased awareness of the effects of tokenism, the knowledge that women are being exploited by many of the multinational conglomerates, the provision of information concerning women's utilization in the world of work, and the realization that most women have the energy to perform double and triple duty. Of greatest importance is the sharing of information on these and all other issues addressed in this volume which will enhance feelings of compassion needed to promote equality in the workplace and in the home.

In closing, as parents and educators, we should encourage girls (and boys) to achieve to their fullest in school and to learn marketable skills. As scientists, we should engage in action-oriented research that will produce results needed to form the basis for the development of programs and laws designed for the betterment of womankind in particular and humankind in general. As citizens, we should endeavor to enforce these laws and support these programs. Further, we should promote values in our society that encourage the growth of boys and girls into responsible adults and parents. The fruits of this dedication to improving conditions specifically associated with women should lead to a much improved society. It is to be hoped that the ideas generated at this symposium and the research reported in this volume will contribute in some way to this grand endeavor.

A.H.

BIBLIOGRAPHY

ARCHIBALD, K. (1970) Sex and the Public Service. Ottawa: Queen's Printer.

ARMSTRONG, H. (1979) "Job creation and unemployment in postwar Canada," in R.M. Novick (ed.) Full Employment: Social Questions for Public Policy (Urban Seminar Six). Toronto: Social Planning Council of Metropolitan Canada.

ARMSTRONG, H. and P. ARMSTRONG (1975) "The segregated participation of women in the Canadian labor force, 1941-71." The Canadian Review of Sociology and Anthropology 12 (November): 370-384.

ARMSTRONG, P. (1979) "Women and unemployment," in R.M. Novick (ed.) Full Employment: Social Questions for Public Policy (Urban Seminar Six) Toronto: Social Planning Council of Metropolitan Toronto.

ARMSTRONG, P. (1980a) "Attacking the unemployed—UIC: reform or revolution?" Perception: A Canadian Journal of Social Comment 3 (March-April): 31-33.

ARMSTRONG, P. (1980b) "Women and unemployment." Presented at Mount St. Vincent University, Halifax, N.S., January.

ARMSTRONG, P. and H. ARMSTRONG (1978) The Double Ghetto: Canadian Women and their Segregated Work. Toronto: McClelland and Stewart.

ATHANASSIADIS, C. (1951) Mathematical Models in Economics. Athens: Papazissis Publ. Co.

ATHANASSIOU, S. (1975) Manpower Planning in Greece. London: English Univ. Press.

ATHANASSIOU, S. (1978) Some Economic-Demographic Interactions Affecting Female Manpower Supply: The Greek Experience. New York: Plenum.

BAKER, E. (1964) Technology and Women's Work. New York: Columbia Univ. Press.

BARNES, W.E. and E.B. JONES (1974) "Differences in male and female quitting." Journal of Human Resources 9 (March): 438-451.

BARRETT, N.S. (1979a) "Statement," pp. 1-17 in U.S. Senate, Committee on Human Resources, Hearings on The Coming Decade: American Women and Human Resources, Policies, and Programs (February 9). Washington, DC: U.S. Government Printing Office.

BARRETT, N.S. (1979b) "Women in the job market: occupations, earnings, and career opportunities," pp. 31-62 in R.E. Smith (ed.) The Subtle Revolution: Women at Work. Washington, DC: The Urban Institute.

BARUCH, G.K. and R.C. BARNETT (1978) "Employment status and well-being, a study of mothers of preschool age children." Presented to the Vermont Conference on the Primary Prevention of Psycho-pathology.

BECKER, G. and B. CHISWICK (1966) "Education and the distribution of earnings." American Economic Review 56 (May): 358-378.

BEM, S.L. (1974) "The measurement of psychological androgyny." Journal of Consulting and Clinical Psychology 42: 155-162.

BERGMAN, L.R. and A. DUNÉR (1975) "Vart tar toppbegåvningarna vägen?" (Choice of career for gifted high school adolescents.) Rapporter, No. 5, Department of Psychology, University of Stockholm. (Includes an English summary.)

BERLINGSKE TIDENDE (1979) "Results of a Gallup Poll conducted in Denmark on child care." (September 16).

BERTELSEN, O. and J. USSING (1974) Familiestørrelse og Livsstil. (Family Size and Style of Life.) Copenhagen: The Danish National Institute of Social Research.

BLAU, P. and O.D. DUNCAN (1967) The American Occupational Structure. New York: John Wiley & Sons.

BLAXALL, M. and B. REAGAN [eds.] (1976) Women and the Workplace: The Implications of Occupational Segregation. Chicago: Univ. of Chicago Press.

BLOCK, J.H. (1973) "Conceptions of sex-role: some cross-cultural and longitudinal perspectives." American Psychologist 28 (June): 512-526.

BLUMROSEN, R.G. (1979) "Wage discrimination, job segregation, and Title VII of the Civil Rights Act of 1964." University of Michigan Journal of Law Reform 12 (Spring): 399-400.

BOND, J.R. and W.E. VINACKE (1961) "Coalitions in mixed-sex triads." Sociometry 24: 61-75.

BOULDING, E. (1977) Women in the Twentieth Century World. New York: John Wiley & Sons.

BOYD, M. (1977) "The forgotten minority: the socio-economic status of divorced and separated women," in P. Marchak (ed.) The Working Sexes. Vancouver: Institute of Industrial Relations, University of British Columbia.

BRAVERMAN, H. (1974) Labor and Monopoly Capital: The Degradation of Work in the Twentieth Century. New York: Monthly Review Press.

BRECHER, J. (1979) "Roots of power: employers and workers in the electrical products industry," in A. Zimbalist (ed.) Case Studies on the Labor Process. New York: Monthly Review Press.

BRINKMANN, C. (1979) "Erwerbsbeteiligung und Arbeitsmarktverhält-nisse: Neue empirische Ergebnisse zur 'Entmutigung' und zusätz-lichen 'Ermutigung' von weiblichen Erwerbspersonen." Mimeo-graphed.

BROWN, G.W., M.N. BHROLCHAIN, and T. HARRIS (1975) "Social class and psychiatric disturbance among women in an urban population." Sociology 9 (May): 225-254.

BROWN, M. and M. MANSER (1978) "Neoclassical vs. bargaining approaches to the joint estimation of household labor supply equations." Discussion Paper No. 401, Department of Economics, State University of New York at Buffalo, November.

BUCKLEY, H. (1973) "Interpreting the unemployment statistics." Notes on Labour Statistics, 1972. Statistics Canada Cat. No. 72-207 (February): 5-8.

BUCKLEY, J.W. (1947) "Wages." Trades and Labour Congress Journal 26 (July): 14.

BURSTEIN, M., N. TIENHAARA, P. HEWSON, and B. WARRANDER (1975) Canadian Work Values. Ottawa: Information Canada.

CANADA, DEPARTMENT OF FINANCE (1980) Economic Review; April 1980: A Perspective on the Decade. Ottawa: Supply and Services Canada.

CANADA, DEPARTMENT OF LABOUR (1958) A Survey of Married Women Working for Pay in Eight Canadian Cities. Ottawa: Queen's Printer.

CANADA, DEPARTMENT OF LABOUR (1960) Occupational Histories of Married Women Working for Pay. Ottawa: Queen's Printer.

CANADA, DEPARTMENT OF LABOUR (1965) Women at Work in Canada: A Fact Book on the Female Labour Force, 1964. Ottawa: Queen's Printer.

CANADA, DOMINION BUREAU OF STATISTICS (1942) Reserve of Labour among Canadian Women. Statistics Canada No. 71-D-51. Ottawa: Queen's Printer.

CANADA, DOMINION BUREAU OF STATISTICS (1963) Annual Report on Benefits Established and Terminated under the Unemployment Insurance Act. Cat. No. 73-201. Ottawa: Queen's Printer.

CANADIAN COUNCIL ON SOCIAL DEVELOPMENT (1976) Women in Need. Ottawa: Canadian Council on Social Development.

CANDEA, L. and G. CHAPMAN (1981) Harassment and Discrimination of Women in Employment. Washington, DC: Center for Women Policy Studies.

CASSETTY, J. (1978) Child Support and Public Policy. Lexington, MA: Lexington Books.

CENTER FOR THE AMERICAN WOMAN AND POLITICS (1978) Women in Public Life. New Brunswick, NJ: Center for the American Woman and Politics.

CHAPMAN, J.R. (1980) Economic Realities and the Female Offender. Lexington, MA: Lexington Books, D.C. Heath.

CLEMENT, W. (1975) The Canadian Corporate Elite, An Analysis of Economic Power. Toronto: McClelland and Stewart.

COGAN, J.F. (1980) "Labor supply with costs of labor market entry," pp. 327-365 in J.P. Smith (ed.) Female Labor Supply: Theory and Estimation. Princeton, NJ: Princeton Univ. Press.

COLLINS, K. (1978) Women and Pensions. Ottawa: Canadian Council on Social Development.

COLLVER, A. and E. LANGLOIS (1962) "The female labour force in metropolitan areas: an international comparison." Economic Development and Cultural Change 10 (July): 367-385.

CONNELLY, P. (1978) Last Hired, First Fired: Women and the Canadian Work Force. Toronto: The Women's Press.

COPP, T. (1974) The Anatomy of Poverty: The Condition of the Working Class in Montreal, 1897-1929. Toronto: McClelland and Stewart.

CROSBY, F. (1976) "A model of egoistical relative deprivation." Psychological Review 83: 85-113.

CROSS, M.S. [ed.] (1974) The Workingman in the Nineteenth Century. Toronto: Oxford Univ. Press.

CROSS, S.D. (1977) "The neglected majority: the changing role of women in 19th century Montreal," pp. 66-86 in S.M. Trofimenkoff and A. Prentice (eds.) The Neglected Majority: Essays in Canadian Women's History. Toronto: McClelland and Stewart.

CUNEO, C. (1980) "Class contradictions in Canada's international setting," pp. 83-102 in J.P. Grayson (ed.) Class, State, Ideology, and Change. Toronto: Holt, Rinehart & Winston of Canada.

CURREY, V. (1977) "Comparative theory and practice," in M. Githens and J. Prestage (eds.) A Portrait of Marginality: The Political Behavior of the American Woman. New York: David McKay.

CYERT, R.M. and J.C. March (1963) A Behavioral Theory of the Firm. Englewood Cliffs, NJ: Prentice-Hall.

DEAUX, K. and T. EMSWILLER (1974) "Explanation of successful performance on sex-linked tasks: what is skill for the male is luck for the female." Journal of Personality and Social Psychology 29: 80-85.

DENTON, F. (1970) The Growth of Manpower in Canada. Ottawa: Queen's Printer.

DERNBERG, T. and K. STRAND (1966) "Hidden unemployment 1953-62: a quantitative analysis by age and sex." American Economic Review 56 (March): 71-95.

DICKINSON, K. (1975) "Child care," in G.J. Duncan and J.N. Morgan (eds.) Five Thousand American Families--Patterns of Economic Progress, Vol. 3. Ann Arbor, MI: Institute for Social Research, University of Michigan.

DUNCAN, G. and C.R. HILL (1975) "Model choice in child care arrangements," in G.J. Duncan and J.N. Morgan (eds.) Five Thousand American Families—Patterns of Economic Progress, Vol. 3. Ann Arbor, MI: Institute for Social Research, University of Michigan.

DUNCAN, O.D. (1979) "Indicators of sex typing: traditional and egalitarian, situational, and ideological responses." American Journal of Sociology 85 (March-April): 251-260.

EASTERLIN, R. (1973) "Does money buy happiness?" The Public Interest 30 (Winter): 3-10.

ECONOMIC COUNCIL OF CANADA (1976) People and Jobs. Ottawa: Information Canada.

ECONOMIC REPORT OF THE PRESIDENT OF THE UNITED STATES (1973) Washington, DC: U.S. Government Printing Office.

EGLE, F. (1979) "Ökonometrische Ansätze und Sättigungsfunktionen zur Erkärung und Projektion des Erwerbspersonen-potentials." Mimeographed.

EICHLER, M. (1980) "Family income: a critical look at the concept." Status of Women 6, 2: 20-24.

EMPLOYMENT AND IMMIGRATION CANADA (1979) "Notes on persons who voluntarily quit their jobs without just cause." Mimeographed, Employment and Immigration Canada, Ottawa, December.

EPSTEIN, C. (1978) "The positive effects of the double minority." American Journal of Sociology 78 (January): 912-935.

FELDMAN-SUMNERS, S. and S.B. KIESLER (1974) "Those who are number two try harder: the effect of sex on attributions of causality." Journal of Personality and Social Psychology 30: 846-855.

FELDMAN, S.D. (1974) Escape from the Doll's House: Women in Graduate and Professional School Education. New York: McGraw-Hill.

FERREE, M.M. (1976) "The confused American housewife." Psychology Today 10: 76-80.

FISHBEIN, M. [ed.] (1967) Readings in Attitude Theory and Measurement. New York: John Wiley & Sons.

FLYNN, E.G. (1942) Women in the War. New York: Workers Library Publishers, Inc.

FRANZ, W. (1974) "Ein makroökonometrisches vierteljahresmodell des Arbeitsmarktes der Bundesrepublik Deutschland 1960-1971." Mannheim.

FRANZ, W. (1977) "International factor mobility and the labor market: a macroeconometric analysis of the German labor market." Empirical Economics 2: 11-30.

FRANZ, W. (1979) "Youth unemployment: the German experience." Discussion Paper No. 116/79, Institut für Volkswirtschaft und Statistik der Universität Mannheim, Mannheim.

FRIEDLANDER, S. (1965) Labor Migration and Economic Growth. Cambridge, MA: Massachusetts Institute of Technology Press.

GARLAND, H. and K.B. PRICE (1977) "Attitudes toward women in management and attributions for their success and failure in a managerial position." Journal of Applied Psychology 62: 29-33.

GATES, M. (1977) "Homemakers into widows and divorcees: can the law provide economic protection?" in J.R. Chapman and M. Gates (eds.) Women into Wives: The Legal and Economic Impact of Marriage. Beverly Hills, CA: Sage Publications.

GELBER, S. (1975) "The compensation of women," pp. 7-16 in S. Gelber (ed.) Women's Bureau '74. Ottawa: Information Canada.

GEORGE, S. (1977) How the Other Half Dies: The Real Reasons for World Hunger. New York: Hammondsworth.

GILMAN, C.P. (1966) "Woman and economics," in C. Degler (ed.) Pivotal Interpretation of American History. New York: Harper & Row.

GLOVER, K. (1943) Women at Work in Wartime. New York: Public Affairs Committee, Inc.

GONICK, C. (1978) Out of Work. Toronto: Lorimer.

GORDON, N.M. (1979) "Institutional responses: the federal income tax system," in R.E. Smith (ed.) The Subtle Revolution: Women at Work. Washington, DC: The Urban Institute.

GREEN, C. and J.M. COUSINEAU (1976) Unemployment in Canada. Ottawa: Ministry of Supply and Services.

GREENBLATT, B. (1977) Responsibility for Child Care. San Francisco: Jossey-Bass.

GREGORY, C.W. (1974) Women in Defense Work during World War II: An Analysis of the Labor Problem and Women's Rights. New York: Exposition Press.

GRONAU, R. (1973) "The intrafamily allocation of time: the value of the housewives' time." American Economic Review 63 (September): 634-651.

GRONAU, R. (1977) "Leisure, home production, and work—the theory of the allocation of time revisited." Journal of Political Economy 85 (December): 1099-1123.

GUNDERSON, M. (1976) "Work patterns," pp. 93-142 in G.C.A. Cook (ed.) Opportunity for Choice: A Goal for Women in Canada. Ottawa: Statistics Canada and C.D. Howe Research Institute.

HALL, R.E. (1979) "Labor Supply and Aggregate Fluctuations." Working Paper No. 385, National Bureau of Economic Research, Cambridge, MA.

HANDL, J. (1978) "Ausma β und Determinanten der Erwerbsbeteiligung von Frauen," pp. 189-256 in Institut für Arbeitsmarkt- und Berufsforschung, Probleme bei der Konstruktion sozioökonomischer Modelle. Beiträge zur Arbeitsmarkt-und Berufsforschung, Vol. 31, Nürnberg.

HANDL, J, K.U. MAYER, W. MÜLLER, and A. WILLMS (1979) "Prozesse sozialstrukturellen Wandels am Beispiel der Entwicklung von Qualifikations—und Erwerbsstruktur der Frauen im Deutschen Reich und der Bundesrepublik Deutschland." Arbeitspapier Nr. 6, VASMA-Universität Mannheim, Mannheim.

HANOCK, G. (1980) "A multivariate model of labor supply: methodology and estimation," pp. 249-326 in J.P. Smith (ed.) Female Labor Supply: Theory and Estimation. Princeton, NJ: Princeton Univ. Press.

HANSEN, P. (1974) "Sex differences and supervision." Presented at the annual meetings of the American Psychological Association, New Orleans, September.

HAYNES, S.G. and M. FEINLEIB (1980) "Women, work, and coronary heart disease: prospective findings from the Framingham heart study." American Journal of Public Health 70 (March-April): 133-140.

HECKMAN, J.J. (1974) "Shadow prices, market wages and labor supply." Econometrica 42 (July): 679-694.

HECKMAN, J.J. (1978) "A partial survey of recent research on the labor supply of women." American Economic Review 68 (May): 200-206.

HECKMAN, J.J., M. KILLINGSWORTH, and T. MACURDY (1979) "Recent theoretical and empirical studies of labor supply: a partial survey." Prepared for a conference on the labor market sponsored by H.M. Treasury, Department of Employment and Manpower Services Commission, Magdalen College, Oxford, September 10-12.

HECKMAN, J.J. and T.E. MACURDY (1980) "A life cycle model of female labor supply." Review of Economic Studies 47 (January): 47-74.

HECKMAN, J.J. and R.J. WILLIS (1977) "A beta-logistic model for the analysis of sequential labor force participation by married women." Journal of Political Economy 85: 27-58.

HECKMAN, J.J. and R.J. WILLIS (1979) "Reply to Mincer and Ofek." Journal of Political Economy 87: 203-211.

HENNIG, M. and A. JARDIN (1978) The Managerial Woman. New York: Pocket Books.

HEWITT, L.L. (1974) Women Marines in World War I. Washington, DC: History and Museums Division, Headquarters, U.S. Marine Corps.

HILL, M. (1979) "The wage effects of marital status and children." Journal of Human Resources 14 (Fall): 579-594.

HOFFERTH, S.L. (1979) "Day care in the next decade: 1980-1990." Journal of Marriage and the Family. Special Issue: "Family Policy" 41 (August): 649-658.

HOFFERTH, S.L. (1980) "High school experience on the attainment process of noncollege boys and girls: when and why do their paths diverge?" Urban Institute Working Paper, February.

HOFFERTH, S.L. and K.A. MOORE (1979) "Women's employment and marriage," pp. 99-124 in R.E. Smith (ed.) The Subtle Revolution: Women at Work. Washington, DC: The Urban Institute.

HOIBERG, A. (1980a) "Military staying power," pp. 212-243 in S. Sarkesian (ed.) Combat Effectiveness: Cohesion, Stress, and the Volunteer Military. Beverly Hills, CA: Sage Publications.

HOIBERG, A. (1980b) "Military occupations: the cutting edge for women?" Presented at the annual meetings of the American Psychological Association, Montreal, September.

HOIBERG, A. and J. ERNST (1980) "Motherhood in the military: conflicting roles for Navy women?" International Journal of Sociology of the Family 10 (July-December): 265-280.

HOOPER, M. (1976) "An empirical analysis of the social identities of women: validation of a proposed measurement procedure." Unpublished paper, Department of Sociology, Temple University, Philadelphia, PA.

HORNER, M.S. (1972) "Toward an understanding of achievement-related conflicts in women." Journal of Social Issues 28 (Spring): 157-176.

HOUSE, J.D. (1980) The Last of the Free Enterprisers: The Oil Men of Calgary. Toronto: Macmillan.

HUSÉN, T. (1969) Talent, Opportunity, and Career. Stockholm: Almqvist & Wiksell.

INSTITUTO Nacional de Estadística (1976) "Características de la Población Española deducida del Padrón Municipal de Habitantes según la inscripción realizada en diciembre de 1975." Andalucía, Tomo I-Vol. 8. Madrid: Instituto Nacional de Estadística.

IZREALI, D.N. (1975) "The middle manager and the tactics of power expansion: a case study." Sloan Management Review 16: 57-60.

JOHNSON, A. (1977) "Sex differentials in coronary heart disease: the explanatory role of primary risk factors." Journal of Health and Social Behavior 18: 46-54.

JOHNSON, L.A. (1972) "The development of class in Canada in the twentieth century," pp. 141-183 in G. Teeple (ed.) Capitalism and the National Question in Canada. Toronto: Univ. of Toronto Press.

JOHNSON, L.A. (1974a) "The political economy of Ontario women in the nineteenth century," pp. 13-31 in J. Acton, P. Goldsmith, and B. Shepard (eds.) Women at Work: Ontario, 1850-1930. Toronto: Canadian Women's Educational Press.

JOHNSON, L.A. (1974b) Poverty in Wealth: The Capitalist Labour Market and Income Distribution in Canada. Toronto: New Hogtown Press.

JONES, C.A., N.M. GORDON, and I.V. SAWHILL (1976) Child Support
    Payments in the United States. Washington DC: The Urban Institute.
JONES, E. and J. LONG (1979) "Part-week work and human capital
    investment by married women." Journal of Human Resources 14
    (Fall): 563-578.
KANDIYOTI, D. (1974) "Social change and social stratification in a
    Turkish village." Journal of Peasant Studies 2: 206-219.
KANDIYOTI, D. (1977) "Sex roles and social change: a comparative
    appraisal of Turkey's women." Signs: Journal of Women in Culture
    and Society 3: 57-73.
KANTER, R.M. (1977a) Men and Women of the Corporation. New York:
    Basic Books.
KANTER, R.M. (1977b) Work and Family in the United States: A Critical
    Review and Agenda for Research and Policy. New York: Russell
    Sage Foundation.
KEALEY, G. [ed.] (1973) Canada Investigates Industrialism.   Toronto:
    Univ. of Toronto Press.
KEYFITZ, N. (1971) Changes in the Birth and Death Rates and their
    Demographic Effects. Baltimore, MD: Johns Hopkins Univ. Press.
KIDD, D. (1974) "Women's organization: learning from yesterday," pp.
    331-361 in J. Acton, P. Goldsmith, and B. Shepard (eds.) Women at
    Work: Ontario, 1850-1930. Toronto: Canadian Women's Educational
    Press.
KJERKEGAARD, E.M. [ed.] (1976) Levevilkår i Danmark.   Statistisk
    oversigt 1976. (Living conditions in Denmark.   Compendium of
    Statistics 1976.) Copenhagen: Danmarks Statistik and The Danish
    National Institute of Social Research.
KLEIN, A. and W. ROBERTS (1974) "Besieged innocence: the 'problem'
    and problems of working women—Toronto, 1896-1914," pp. 211-260
    in J. Acton, P. Goldsmith, and B. Shepard (eds.) Women at Work:
    Ontario, 1850-1930. Toronto: Canadian Women's Educational Press.
KLEIN, V. (1965) Britain's Married Women Workers. London: Routledge
    and Kegan Paul.
KLERMAN, G. (1979) "Is this the age of melancholy?" Psychology Today
    12 (April): 36-42.
KOCHER, J. (1973) Rural Development, Income and Fertility Decline.
    New York: The Population Council.
KOLLER, A. (1949) "Die Verheirate Berufsträtige Frau in der Schweiz."
    Schweizerische Zaitschrift für Volkswirtschaft und Statistic, No.
    4/5, August-October.
KOLLER, M. and I. KONIG (1977) "Internationaler Vergleich der Arbeit-
    slosenquoten." Institut für Arbeitsmarkt—und Berufsforschung der
    Bundesanstalt fur Arbeit, Beiträge zur Arbeitsmarkt—und Berufsfor-
    schung, Vol. 29, Nürnberg.
KOUTSOYANNIS-KOKOVA, A.   (1964) Production Functions in Greek
    Industry. Athens: Center of Planning and Economic Research.
KURZ, M., P. ROBINS, and R. SPIEGELMAN (1975) "A study of the
    demand for child care by working mothers." Research Memorandum
    27, Stanford Research Institute, Menlo Park, CA.

LAJEWSKI, H.C. (1959) Child Care Arrangements of Full-Time Working Mothers. Children's Bureau Publication No. 278, U.S. Department of Health, Education, and Welfare. Washington, DC: U.S. Government Printing Office.

LANDES, E. (1974) "Male-female differences in wages and employment: a specific human capital model." Unpublished Ph.D. dissertation, Department of Economics, University of Chicago.

LAUTARD, H.E. (1976) "The segregated labour force participation of men and women in Canada: long run trends in occupational segregation by sex, 1891-1961." Presented at the annual meetings of the Western Association of Sociology and Anthropology.

LAVE, J. and S.S. ANGRIST (1974) "Child care arrangements of working mothers: social and economic aspects." Unpublished manuscript.

LAWS, J.L. and S.S. TANGRI (1979) "Institutional sexism: why there is no conspiracy." Presented at the annual meetings of the American Psychological Association, New York City, August.

LAZEAR, E.P. and R.T. MICHAEL (1980) "Real income equivalence among one-earner and two-earner families." American Economic Review 70 (May): 203-208.

LESLIE, G. (1974) "Domestic service in Canada, 1880-1920," pp. 71-126 in J. Acton, P. Goldsmith, and B. Shepard (eds.) Women at Work: Ontario, 1850-1930. Toronto: Canadian Women's Educational Press.

LESTER, R.A. (1945) Economics of Labor. New York: Macmillan.

LIANOS, T. and K. PRODROMIDIS (1974) Aspects of Income Distribution in Greece. Athens: Center of Planning and Economic Research.

LIEBERMAN, A.F. (1978) "Psychology and day care." Social Research 45: 416-451.

LILJESTRÖM, R., G. FÜRST MELLSTRÖM, and G. LILJESTRÖM (1978) Roles in Transition. Stockholm: Liber.

LIPTON, C. (1967) The Trade Union Movement of Canada, 1827-1959. Toronto: NC Press.

LIRTZMAN, S.I. and M.A. WAHBA (1972) "Determinants of coalition behavior of men and women: sex roles or situational requirements?" Journal of Applied Psychology 56: 406-411.

LIVI-BACCI, M. and H. HAYMANN (1971) "Report on the demographic and social pattern of migrants in Europe especially with regard to international migration." Presented at the Second European Conference, Strasbourg, August-September.

LOEVINGER, J., R. WESSLER, and C. REDMORE (1970) Measuring Ego Development. San Francisco: Jossey-Bass.

LONG, C. (1958) The Labor Force under Changing Income and Employment. Princeton, NJ: Princeton Univ. Press.

LOVE, R. (1979) Income Distribution and Inequality in Canada. Statistics Canada Census Analytical Study. Ottawa: Minister of Supply and Services.

LOW, S. and P.G. SPINDLER (1968) Child Care Arrangements of Working Mothers in the United States. U.S. Department of Health, Education, and Welfare, Children's Bureau; U.S. Department of Labor, Women's Bureau. Washington, DC: U.S. Government Printing Office.

LOWE, G. (1978) "The Canadian union of bank employees: a case study."
    Working Paper No. 78-05, Centre for Industrial Relations, University
    of Toronto, October.
LOWE, G. (1979) "The administrative revolution: the growth of clerical
    occupations and the development of the modern office in Canada,
    1911-1931." Unpublished Ph.D. dissertation, Department of Soci-
    ology, University of Toronto.
LUCAS, R.E., JR. and L.A. RAPPING (1970) "Real wages, employment,
    and inflation," pp. 257-305 in E.S. Phelps et al. (eds.) Microeconomic
    Foundations of Employment and Inflation Theory. New York: W.W.
    Norton.
MACDONALD, B.M. (1978) "Flows into unemployment." Research Paper
    No. 17, Statistics Canada Labour Force Survey Division, Ottawa.
MACDONALD, L. (1980) "Equal pay—how far off?" Canadian Dimension
    14 (May): 21-24.
MACHIAVELLI, N. (1964) (1513) "Il principe," in T.G. Geigin (ed.) The
    Prince. New York: Appleton-Century-Crofts.
MACLEOD, C. (1974) "Women in production: the Toronto dressmakers'
    strike of 1931," pp. 309-330 in J. Acton, P. Goldsmith, and B. Shepard
    (eds.) Women at Work: Ontario, 1850-1930. Toronto: Canadian
    Women's Educational Press.
MACLEOD, N. and K. HORNER (1980) "Analyzing postwar changes in
    Canadian income distribution," in Economic Council of Canada,
    Reflections on Canadian Incomes. Ottawa: Supply and Services
    Canada.
MACURDY, T.E. (1980) "An empirical model of labor supply in a life
    cycle setting." Working Paper No. 421, National Bureau of Economic
    Research, Cambridge, MA.
MADDEN, J.F. (1980) "Urban land use and the growth in two-earner
    households." American Economic Review 70 (July): 191-198.
MADDEN, J.F. (forthcoming) "Why women work closer to home." Urban
    Studies.
MADDEN, J.F. and M.J. WHITE (1980) "Spatial implications of increases
    in the female labor force: a theoretical and empirical synthesis."
    Land Economics 5 (November): 432-446.
MANIS, J. and H. MARKUS (1978) "Careers and career attitudes: age,
    education, and timing effects." Presented at the annual meetings of
    the American Psychological Association, Toronto, September.
MARCHAK, P. (1973) "The Canadian labour force: jobs for women," in M.
    Stephenson (ed.) Women in Canada. Toronto: New Press.
MARCUS, A.C. and T.E. SEEMAN (1981) "Sex differences in reports of
    illness and disability: a preliminary test of the 'fixed role obligations'
    hypothesis." Journal of Health and Social Behavior 22 (June): 174-
    182.
MARKUSH, R. and R. FAVERO (1974) "Epidemiologic assessment of
    stressful life events, depressed mood, and psychophysiological symp-
    toms—a preliminary report," in B.P. Dohrenwend and B.S. Dohren-
    wend (eds.) Stressful Life Events: Their Nature and Effects. New
    York: John Wiley & Sons.

MARSDEN, L. (1980) "The role of the national action committee on the status of women in facilitating equal pay policy in Canada," pp. 242-260 in R.S. Ratner (ed.) Equal Employment Policy for Women. Philadelphia: Temple Univ. Press.

MARSDEN, L.R. and E.B. HARVEY (1979) The Fragile Federation: Social Change in Canada. Toronto: McGraw-Hill-Ryerson.

MARTIN, N.H. and J.H. SIMS (1956) "Thinking ahead, power tactics." Harvard Business Review 34: 25-26.

MASON, K.O. and L.L. BUMPASS (1975) "U.S. women's sex-role ideology, 1970." American Journal of Sociology 80: 1212-1219.

MASON, K.O., J.L. CZAJKA, and S. ARBER (1976) "Change in U.S. women's sex-role attitudes, 1964-1974." American Sociological Review, 41: 573-596.

MATTHAEI, J.A. (1980) "Consequences of the rise of the two-earner family: the breakdown of the sexual division of labor." American Economic Review 70 (May): 198-202.

MCCARTHY, A. (1972) Private Faces/Public Places. Garden City, NY: Doubleday.

MCCARTHY, A. (1977) Circles: A Washington Story. Garden City, NY: Doubleday.

MCDIARMID, O. (1977) Unskilled Labor for Development. Baltimore, MD: Johns Hopkins Univ. Press.

MCGIBBON, P.M. (1975) "Women's year: a hopeful milestone." Canadian Business Review 2: 1.

MCILVEEN, N. and H. SIMS (1978) The Flow Components of Unemployment in Canada. Statistics Canada Cat. No. 71-527E, Series A, No. 11. Ottawa: Information Canada.

MELTZ, N. and D. STAGER (1979) The Occupational Structure of Earnings in Canada. Ottawa: Supply and Services Canada.

MINCER, J. (1960a) "Employment and consumption." Review of Economics and Statistics 42: 20-26.

MINCER, J. (1960b) "Labor supply, family income, and consumption." American Economic Review, 50: 475-483.

MINCER, J. (1962) "Labor force participation of married women: a study of labor supply," pp. 63-105 in H.G. Lewis (ed.) Aspects of Labor Economics. National Bureau of Economic Research. Princeton, NJ: Princeton Univ. Press.

MINCER, J. (1974) Schooling, Experience, and Earnings. New York: Columbia Univ. Press.

MINCER, J. and H. OFEK (1979) "The distribution of lifetime labor force participation of married women: comment." Journal of Political Economy 87: 197-201.

MINCER, J. and S. POLACHEK (1974) "Family investments in human capital: earnings of women." Journal of Political Economy 82 (March-April): 76-108.

MINISTRY OF COORDINATION AND CENTER OF PLANNING AND ECONOMIC RESEARCH (1979) Economic and Social Development Plan 1978-82: Preliminary Guidelines. Athens: Ministry of Coordination and Center of Planning and Economic Research.

MINTZBERG, H. (1973) The Nature of Managerial Work. New York:
    Harper & Row.
MOORE, K.A. and S.L. HOFFERTH (1979) "Women and their children," in
    R.E. Smith (ed.) The Subtle Revolution: Women at Work. Wash-
    ington, DC: The Urban Institute.
MOORE, K.A., S.L. HOFFERTH, S.B. CALDWELL, and L.J. WAITE (1979)
    Teen-age Motherhood: Social and Economic Consequences. Washing-
    ton, DC: The Urban Institute.
MYRDAL, A. and V. KLEIN (1956) Die Doppelrolle der Frau in Familie
    und Beruf. Köln.
NAKAMURA, M., A. NAKAMURA, and D. CULLEN (1979a)   "Job
    opportunities, the offered wage, and the labor supply of married
    women." American Economic Review 69 (December): 787-805.
NAKAMURA, A., M. NAKAMURA, and D. CULLEN in collaboration with
    D. Grant and H. Orcutt (1979b) Employment and Earnings of Married
    Females. Ottawa: Statistics Canada, December.
NATHANSON, C. (1980) "Social roles and health status among women:
    the significance of employment." Social Science and Medicine 14A:
    463-471.
NATIONAL ACADEMY OF SCIENCES (1979) Climbing the Academic
    Ladder: Doctoral Women Scientists in Academe. Washington, DC:
    Committee on the Education and Employment of Women in Science
    and Engineering, Commission on Human Resources.
NATIONAL ACADEMY OF SCIENCES (1980a) Science, Engineering, and
    Humanities Doctorates in the United States: 1979 Profile. Washing-
    ton, DC: Commission on Human Resources.
NATIONAL ACADEMY OF SCIENCES (1980b)   Summary Report 1979:
    Doctorate Recipients from United States Universities. Washington,
    DC: Commission on Human Resources.
NATIONAL ACADEMY OF SCIENCES (1980c)   Women Scientists in
    Industry and Government: How Much Progress in the 1970s? Wash-
    ington, DC: Committee on the Education and Employment of Women
    in Science and Engineering, Commission on Human Resources.
NATIONAL ADVISORY COUNCIL ON ECONOMIC OPPORTUNITY (1980)
    Twelfth Annual Report of the National Advisory Council on Economic
    Opportunity. Washington, DC: U.S. Government Printing Office.
NATIONAL COMMISSION ON WORKING WOMEN (1979) National Survey
    of Working Women: Perceptions, Problems, and Prospects. Washing-
    ton, DC: National Commission on Working Women.
NATIONAL COUNCIL OF WELFARE (1976) One in a World of Two's.
    Ottawa: National Council of Welfare.
NATIONAL COUNCIL OF WELFARE (1979)   Women and Poverty.
    Ottawa: National Council of Welfare.
NATIONAL COUNCIL OF WOMEN OF CANADA (1975)   Women of
    Canada: Their Life and Work. Distributed at the Paris International
    Exhibition, 1900.   Reprinted by National Council of Women of
    Canada.
NATIONAL NETWORK OF BUSINESS SCHOOL WOMEN REPORT (1979)
    Philadelphia: The Wharton School, University of Pennsylvania.

NEWLAND, K. (1979) Global Employment and Economic Justice: The Policy Challenge. Washington, DC: Worldwatch Institute.

NOELLE-NEUMANN, E. (1977) "Turbulences in the climate of opinion: methodological applications of the spiral of silence theory." Public Opinion Quarterly 41: 143-158.

OLSEN, D. (1980) The State Elite. Toronto: McClelland and Stewart.

O'NEILL, J. (1976) "Returns to Social Security." Presented at the annual meetings of the American Economics Association, Atlantic City, NJ, September.

O'NEILL, J. (1980) "The tax treatment of married and single taxpayers." Statement before hearings of the U.S. House of Representatives, Committee on Ways and Means, April 2.

OSGOOD, C.E., G.J. SUCI, and P.H. TANNENBAUM (1957) The Measurement of Meaning. Urbana, IL: Univ. of Illinois Press.

OSTROM, T.M. and H.S. UPSHAW (1968) "Psychological perspective and attitude change," in A.G. Greenwald et al. (eds.) Psychological Foundation of Attitudes. New York: Academic Press.

OSTRORICH, V. (1970) Características Y. Evolución de la Población Económicamente Activa. Santiago: Universidad de Chile.

OSTRY, S. (1967) The Occupational Composition of the Canadian Labour Force. Ottawa: Dominion Bureau of Statistics.

OSTRY, S. (1968a) The Female Worker in Canada. Ottawa: Queen's Printer.

OSTRY, S. (1968b) Unemployment in Canada. Ottawa: Queen's Printer.

OSTRY, S. and M.A. ZAIDI (1979) Labour Economics in Canada. Toronto: Macmillan Co. of Canada.

PAPANEK, H. (1979) "Family status production: the 'work' and 'non-work' of women." Signs: Journal of Women in Culture and Society 4 (Summer): 775-781.

PATEL, K. (1980) "Women at work." Report of the World Conference on the United Nations Decade for Women: Equality, Development, and Peace. Copenhagen, July 14-30.

PAULY, P. (1978) "Theorie und Empirie des Arbeitsmarktes: Eine ökonometrische Analyse für die Bundesrepublick Deutschland, 1960-1974." Schriftenreihe des Sozialökonomischen Seminars der Universität Hamburg, Vol. 8, Frankfurt.

PEARCE, D. (1979) "Women, work, and welfare: the feminization of poverty," in K.W. Feinstein (ed.) Working Women and Families. Beverly Hills, CA: Sage Publications.

PEARLIN, L. (1978) "Sex roles and depression," in N. Datan and L. Ginsberg (eds.) Life-Span Developmental Psychology: Normative Life Crises. New York: Academic Press.

PEARLIN, L. and J. JOHNSON (1977) "Marital status, life-strains, and depression." American Sociological Review 42: 704-715.

PENTLAND, H.C. (1950) "The role of capital in Canadian economic development before 1875." The Canadian Journal of Economics and Political Science 16 (November): 457-474.

PENTLAND, H.C. (1959) "The development of a capitalistic labour market in Canada." Canadian Journal of Economics and Political Science 25 (November): 450-461.

PEOPLE'S COMMISSION ON UNEMPLOYMENT IN NEWFOUNDLAND AND LABRADOR (1978) "Now That We've Burned Our Boats. . ." Ottawa: Mutual Press.

PERRUCCI, C. (1975) "Sex-based professional socialization among graduate students in science," pp. 83-123 in National Research Council, Research Issues in the Employment of Women: Proceedings of a Workshop. Washington, DC: National Academy of Sciences.

PETTIGREW, A.M. (1975) "Toward a political theory of organization intervention." Human Relations 28: 191-208.

PETTY, M.M. and G.K. LEE (1975) "Moderating effects of sex of supervisor and subordinate on relationships between supervisor behavior and subordinate satisfaction." Journal of Applied Psychology 60: 624-628.

PETTY, M.M. and R.H. MILES (1976) "Leader sex-role stereotyping in a female dominated work culture." Personnel Psychology 29: 393-404.

PIERSON, R. (1977) "Women's emancipation and the recruitment of women into the labour force in World War II," pp. 125-145 in S.M. Trofimenkoff and A. Prentice (eds.) The Neglected Majority: Essays in Canadian Women's History.   Toronto: McClelland and Stewart.

PINCHBECK, I. (1930) Women Workers and the Industrial Revolution, 1750-1850. London: George Routledge & Sons.

POLACHEK, S. (1973) "Work experience and the difference between male and female wages." Unpublished Ph.D. dissertation, Department of Economics, Columbia University, New York, NY.

PORTER, J. (1965) The Vertical Mosaic. Toronto: Univ. of Toronto Press.

PORTER, L. (1976) "Organizations as political animals." Presented at the annual meetings of the American Psychological Association, Washington, DC, September.

PRESSAT, R. (1970) Population. London: C.A. Watts Co.

PRESSER, H.B. (1978) "Childrearing, work, and welfare: research issues." Journal of Population 1 (Summer): 167-180.

PRESSER, H.B. and W. BALDWIN (1980) "Child care as a constraint on employment: prevalence, correlates, and bearing on the work and fertility nexus." American Journal of Sociology 85 (May): 1202-1213.

RADLOFF, L.S. (1975) "Sex differences in depression: the effects of occupation and marital status." Sex Roles 1, 3: 249-265.

RADLOFF, L.S. (1977) "The CES-D Scale: a self-report depression scale for research in the general population."   Applied Psychological Measurement 1, 3: 385-401.

REAGAN, B.B. (1979a) "In retrospect: summary and issues." in B.B. Reagan (ed.) Issues in Federal Statistical Needs Relating to Women. Current Population Reports, Special Studies, Series P-23, No. 83, July. Washington, DC: U.S. Government Printing Office.

REAGAN, B.B. (1979b) "De facto job segregation," pp. 90-102 in Ann Foote Cahn (ed.) Women in the U.S. Labor Force. New York: Praeger.

REBECCA, M., R. HEFNER, and B. OLEHANSKY (1976) "A model of sex-role transcendence." Journal of Social Issues 32: 197-206.

REES, A. (1968) The Economics of Trade Unions. Chicago: Univ. of Chicago Press.

RIBOUD, M. (1977) "An analysis of earnings distribution in France." Unpublished Ph.D. dissertation, Department of Economics, University of Chicago.

ROBB, L.A. and B.G. SPENCER (1976) "Education: enrollment and attainment," pp. 53-88 in G.C.A. Cook (ed.) Opportunity for Choice: A Goal for Women in Canada. Ottawa: Statistics Canada and C.D. Howe Research Institute.

ROBERTS, W. (1976) Honest Womanhood: Feminism, Femininity, and Class Consciousness among Toronto Working Women, 1893 to 1914. Toronto: New Hogtoroso Press.

RODES, T.W. and J.C. MOORE (1975) National Child Care Consumer Study 1975, Vol. 3. Prepared for the U.S. Department of Health, Education, and Welfare. Washington, DC: U.S. Government Printing Office.

ROKEACH, M. (1973) The Nature of Human Values. New York: Free Press.

ROSS, H.L. and I. SAWHILL (1975) Time of Transition: The Growth of Families Headed by Women. Washington, DC: The Urban Institute.

ROTENBERG, L. (1974) "The wayward worker: Toronto's prostitute at the turn of the century," pp. 33-70 in J. Acton, P. Goldsmith, and B. Shepard (eds.) Women at Work: Ontario, 1850-1930. Toronto: Canadian Women's Educational Press.

ROUSELL, C. (1974) "Relationship of sex of department head to department climate." Administrative Science Quarterly 19: 211-220.

ROYAL BANK OF CANADA (1978) Between Ourselves, Task Force on the Status of Women. Toronto: Royal Bank of Canada.

ROYAL COMMISSION ON THE RELATIONS OF LABOUR AND CAPITAL (1889) Report of the Royal Commission on the Relations of Labour and Capital. Ottawa: Queen's Printer.

ROYAL COMMISSION ON THE STATUS OF WOMEN IN CANADA (1970) Report of the Royal Commission on the Status of Women in Canada. Ottawa: Information Canada.

RUDEBRANT, S. and S. THÖRN (1979) "Barn på daghem, familje-daghem och hemma--en uppföljningsstudie." Institutionen för pedagogik. Högskolan för lärerutbildning i Stockholm.

RUDERMAN, A. (1968) Child Care and Working Mothers. New York: Child Welfare League of America, Inc.

RYTEN, E. (1975) "Our best educated women." Canadian Business Review 2: 12-16.

SANGSTER, D. (1973) The Role of Women in the Economy. Ottawa: Manpower and Immigration.

SAWHILL, I. (1976) "Women with low-incomes," in M. Blaxall and B. Reagan (eds.) Women and the Workplace: The Implications of Occupational Segregation. Chicago: Univ. of Chicago Press.

SAYLES, L. (1964) Managerial behavior. New York: McGraw-Hill.

SCB. (1980) Barnomsorgsundersökningen 1980. Del 2 Förskolebarn, 0-6 år. (Survey of children care needs 1980. Part 2, Pre-school Children, 0-6 years old.) Stockholm: National Central Bureau of Statistics.

SCHEIN, V.E. (1973) "The relationship between sex role stereotypes and requisite management characteristics." Journal of Applied Psychology 57: 95-100.

SCHEIN, V.E. (1975) "The relationship between sex role stereotypes and requisite management characteristics among female managers." Journal of Applied Psychology 60: 340-344.

SCHEIN, V.E. (1977) "Individual power and political behaviors in organizations: an inadequately explored reality." Academy of Management Review 2: 64-72.

SCHEIN, V.E. (1979) "Examining an illusion: the role of deceptive behavior in organizations." Human Relations 32: 287-295.

SCHWARTZMAN, V. (1978) I quit the UIC and found happiness." Ontario Report 3 (September): 20-24.

SCHWARZ, K. (1978) "Erwerbstätigkeit verheirateter Frauen. Ergebnisse des Mikrozensus." Wirtschaft und Statistik: 473-480.

SCOTT, J. (THOMAS) (1976) "Conditions of female labour in Ontario," pp. 172-181 in R. Cook and W. Mitchinson (eds.) The Proper Sphere. Toronto: Oxford Univ. Press.

SEARS, P.S. and A. BARBEE (1977) "Career and life satisfactions among Terman's gifted women," in J. Stanley, W. George, and C. Solano (eds.) The Gifted and the Creative: Fifty-Year Perspective. Baltimore, MD: Johns Hopkins Univ. Press.

SECUNDA, S.K. (1973) Special Report: 1973. The Depressive Disorders. Rockville, MD: Institute of Mental Health.

SEKSCENSKI, E.S. (1980) "Women's share of moonlighting nearly doubles during 1969-79." Monthly Labor Review 103 (May): 36-39.

SHAEFFER, R.B. and H. AXEL (1978) Improving job opportunities for women. New York: The Conference Board.

SHAW, P. (1979) Canada's Farm Population. Statistics Canada Census Analytical Study. Ottawa: Minister of Supply and Services.

SHORTLIDGE, R.L. and P. BRITO (1977) "How women arrange for the care of their children while they work: study of child care arrangements, costs, and preferences in 1971." Unpublished manuscript.

SKOULAS, N. (1974) Determinants of the Participation Rate of Married Women in the Canadian Labour Force: An Econometric Analysis. Statistics Canada Cat. No. 71-522. Ottawa: Information Canada.

SLABY, R.G. and K.S. FREY (1975) "Development of gender constancy and selective attention to same-sex models." Child Development 46: 339-347.

SMITH, J.P. [ed.] (1980) Female Labor Supply: Theory and Estimation. Princeton, NJ: Princeton Univ. Press.

SMITH, P. (1978) The Treasure-Seekers. The Men Who Built Home Oil. Toronto: Macmillan.

SMITH, R.E. (1979a) Women in the Labor Force in 1990. Washington, DC: The Urban Intititute.

SMITH, R.E. [ed.] (1979b) The Subtle Revolution: Women at Work. Washington, DC: The Urban Institute.

SOCIAL PLANNING COUNCIL OF METROPOLITAN TORONTO (1978) The Problem is Jobs. . . Not People. Mimeographed, Social Planning Council of Metropolitan Toronto, Toronto.

SONQUIST, J.A., E. BAKER, and J.N. MORGAN (1971) Searching for Structure. Ann Arbor, MI: Institute for Social Research, University of Michigan.

SOU. (1979) Barnomsorg—Behov, Efterfrågan, Planeringsunderlag. Huvudbetänkande af Planeringsgruppen för Barnomsorg. (Child Care. Main Commission Report by the Planning Group for Child Care.) Stockholm: Liber.

SPENCE, J.T. and R.L. HELMREICH (1978) Masculinity and Femininity: Their Psychological Dimensions, Correlates, and Antecedents. Austin: Univ. of Texas Press.

SPENCER, B.C. and D.C. FEATHERSTONE (1970) Married Female Labour Force Participation: A Micro Study. Statistics Canada Cat. No. 71-516. Ottawa: Queen's Printer.

STATISTICS CANADA (formerly Dominion Bureau of Statistics) (1921-71) Census of Canada: 1921, 1931, 1941, 1951, 1961, 1971. Ottawa: Information Canada.

STATISTICS CANADA (1979a) The Labour Force. Cat. No. 71-001. Ottawa: Information Canada.

STATISTICS CANADA (1979b) Labour Force Annual Averages 1975-1978. Cat. No. 71-529. Ottawa: Information Canada.

STATISTICS CANADA (1979c) Historical Labour Force Statistics—Actual Data, Seasonal Factors, Seasonally Adjusted Data. Cat. No. 71-201. Ottawa: Information Canada.

STATISTICS CANADA (1980a) Canada's Female Labour Force. Cat. No. 98-804E. Ottawa: Information Canada.

STATISTICS CANADA (1980b) Income Distribution by Size in Canada 1978. Cat. No. 12-307. Ottawa: Information Canada.

STEINER, G.Y. (1971) The State of Welfare. Washington, DC: The Brookings Institution.

STIEHM, J. (1976a) "Invidious intimacy." Social Policy 6 (March-April): 12-16.

STIEHM, J. (1976b) "The pursuit of equality: a comparison of German and American women." Frontiers 1 (Winter): 63-70.

STIPAK, B. (1977) "Attitudes and belief systems concerning urban services." Public Opinion Quarterly 41: 41-55.

STIRLING, R. and D. KOVRI (1979) "Unemployment indexes—the Canadian context," in J.A. Fry (ed.) Economy, Class, and Social Reality. Toronto: Butterworths.

STROBER, M.H. (1977) "Wives' labor force behavior and family consumption patterns." American Economic Review 67 (February): 410-417.

STROBER, M.H. (1979) "Should separate family budgets be constructed for husband-wife-earner (HWE) and husband-only-earner (HOE) families at various income levels?" Statement for Bureau of Labor Statistics Expert Panel, November.

SWEET, J.A. (1970) "Family composition and the labor force activity of American wives." Demography (May): 195-209.

TERBORG, J.R. (1977) "Women in management: a research review." Journal of Applied Psychology 62: 647-664.

TILLY, L.I. and J.W. SCOTT (1978) Women, Work, and Family. Toronto: Holt, Rinehart & Winston.

TRANSGAARD, H. (forthcoming) An Analysis of Danish Sex-Linked Attitudes. Copenhagen: The Danish National Institute of Social Research.

TUSHMAN, M.L. (1977) "A political approach to organizations: a review and rationale." Academy of Management Review 2: 206-216.

UNDERWOOD, L. (1979) Women in Federal Employment Programs. Washington, DC: The Urban Institute.

UNEMPLOYMENT INSURANCE CANADA (1977) Comprehensive Review of the Unemployment Insurance Program in Canada. Ottawa: Unemployment Insurance Commission.

UNITED NATIONS (1975a) The Population Debate: Dimensions and Perspectives. World Population Conference, Bucharest, 1974. New York: United Nations.

UNITED NATIONS (1975b) Application of International Standards to Census Data on the Economically Active Population. New York: United Nations.

UNITED NATIONS (1979) "Earnings: the pay differential for women: some comparisons for selected ECE countries," pp. 13-15 in Proceedings of the Seminar on the Participation of Women in the Economic Evolution of the ECE Region. New York: United Nations.

U.S. BUREAU OF THE CENSUS (1950) "Children and youth: 1950" Current Population Reports, Series P-20, No. 32, December. Washington, DC: U.S. Government Printing Office.

U.S. BUREAU OF THE CENSUS (1951) "Marital status and household characteristics." Current Population Reports, Series P-20, No. 33, February. Washington, DC: U.S. Government Printing Office.

U.S. BUREAU OF THE CENSUS (1974) "Female family heads, growth, structure, and composition, other demographic characteristics, economic characteristics, housing, primary individuals and subfamilies." Current Population Reports, Series P-23, No. 50, July. Washington, DC: U.S. Government Printing Office.

U.S. BUREAU OF THE CENSUS (1976a) "Money income in 1974 of families and persons in the United States." Current Population Reports, Series P-60, No. 101, January. Washington, DC: U.S. Government Printing Office.

U.S. BUREAU OF THE CENSUS (1976b) "A statistical portrait of women." Current Population Reports, Series P-20, No. 58. Washington, DC: U.S. Government Printing Office.

U.S. BUREAU OF THE CENSUS (1976c) "Daytime care of children: October 1975 and February 1976." Current Population Reports, Series P-20, No. 298. Washington, DC: U.S. Government Printing Office.

U.S. BUREAU OF THE CENSUS (1977) "Household and family characteristics: March 1976." Current Population Reports, Series P-20, No. 311, August. Washington, DC: U.S. Government Printing Office.

U.S. BUREAU OF THE CENSUS (1978a) "Nursery school and kindergarten enrollment of children and labor force status of their mothers: October 1967 to October 1976." Current Population Reports, Series P-20, No. 318. Washington, DC: U.S. Government Printing Office.

U.S. BUREAU OF THE CENSUS (1978b) "Money income in 1976 of families and persons in the United States." Current Population Reports, Series P-60, No. 114, July. Washington, DC: U.S. Government Printing Office.

U.S. BUREAU OF THE CENSUS (1979a) "Money Income in 1977 of Families and Persons in the United States." Current Population Reports, Series P-60, No. 118, March. Washington, DC: U.S. Government Printing Office.

U.S. BUREAU OF THE CENSUS (1979b) "Characteristics of population below the poverty level: 1977." Current Population Reports, Series P-60, No. 119, March. Washington, DC: U.S. Government Printing Office.

U.S. BUREAU OF THE CENSUS (1979c) "Population profile of the United States: 1978." Current Population Reports, Series P-20, No. 336, April. Washington, DC: U.S. Government Printing Office.

U.S. BUREAU OF THE CENSUS (1979d) "Marital status and living arrangements: March 1978." Current Population Reports, Series P-20, No. 338, May. Washington, DC: U.S. Government Printing Office.

U.S. BUREAU OF THE CENSUS (1979e) "Divorce, child custody, and child support." Current Population Reports: Special Studies, Series P-23, No. 84, June. Washington, DC: U.S. Government Printing Office.

U.S. BUREAU OF THE CENSUS (1979f) "Household and families by type: March 1979 (Advance Report)." Current Population Reports, Series P-20, No. 345, October. Washington, DC: U.S. Government Printing Office.

U.S. CONGRESS, CONGRESSIONAL BUDGET OFFICE (1978) Child Care and Preschool Options for Federal Support. Washington, DC: U.S. Government Printing Office.

U.S. DEPARTMENT OF COMMERCE (1977) Statistical Abstract of the United States. Washington, DC: U.S. Government Printing Office.

U.S. DEPARTMENT OF COMMERCE (1979) Statistical Abstract of the United States. Washington, DC: U.S. Government Printing Office.

U.S. DEPARTMENT OF HEALTH, EDUCATION, AND WELFARE (1976) Statistical Highlights from the National Child Care Consumer Study. DHEW Publication No. (OHD) 76-31096. Washington, DC: U.S. Government Printing Office.

U.S. DEPARTMENT OF HEALTH, EDUCATION, AND WELFARE, SOCIAL AND REHABILITATION SERVICE (1974) Findings of the 1973 AFDC Study, Part I. SRS 74-03764. Washington, DC: U.S. Government Printing Office.

U.S. DEPARTMENT OF LABOR, BUREAU OF LABOR STATISTICS (1977) U.S. Working Women: A Data Book. Washington, DC: U.S. Government Printing Office.

U.S. DEPARTMENT OF LABOR, BUREAU OF LABOR STATISTICS (1979a) Women in the Labor Force: Some New Data Series. Washington, DC: U.S. Government Printing Office.

U.S. DEPARTMENT OF LABOR, BUREAU OF LABOR STATISTICS (1979b) International Comparisons of Unemployment. Washington, DC: U.S. Government Printing Office.

U.S. DEPARTMENT OF LABOR, WOMEN'S BUREAU (1980a) Employment Goals of the World Plan of Action: Developments and Issues in the U.S. Prepared for the World Conference on the United Nations Decade for Women, Copenhagen, Denmark, July. Washington, DC: U.S. Government Printing Office.

U.S. DEPARTMENT OF LABOR, WOMEN'S BUREAU (1980b) The Employment of Women: General Diagnosis of Developments and Issues. U.S. Report for OECD High Level Conference on the Employment of Women, Washington, DC, April. Washington, DC: U.S. Government Printing Office.

U.S. NATIONAL CENTER FOR EDUCATION STATISTICS (1979a) Fall Enrollment in Higher Education 1977. Washington, DC: U.S. Government Printing Office.

U.S. NATIONAL CENTER FOR EDUCATION STATISTICS (1979b) Digest of Education Statistics. Washington, DC: U.S. Government Printing Office.

VETTER, B.M. (1980) "Working women scientists and engineers." Science 207 (January 4): 28-34.

VICKERY, C. (1977) "The time poor: a new look at poverty." Journal of Human Resources 12 (Winter): 27-48.

VICKERY, C. (1978) "The changing household: implications for devising an income support program." Public Policy 26 (Spring): 121-151.

VICKERY, C. (1979) "Women's economic contribution to the family," pp. 159-200 in R.E. Smith (ed.) The Subltle Revolution: Women at Work. Washington, DC: The Urban Institute.

VINACKE, W.E. (1959) "Sex roles in a three-person game." Sociometry 22: 343-360.

WABE, S. (1932) "Labour force participation rates in the labour metropolitan region." Journal of Labour Statistics, Series A 1932, Part 2.

WAHBA, M.A. and S.I. LIRTZMAN (1971) "Toward a general theory of coalition behavior." Unpublished working paper. Baruch College, City University of New York.

WAITE, L.J., R. SHORTLIDGE, and L.E. SUTER (1974) "Child-care arrangements of working mothers in 1971." Presented at the annual meetings of the American Statistical Association, St. Louis, MO.

WALDMAN, E. and K.R. GOVER (1971) "Children of women in the labor force." Monthly Labor Review (July): 19-25.

WALDRON, I. (1976) "Why do women live longer than men?" Social Science and Medicine 10: 349-367.

WALDRON, I. and S. JOHNSTON (1976) "Why do women live longer than men? II: accidents, alcohol, and cirrhosis." Journal of Human Stress 2 (June): 19-30.

WALES, T.J. and A.D. WOODLAND (1977) "Estimation of the allocation of time for work, leisure, and housework." Econometrics 45 (January): 115-132.

WATTS, H.W. and F. SKIDMORE (1979) "Household structure: necessary changes in categorization and data collection, and postscript," pp. 62-69 in B.B. Reagan (ed.) Issues in Federal Statistical Needs Relating to Women. Current Population Reports, Special Studies, Series P-23, No. 83. Washington, DC: U.S. Government Printing Office.

WEISS, R.S. and N.M. SAMELSON (1958) "Social roles of American women: their contribution to a sense of usefulness and importance." Marriage and Family Living 20 (November): 358-366.

WEISSMAN, M. and G. KLERMAN (1977) "Sex differences and the epidemiology of depression." Archives of General Psychiatry 34: 98-111.

WEISSMAN, M. and E.S. PAYKEL (1974) The Depressed Woman: A Study in Social Relationships. Chicago: Univ. of Chicago Press.

WEITZMAN, L.J. and R.B. DIXON (1980) "The alimony myth: does no-fault divorce make a difference?" Family Law Quarterly (Fall).

WELCH, S. and A. BOOTH (1977) "Employment and health among married women." Sex Roles 3 (August): 385-396.

WESTINGHOUSE LEARNING CORPORATION (1971) Day Care Survey 1970: Summary Report and Basic Analysis. Report presented to the Office of Economic Opportunity. Oak Lawn, IL: Westinghouse Learning Corp.

WHITE, J. (1980) Women and Unions. Ottawa: Canadian Government Publishing Centre.

WIGGINS, J.S. (1978) "A psychological taxonomy of trait-descriptive terms: the interpersonal domain." Journal of Personality and Social Psychology 37: 395-412.

WIGGINS, J.S. and A. HOLZMULLER (1978) "Psychological androgyny and interpersonal behavior." Journal of Consulting and Clinical Psychology 46: 40-52.

WOLF-SEIBEL, H.R. (1980) Sex differences in determinants of educational and occupational goal orientation: a study among 12,000 Swedish sixth graders. Research Report No. 13, Project Metropolitan, Stockholm.

"WOMEN IN THE LABOUR FORCE" (1947) Trades and Labour Congress Journal 26 (May): 18.

WOMEN'S BUREAU, LABOUR CANADA (1978) Women in the Labour Force, Facts and Figures. Labour Force Activity. Ottawa: Minister of Supply and Services.

WOOLSEY, S. and D. NIGHTINGALE (1977) "Day care utilization trends: socioeconomic status, ethnicity, and mobility." Photocopied, The Urban Institute, Washington, DC.

ZARETSKY, E. (1976) Capitalism, the Family, and Personal Life. New York: Harper & Row.

ZIMBALIST, A.S. [ed.] (1979) Case Studies on the Labor Process. New York: Monthly Review Press.

ZISSIMOPOULOS, A. (1975) Determinative Factors of Female Employment, Measures—Incentives for Further Participation. Athens: Center of Planning and Economic Research.

ABOUT THE AUTHORS

Hugh Armstrong teaches sociology at Vanier College, Montreal, and is the coauthor of the book, <u>The Double Ghetto: Canadian Women and their Segregated Work</u>. He has presented his research findings on women's employment and unemployment at professional conferences and has published several articles on women, work, and the state.

Pat Armstrong is coauthor of the book, <u>The Double Ghetto: Canadian Women and their Segregated Work</u>. She teaches sociology at Vanier College, Montreal, and is a member of the Canadian Sociology and Anthropology Association. She has conducted research for the Canadian Research Institute for the Advancement of Women and has published several articles on women in the work force and women's unemployment.

Stylianos K. Athanassiou is an Economist and Statistician at the Center for Planning and Economic Research, Athens, Greece. He received his doctorate from the University of Uppsala, Sweden, and served on the United Nations Economic Commission for Latin America from 1972 to 1976. Dr. Athanassiou has published numerous books and articles in the areas of economic quantitative analysis and manpower and demographics.

Deborah Belle is Research Associate and Lecturer at the Harvard University Graduate School of Education. She also is Director of the Stress and Families Project. She has been a member of the faculties of Northwestern University, Wellesley College, and the Massachusetts Institute of Technology. Dr. Belle's research has been conducted in the areas of children's social behavior as well as families and stress.

Barbara R. Bergmann is Professor of Economics at the University of Maryland. She has served on the staffs of the Council of Economic Advisors, the Brookings Institution, and the Agency for International Development. Dr. Bergmann's specialties include the economics of gender roles and the computer simulation of eocnomic systems.

Clair (Vickery) Brown is a Labor Economist at the University of California, Berkeley. Her areas of research include women's work, income support programs, and unemployment. Dr. Brown is currently writing a book on the interelationship between the changes in women's work and family life styles in twentieth-century America.

369

Gordon R. Chapman is an Economist and Senior Associate at the Center for Women Policy Studies. He received a baccalaureate degree in philosophy at Stanford University and a Master of Arts degree in economics from Oklahoma University. His work has focused on studies of technological development and social change with the U.S. government and international organizations. His most recent research endeavors included contributions to Economic Realities and the Female Offenders and "Harassment and Discrimination of Women in Employment."

Jane Roberts Chapman is Director of the Center of Women Policy Studies which she and Margaret J. Gates established in 1972 for the purpose of helping to meet the growing needs for action-oriented policy studies on the economic, social, and legal status of women. As a social scientist, she has investigated problems relating to the social and economic status of women and has published major studies, testified before Congress, and developed specific programs to provide technical assistance to organizations serving the special needs of women. She edited Economic Independence for Women: The Foundation for Equal Rights and was coeditor of Women Into Wives: The Legal and Economic Impact of Marriage and The Victimization of Women. She also is the author of Economic Realities and the Female Offender.

M. Patricia Connelly is Associate Professor and Chairperson of the Department of Sociology at Saint Mary's University in Halifax, Nova Scotia, where she teaches courses on women in the economy and the sociology of labor. She is on the editorial board of Atlantis: A Women's Studies Journal and is an associate editor of The Canadian Review of Sociology and Anthropology. She is the author of Last Hired, First Fired: Women in the Canadian Work Force which develops theoretically and then traces empirically the position of women as a reserve labor force. Currently, she is investigating the relationship between women's wage and domestic labor through a series of case studies on Atlantic Canada Communities and is coauthoring a nonsexist introductory sociology textbook.

Camille Kim Cook is a Research Analyst at the Naval Health Research Center, San Diego, California. Currently enrolled in law school, she plans to focus on civil rights and women's issues. As an undergraduate student, her studies and activities centered on women's issues. Recently, she served as a delegate to the first annual conference on women students in leadership, and she was a member of the Advisory Board for the conference, "Future, Technology and Women."

Wolfgang Franz teaches economics at the University of Mannheim, Federal Republic of Germany. He recently spent his sabbatical year (1979-80) at the National Bureau of Economic Research in Cambridge, Massachusetts. His main interest is labor economics, and he has written several articles on a macroeconomic analysis of the German labor market and on consumer demand. Dr. Franz is currently writing a book on youth unemployment in the Federal Republic of Germany.

Anne Hoiberg is a Research Psychologist and the Manager of the Longitudinal Studies Program at the Naval Health Research Center, San Diego, California. She served as editor for a special issue of <u>Armed Forces and Society</u>, titled, "Women as New 'Manpower'." She has published numerous articles on women in the military and has presented her research findings at professional conferences. At present, she directs research on occupational health, the health and performance effectiveness of Navy women, and the longitudinal health patterns of Navy personnel.

Lilli S. Hornig is the Executive Director of Higher Education Resource Services (H.E.R.S.) at Wellesley College. H.E.R.S. conducts research on women in higher education and on programs designed to improve the status of academic professional women. Dr. Hornig, a graduate of Bryn Mawr College, holds a doctorate in chemistry from Harvard University and has been a faculty member at Brown University and at Trinity College (Washington, D.C.) where she chaired the chemistry department. Her chemical research interests lie in the area of mechanisms by which chemical substances produce cancer. Her longstanding concern for providing better opportunities for academic women has led her to a more general interest in the utilization of human resources in academe. She is a member of the National Academy of Sciences' Commission on Human Resources and serves as the Chairperson of its Committee on the Education and Employment of Women in Science and Engineering.

Lorna R. Marsden is Professor of Sociology and Associate Dean of the School of Graduate Studies at the University of Toronto. A native of Canada, she received her doctorate from Princeton University. Active as a feminist in Canada, Dr. Marsden was president of the National Action Committee on the Status of Women from 1975 to 1977 and combines her research interests in the sociology of work and social change in Canadian society with her community interests in the status of women.

Lawrence A. Palinkas is a Research Analyst at the Naval Health Research Center, San Diego, California. He received his doctorate in anthropology from the University of California, San Diego, where he currently is Visiting Lecturer. He also is the Associate Director of Impact Assessment, Inc., a research firm specializing in issues of development and social change. His research interests include the political and psychological implications of social change, social stress and mental health, and ethnic group identity.

Maria de Lourdes Pintasilgo, the Keynote Speaker, served as Prime Minister of Portugal during 1979 and early 1980. She was President of the Commission on the Status of Women in Portugal from 1970 to 1974. As Portugal's Minister of Social Affairs (1974-75), she was instrumental in introducing equal rights for women into the country's new constitution. Since 1975, she has served as Portugal's Ambassador to UNESCO; she also is a member of the Executive Committee of UNESCO.

Harriet B. Presser is currently Professor of Sociology at the University of Maryland. She received her Ph.D. from the University of California, Berkeley, in 1969, and has taught at the University of Sussex (England) and Columbia University. She also was a Research Associate of the Population Council. She has done research on sterilization in Puerto Rico, the United States, and world-wide as well as on teen-age fertility in the United States. At present, she is conducting research on child care in relation to female employment.

Barbara B. Reagan is Professor of Economics at Southern Methodist University, Dallas, Texas. Recently, she completed three years as an editor of the Journal of Economic Literature, and is a past president of the Southwestern Social Science Association. In addition to teaching, publishing, and speaking, Dr. Reagan currently is a member of the Board of Directors of the Federal Home Loan Bank Board of Little Rock which covers five states.

Michelle Riboud teaches economics at the University of Orleans, France. She recieved a doctorate in economics from the University of Paris (1974) and a Ph.D. in economics from the University of Chicago (1977). She held the position of Lecturer at the University of Illinois (1972-73) and taught economics at the University of Abidjan, Ivory Coast (1975-77). Dr. Riboud also served as a member of the French research institute "Casa de Velazquez" in Madrid from 1977 to 1980. Her research has focused mainly on the fields of human capital and labor economics.

Anne-Sofie Rosén teaches personality and social psychology at the University of Stockholm, Sweden. Her research interest is in human socialization and value changes. The research reported in her chapter has been supported by grants from the Swedish Council of Research in the Humanities and Social Sciences, the Bank of Sweden Tercenteneray Foundation, and the World Health Organization.

Virginia E. Schein is an Organizational Psychologist and Consultant to business and industry in human resource management. She received her doctorate in industrial psychology from New York University and is a former Associate Professor of Management at the Wharton School, University of Pennsylvania. The author of over 20 scientific publications, she is an elected member of the Council of Representatives of the American Psychological Association, past president of the Metropolitan New York Association of Applied Psychology, and listed in Who's Who of American Women. Dr. Schein's combined business and academic experiences include faculty positions at Yale University's School of Organization and Management and Case Western Reserve University and Director of Personnel Research for Metropolitan Life Insurance Company.

Judith Stiehm is Associate Professor of Political Science at the University of Southern California where she also is Chairperson of the Program for the Study of Women and Men in Society. She is author of Nonviolent Power: Active and Passive Resistance in America and Bring Me Men and Women: Mandated Change at the U.S. Air Force Academy.

Sandra S. Tangri is Senior Research Associate at the Urban Institute, Washington, D.C. She received her doctorate in social psychology from the University of Michigan and has taught at Wayne State University, the University of Michigan, Douglass College at Rutgers University, and Richmond College at the City University of New York. For five years, Dr. Tangri was Director of Research for the U.S. Commission on Civil Rights. Her research has been on women's roles, particularly how these are affected by women's increased entry into the labor force and into nontraditional occupations. Related areas of her research are the personality dynamics of achievement in women, population programs and the role of women in them, and institutional discrimination.

Ruth Tebbets is a Post-Doctoral Fellow in a program on personality and social structure in the Department of Sociology at the University of California, Berkeley. She received her doctorate in psychology and social relations from Harvard University. Dr. Tebbets' dissertation was a comprehensive analysis of work and family factors related to mental health in a sample of women living in poverty.

Henning Transgaard is a Psychologist at the Danish National Institute for Social Research. For several years, his research interests have focused on attitudes, attitude structure, and attitude change. He recently published in English a book-length study of Danish sex-linked attitudes toward married women's employment and forms of child care.

Mary P. Wagner is a Research Analyst at the Naval Health Research Center, San Diego, California. She also is a graduate student at San Diego State University where she received her baccalaureate degree in religious studies. The focus of her scholastic endeavors is the transformative philosophies of both Eastern and Western traditions. Ms. Wagner is a past vice-president and current member of the San Diego State University chapter of the Society for the Academic Study of Religion.

PARTICIPANTS

MARY B. ABU-SABA
Counseling Center
University of North Carolina
Greensboro, North Carolina 27412
U.S.A.

G.M.W. ACDA
Royal Netherlands Navy
Konigin Marialaan 17
The Hague, The Netherlands

MARIA MANUELA AGUIAR
Secretary of State for Emigration
Lisbon, Portugal

HUGH ARMSTRONG
Vanier College
4266 Beaconsfield Avenue
Montreal, Quebec H4A 2H3 Canada

PAT ARMSTRONG
Vanier College
4266 Beaconsfield Avenue
Montreal, Quebec H4A 2H3 Canada

STYLIANOS K. ATHANASSIOU
Centre of Planning and
Economic Research
Marinu Gerulanu 78
Argyrupolis (T12) Athens, Greece

ELISABETH BECK-GERNSHEIM
Institute for Sociology
University of Münster
Piusallee 154b, 44 Münster
Federal Republic of Germany

LENORE BEHAR
Martha Stuart Communications, Inc.
and Child Mental Health Services
Department of Human Resources,
North Carolina, Albemarle Building,
325 North Salisbury Street
Raleigh, North Carolina 27611 U.S.A.

DEBORAH BELLE
Graduate School of Education,
Harvard University, Read House,
Appian Way, Cambridge,
Massachusetts 02138 U.S.A.

BARBARA R. BERGMANN
Department of Economics
University of Maryland
College Park, Maryland 20742 U.S.A.

JENNY BLAKE
11 Redpit
Dilton Marsh
Westbury, Wilts, U.K.

ANNE M. BRISCOE
Department of Medicine
College of Physicians and Surgeons
Columbia University
Harlem Hospital Center
New York, New York 10037 U.S.A.

CLAIR B. (VICKERY) BROWN
Department of Economics
University of California
Berkeley, California 94720 U.S.A.

ANDREA BRÜNJES
Talstrasse 36
7800 Freiburg
Federal Republic of Germany

ANNA CASTRO
Santiago, Chile

GORDON R. CHAPMAN
Center for Women Policy Studies
2000 P Street NW, Suite 508
Washington, D.C. 20036 U.S.A.

JANE ROBERTS CHAPMAN
Center for Women Policy Studies
2000 P Street NW, Suite 508
Washington, D.C. 20036 U.S.A.

JOYCE A. CHELOUCHE
1565 Madison Street, No. 403
Oakland, California 94612 U.S.A.

NICOLA CHERRY
Institute of Occupational Health
London School of Hygiene and
Tropical Medicine, Keppel Street,
London WC1E 7HT U.K.

M. PATRICIA CONNELLY
Department of Sociology
St. Mary's University
Halifax, Nova Scotia B3H 3C3
Canada

ALICE H. COOK
Longhouse, 766 Elm Street
Ithaca, New York 14850 U.S.A.

MARTHA A. DARLING
Office of Senator Bill Bradley
4104 Dirksen, U.S. Senate
Washington, D.C. 20510 U.S.A.

IRENE DEITCH
Department of Psychology
College of Staten Island
City University of New York
715 Ocean Terrace
Staten Island, New York 10301 U.S.A.

JEAN LEONARD ELLIOTT
Department of Sociology and
Anthropology, Dalhousie University
Halifax, Nova Scotia, Canada

HERSCHEL Y. FELDMAN
Jewish Federation of Metropolitan
Chicago, 1 South Franklin Street
Chicago, Illinois 60626 U.S.A.

NATALIE FELDMAN
East Maine School District, No. 63
10150 Dee Road
Des Plaines, Illinois 60016 U.S.A.

MARY L. FISCHER
Graduate School of Management,
Delft, Van Der Helmstraat 270
3067 HL Rotterdam, The Netherlands

NICKI FONDA
Brunel University
Uxbridge, Middlesex UB8 3PH U.K.

MELVIN L. FOULDS
4633 Perham Road
Corona del Mar, California 92625
U.S.A.

WOLFGANG FRANZ
University of Mannheim
Seminargebäude A-5
6800 Mannheim 1
Federal Republic of Germany

DEBORAH S. FREEDMAN
Population Studies Center
University of Michigan
1225 South University Avenue
Ann Arbor, Michigan 48109 U.S.A.

HARRIET S. GERSHENSON
Public Schools, Washington, D.C.
5916 Rossmore Drive
Bethesda, Maryland 20014 U.S.A.

JOAN M. GOODIN
National Commission on Working
Women
1211 Connecticut Avenue NW, Ste. 310
Washington, D.C. 20036 U.S.A.

PATRICIA S. HANNIGAN
School of Human Development and
Community Service
California State University
Fullerton, California 92634 U.S.A.

HORST JÜRGEN HELLE
Institute of Sociology, University
of Munich, Konradstrasse 6
8000 Munich 40
Federal Republic of Germany

FRANCINE HERMAN
Cornell University
Ithaca, New York 14853 U.S.A.

JOAN B. HOELZER
2805 W. Glen Drive, Apt. 23
Falls Church, Virginia 22046 U.S.A.

ANNE HOIBERG
Naval Health Research Center
P.O. Box 85122
San Diego, California 92138 U.S.A.

LILLI S. HORNIG
Higher Education Resources Services
Cheever House, Wellesley College
Wellesley, Massachusetts 02181
U.S.A.

PAULA M. HUDIS
40 Worth Street
San Francisco, California 94114
U.S.A.

PENNY JONES
Henderson Jones Research and
Consultancy
13 New End
London NW3 U.K.

DENIZ KANDIYOTI
Social Sciences Department,
Bogazici University
P.K. 2 Bebek
Istanbul, Turkey

FRANÇOISE LATOUR DE VEIGA PINTO
Women's Centre for Studies in a
Changing Society
55, rue de Varenne
75007 Paris, France

JOYCE B. LAZAR
National Institute of Metal Health
5600 Fishers Lane
Rockville, Maryland 20857 U.S.A.

JEAN LIPMAN-BLUMEN
Center for Women Policy Studies
and University of Maryland
6803 Greyswood Road
Bethesda, Maryland 20034 U.S.A.

HELENA ZNANIECKA LOPATA
Department of Sociology,
Loyola University, 6525 Sheridan
Chicago, Illinois 60626 U.S.A.

LORNA R. MARSDEN
Department of Sociology,
University of Toronto
Toronto, Ontario M5S 1A1 Canada

GERRI MERTENS
Women's Career Development Center
735 East Lexington Drive
Glendale, California 91206 U.S.A.

PRISCILLA MEYER
Veterans Administration
Medical Center, Building 93
Waco, Texas 76703 U.S.A.

MARGUERITE MILKE
Box 2022
APO New York 09283 U.S.A.

JACOB MINCER
Department of Economics
International Affairs Building
Columbia University
New York, New York 10027 U.S.A.

CONSTANCE A. NATHANSON
Department of Population Dynamics
School of Hygiene and Public Health
Johns Hopkins University
Baltimore, Maryland 21205 U.S.A.

RICHARD J. NIEHAUS
Office of Assistant Secretary of the
Navy (Manpower, Reserve Affairs
and Logistics), Room 4E775,
The Pentagon
Washington, D.C. 20350 U.S.A.

VIRGINIA A. NOVARRA
12 St. Edmund's Court
St. Edmund's Terrace
London NW8 7QL U.K.

AYSE ÖNCÜ
Department of Social Sciences
Bogazici University
P.K. 2, Bebek, Istanbul, Turkey

FREDA L. PALTIEL
Status of Women
Health and Welfare Canada
Room 2106, Jeanne Mance Building
Tunney's Pasture
Ottawa, Ontario K1A OK9 Canada

HANNA PAPANEK
Center for Asian Development
Studies, Boston University
Boston, Massachusetts 02215 U.S.A.

MAJOR FRANKLIN C. PINCH
Canadian Forces Personnel
Applied Research Unit
4900 Yonge Street, Suite 600
Willowdale, Ontario M2N 6B7
Canada

MARIA DE LOURDES PINTASILGO
Alameda St° Antonio dos Capuchos
4-5° 1700 Lisbon, Portugal

HARRIET B. PRESSER
Department of Sociology
University of Maryland
College Park, Maryland 20742 U.S.A.

WENDY PRITCHARD
Shell International Petroleum Co., Ltd.
LP/4, Shell Centre, York Road
London SE1 7NA U.K.

BARBARA B. REAGAN
Department of Economics
Southern Methodist University
Dallas, Texas 75275 U.S.A.

HELGA REIMANN
University of Augsburg
Memminger Strasse 14
8900 Augsburg
Federal Republic of Germany

HORST REIMANN
University of Augsburg
Memminger Strasse 14
8900 Augsburg
Federal Republic of Germany

RAQUEL RIBEIRO
Ministry of Social Affairs
Lisbon, Portugal

MICHELLE RIBOUD
Ciudad Jardin Atalaya
Bloque 1, 4° A
Camas, Sevilla, Spain

CARLAN M. ROBINSON
New York University Medical Center
School of Medicine
550 First Avenue
New York, New York 10016 U.S.A.

ANNE-SOFIE ROSÉN
Department of Psychology
University of Stockholm
P.O. Box 6706
113 85 Stockholm, Sweden

VIRGINIA E. SCHEIN
173 Hardesty Road
Stamford, Connecticut 06903 U.S.A.

MIET SMET
Volksvertegenwoordiger
Durmelaan, 4/5
9100 Lokeren, Belgium

JUDITH STIEHM
Department of Political Science
University of Southern California
Von Kleinsmid Center,
University Park
Los Angeles, California 90007 U.S.A.

SANDRA S. TANGRI
Program of Research on Women
and Family Policy
The Urban Institute
2100 M Street NW
Washington, D.C. 20037 U.S.A.

HENNING TRANSGAARD
Danish National Institute of
Social Research
28 Borgergade
DK-1300 Copenhagen K Denmark

JO VAN ROOYEN
Management Studies Division
National Institute for
Personnel Research
P.O. Box 32410
Broomfontein, 2107 South Africa

H.M. IN'T VELD-LANGEVELD
Netherlands Scientific Council
for Government Policy
Post Box 20004
2500 EA The Hague, The Netherlands

ANNA VICENTE
Commission on the Status of Women
Av. Elias Garcia, 12, 1°
Lisbon, Portugal

VAIRA VIKIS-FREIBERGS
Department of Psychology
University of Montreal
Case Postale 6128, Succursale "A"
Montreal, Quebec H3C 33J7
Canada

FLORENCE WILHELM-REZENDE
Labor Ministry
1, Place de Fontenoy
75700 Paris, France

LARAINE T. ZAPPERT
Center for Research on Women
Stanford University
Stanford, California 94305 U.S.A.

INDEX

Abortion, 124, 334, 342

Academic achievement: of girls, 7-8, 16, 18, 37, 87, 91, 187, 291, 321, 329, 344

Achievement: and "fear of success," 14; mood of achievement satisfaction, 13; orientation, 17, 88-89, 91; patterns, sex differences ·in, 329-330

Achieving styles: 13-14, 17, 19; occupational roles, 14; socialization for, 14

Aid to Families with Dependent Children (AFDC), 99, 110, 115, 121-122, 124, 126-127, 181-182, 185

Alimony, 122, 167

Allocation-of-time approach, 255, 262

Andalusia, 199, 255-267

"Androgynous family," 332-333

Androgynous personality, 88, 90-91

Appropriate technology, 30

Attitudes: 305-320; dimensions of, 306-308; effects of climate of opinion on, 308, 311-312, 319; husbands' toward elected women officials, 58; objects of, 306-308, 311, 319; structure of, 306-308, 310, 319; theory of, 306-308, 311, 319; toward child care, 290-291, 305-320; toward work, 185, 313, 322; and value system, 307-308

"Balanced" groups: women in, 55-56, 63-64

Barriers of employment for women. See Labor force participation: barriers of

Beck-Gernsheim, Elisabeth, 288-289

Brandt Commission: report of, 27, 32

Briscoe, Anne, 12, 97-98, 104

California Elected Women for Education and Research (CEWEAR), 56-62

Canada: women in, 10, 65-76, 129-152, 223-237; capitalism in, 224, 227; decision-making sector of, 6, 10, 17, 66, 69, 76; families in, 65, 73, 75-76, 141-142, 148, 223-237; Human Rights Act of, 342; Human Rights Commission in, 10, 72; immigration, effect on women's labor force participation in, 228-229, 232; labor unions in, 10, 71, 225-226, 230-231, 233; married women in labor force of, 140-143, 145, 149, 225, 227-228, 231-232, 234; part-time employment in, 66, 75, 99, 129-131, 134-136, 144, 146-150, 194, 235; political parties in, 10, 72; population distribution of, 65, 228-229, 232, 234; poverty level in, 140-141, 223, 234; unemployment in (includes insurance benefits), 75, 99, 129-134, 138-145, 149-152, 223, 231, 235; urbanization